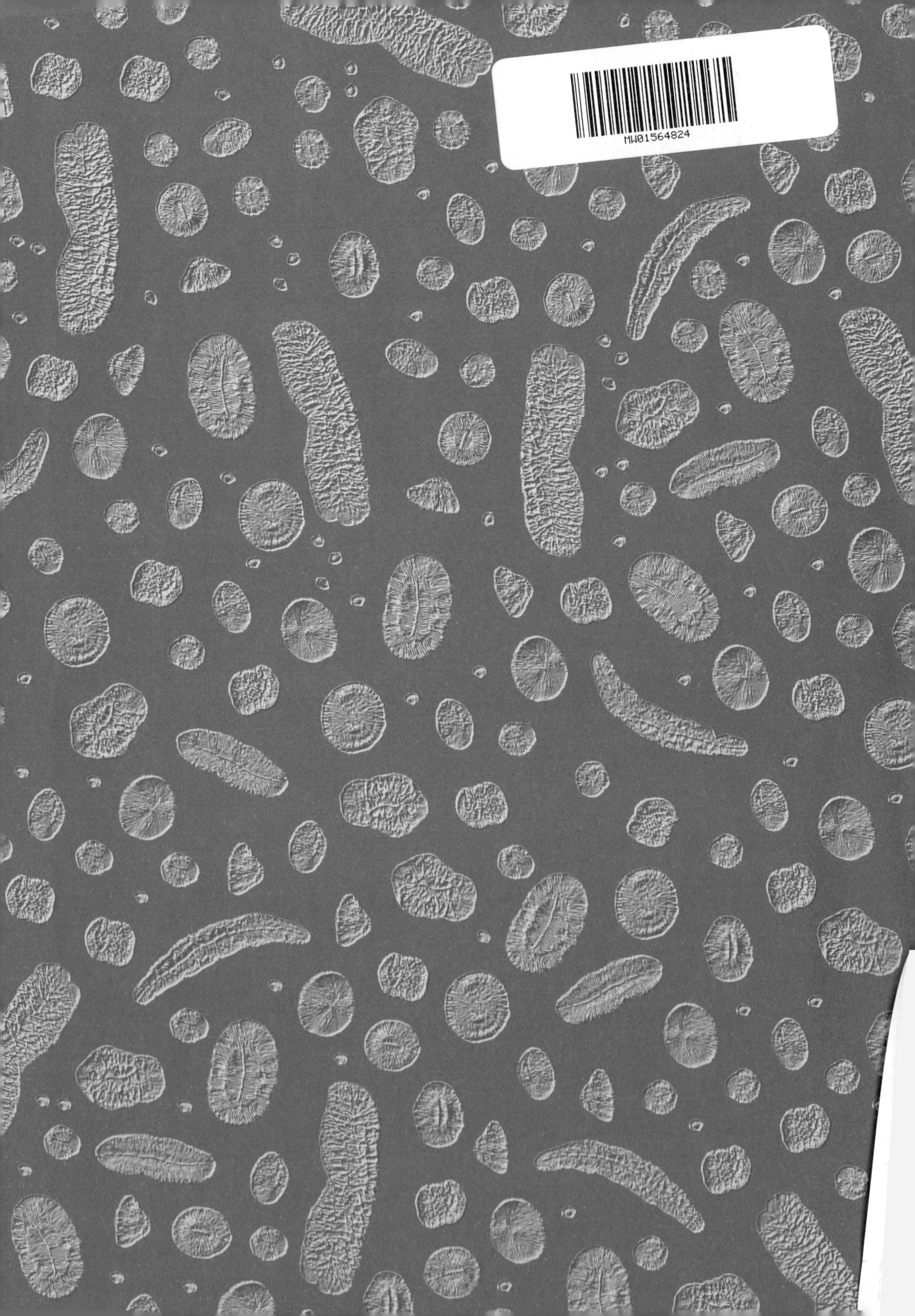

CORALS
of the world

JEN Veron
Author

③

Mary Stafford-Smith
Scientific Editor and Producer

Cover artwork: A compilation of corals from around the world. *Design: Elliot Sprung (aka Paul Elliot)*

Pages iv-v: A typical shallow reef lagoon community of the central Great Barrier Reef, Australia in 1981. At that time, this amount of coral cover was not unusual for reefs of the region. GREAT BARRIER REEF, AUSTRALIA *Photograph: Ed Lovell*

CORALS of the world

JEN Veron
Author

③

Mary Stafford-Smith
Scientific Editor and Producer

Author
Dr JEN Veron

Scientific Editor and Producer
Dr Mary Stafford-Smith

Publisher
Australian Institute of Marine Science,
PMB 3, Townsville MC,
Qld 4810, Australia
email: corals@aims.gov.au
internet: www.aims.gov.au/corals

Design and Production
CRR Qld, 250 Ross River Road,
Townsville, Qld 4814, Australia

Printer
New Litho, Surrey Hills, Melbourne, Australia

Finisher
M & M Binders, Mount Waverley, Melbourne, Australia

First published in 2000

© Australian Institute of Marine Science and CRR Qld Pty Ltd

All rights reserved. No part of this publication may be reproduced, stored in a retrieval system or transmitted in any form by any means without the prior permission of the copyright owners. Please apply in writing to the publisher.

The National Library of Australia Cataloguing-in-Publication data:

> Veron, JEN (John Edward Norwood)
> Corals of the World
>
> Bibliography.
> Includes index.
> ISBN numbers:
> volume 1: 0 642 32236 8
> volume 2: 0 642 32237 6
> volume 3: 0 642 32238 4
> Corals - identification. 2. Corals I. Stafford-Smith, Mary. II Title.

593.6

Contents

Volume 1

Acknowledgments x-xi

About this book 3-7

Observing corals 11-17

Corals and coral reefs 21-29

Geological history 33-43

Structure 47-57

Family Acroporidae 61-447
Genus Montipora 62-167
Genus Anacropora 168-175
Genus Acropora 176-433
Genus Astreopora 434-447

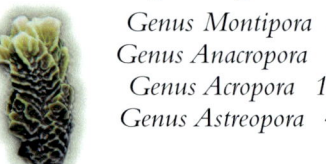

Index 451-463

Volume 2

Family Astrocoeniidae 3-21
Genus Stylocoeniella 4-8
Genus Stephanocoenia 9
Genus Palauastrea 10-11
Genus Madracis 12-21

Family Pocilloporidae 23-65
Genus Pocillopora 24-45
Genus Seriatopora 46-55
Genus Stylophora 56-65

Family Euphyllidae 67-93
Genus Euphyllia 68-81
Genus Catalaphyllia 82-83
Genus Nemenzophyllia 84-85
Genus Plerogyra 86-91
Genus Physogyra 92-93

Family Oculinidae 95-117
Genus Oculina 98-102
Genus Simplastrea 103
Genus Schizoculina 104-105
Genus Galaxea 106-117

Family Meandrinidae 119-131
Genus Meandrina 120-122
Genus Ctenella 123
Genus Dichocoenia 124-125
Genus Dendrogyra 126-127
Genus Gyrosmilia 128-129
Genus Montigyra 129
Genus Eusmilia 130-131

Family Siderastreidae 133-167
Genus Pseudosiderastrea 134-135
Genus Horastrea 136
Genus Anomastraea 137
Genus Siderastrea 138-143
Genus Psammocora 144-157
Genus Coscinaraea 158-167

Family Agariciidae 169-231
Genus Agaricia 170-177
Genus Pavona 178-201
Genus Leptoseris 202-220
Genus Coeloseris 221
Genus Gardineroseris 222-223
Genus Pachyseris 224-231

Family Fungiidae 233-315
Genus Cycloseris 236-247
Genus Diaseris 248-250
Genus Cantharellus 251-253
Genus Heliofungia 254-255
Genus Fungia 256-285
Genus Ctenactis 286-290
Genus Herpolitha 291-293
Genus Polyphyllia 294-295
Genus Sandalolitha 296-299
Genus Halomitra 300-303
Genus Zoopilus 304-305
Genus Lithophyllon 306-309
Genus Podabacia 310-315

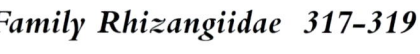

Family Rhizangiidae 317-319
Genus Astrangia 319

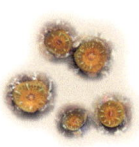

Contents

Family Pectiniidae 321-361
Genus Echinophyllia 322-332
Genus Echinomorpha 333
Genus Oxypora 334-341
Genus Mycedium 342-347
Genus Pectinia 348-361

Family Merulinidae 363-383
Genus Hydnophora 364-373
Genus Paraclavarina 374-375
Genus Merulina 376-381
Genus Boninastrea 382
Genus Scapophyllia 383

Family Dendrophylliidae 385-407
Genus Turbinaria 388-404
Genus Duncanopsammia 405
Genus Balanophyllia 406
Genus Heteropsammia 407

Family Caryophylliidae 409-413
Genus Heterocyathus 412-413

Index 417-429

This Volume

Family Mussidae 3-83
Genus Blastomussa 4-7
Genus Micromussa 8-11
Genus Acanthastrea 12-31
Genus Mussismilia 32-35
Genus Isophyllia 36-37
Genus Lobophyllia 38-51
Genus Symphyllia 52-63
Genus Mussa 64-65
Genus Scolymia 66-71
Genus Mycetophyllia 72-79
Genus Australomussa 80
Genus Indophyllia 81
Genus Cynarina 82-83

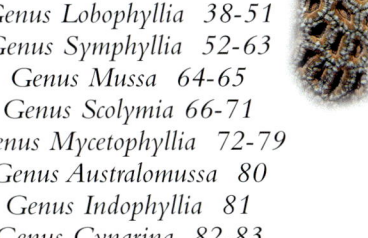

Family Faviidae 85-269
Genus Cladocora 88-90
Genus Caulastrea 91-97
Genus Erythrastrea 98
Genus Manicina 99
Genus Favia 100-131
Genus Barabattoia 132-133
Genus Favites 134-155
Genus Goniastrea 156-175
Genus Platygyra 176-193
Genus Australogyra 194
Genus Oulophyllia 195-201
Genus Leptoria 202-205
Genus Diploria 206-209
Genus Colpophyllia 210-211
Genus Montastrea 212-225
Genus Plesiastrea 226-228
Genus Oulastrea 229
Genus Diploastrea 230-231
Genus Leptastrea 232-238
Genus Parasimplastrea 239
Genus Cyphastrea 240-249
Genus Solenastrea 250-251
Genus Echinopora 252-268
Genus Moseleya 269

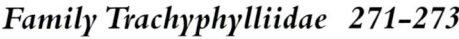

Family Trachyphylliidae 271-273
Genus Trachyphyllia 272-273

Family Poritidae 275-397
Genus Porites 276-345
Genus Stylaraea 346
Genus Poritipora 347
Genus Goniopora 348-379
Genus Alveopora 380-397

Non-scleractinian corals 399-407

Biogeography 411-421

What are species? 425-433

Evolution of species 437-443

Keys to genera and species 447-459

Common names 461

Glossary 463-467

References 469-473

Index 477-489

CORALS of the world

Family Mussidae
Family Faviidae
Family Trachyphylliidae
Family Poritidae

★ ★ ★

Non-Scleractinian Corals

★ ★ ★

Biogeography
What are species?
Evolution of species

★ ★ ★

Keys to genera and species
Common names
Glossary
References

★ ★ ★

Index

Family Mussidae

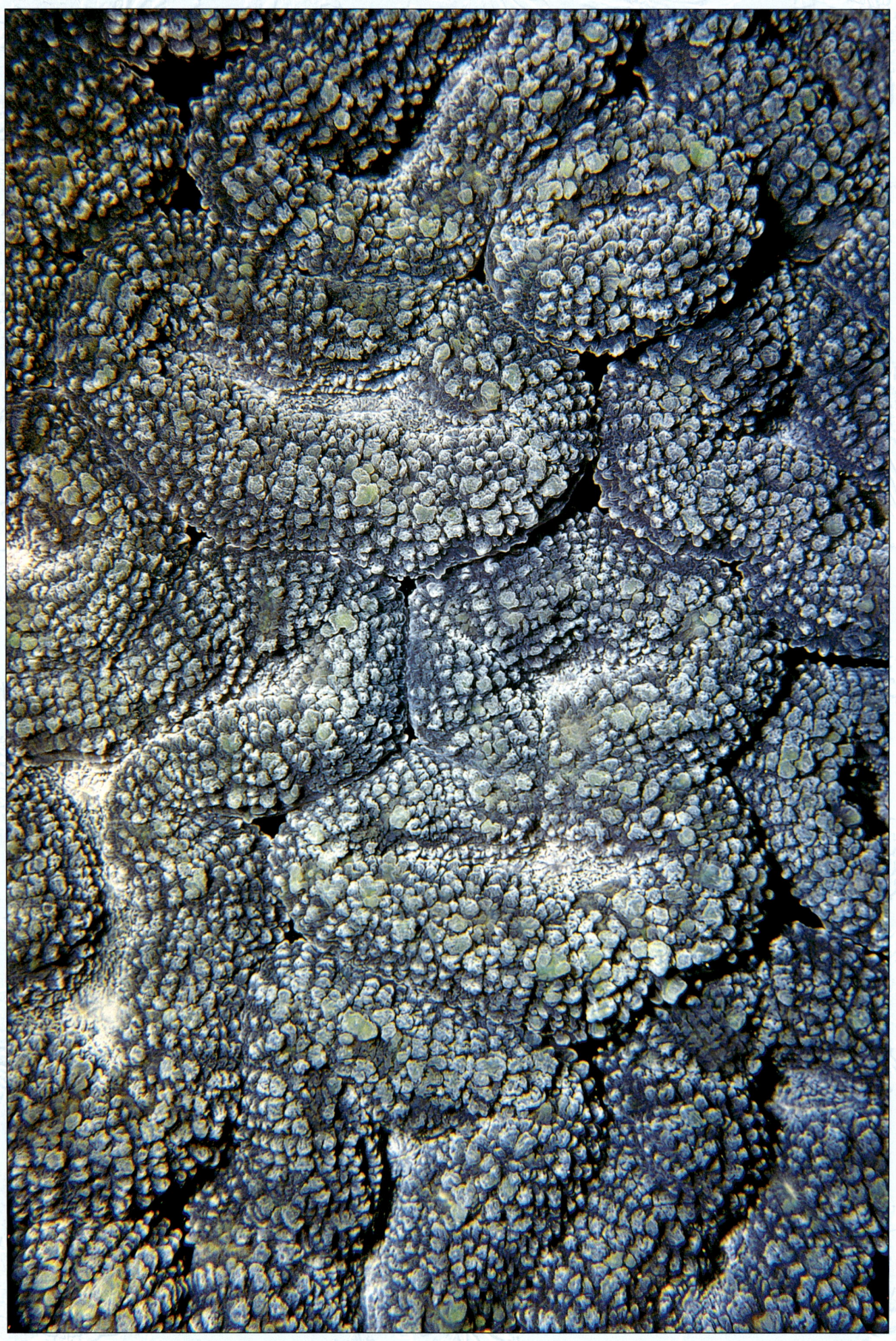

FAMILY
MUSSIDAE
Ortmann, 1890

Characters: All species are zooxanthellate, solitary or colonial. Skeletal structures are solid. Corallites and valleys are large. Septa have large teeth or lobes. Columellae and walls are thick and well developed. **Similar family:** Pectiniidae. **Genera:** Thirteen genera, eight of which are restricted to the Indo-Pacific, four are restricted to the Atlantic, while *Scolymia* occurs in both realms.

Taxonomic references: Chevalier (1975), Veron and Pichon (1980).
Summary key: See p455.

Page viii: Sheltered platform reefs that have irregular edges and good water circulation commonly have a wide range of habitat types and a high species diversity. GREAT BARRIER REEF, AUSTRALIA *Photograph: Roger Steene*

Opposite: The soft parts of most mussids, characterised by this *Lobophyllia hemprichii*, have a coarse texture. The mesoglea layer is also thick and tough. HOUTMAN ABROLHOS ISLANDS, SOUTH-WEST AUSTRALIA *Photograph: author*

GENUS *BLASTOMUSSA*
Wells, 1961

Characters: Colonies are phaceloid to subplocoid. Corallites have weakly developed columellae. Septa slope gently to the corallite centre and have lobed teeth. Corallite walls are enveloped with, and often joined by, epitheca. Tentacles have fleshy mantles extended during the day, often forming a continuous cover, obliterating the colony structure underneath. Tentacles are extended only at night. **Similar genera:** The faviid *Caulastrea* also has phaceloid colonies but corallites do not have fleshy mantles and septa are fine, without lobed teeth. See also the faviid *Parasimplastrea*. **Earliest fossil record:** Pleistocene of the Pacific.

Taxonomic references: Wells (1961), Chevalier (1975), Veron and Pichon (1980).

Blastomussa merleti
Wells, 1961

Characters: Colonies are phaceloid to plocoid, and consist of a few to large numbers of corallites. Corallites are less than 7 millimetres diameter. Septa are mostly in two cycles of which only the first reaches the columella. Septa have slightly serrated margins. Primary septa may be exsert. Columellae are poorly developed. Mantles, but not tentacles, are extended during the day and may form a continuous surface obscuring the underlying growth-form. **Colour:** Commonly dark red with conspicuous green oral discs. May also be pink, orange, brown or uniform dark grey with white margins to primary septa. **Similar species:** *Blastomussa wellsi*, which has much larger corallites with more numerous septa. See also the faviid *Parasimplastrea sheppardi*, which is cerioid and has extensive 'groove and tubercle' formations. **Habitat:** Reef environments, especially where the water is turbid. **Abundance:** Uncommon.

Taxonomic note: Red Sea colonies are mostly plocoid and have been described as a separate species (*Blastomussa loyae* Head, 1978). There are, however, no consistent differences between these colonies and plocoid colonies in other regions, see 'What are species?' p425.
Taxonomic references: Wells (1961), Chevalier (1975), Veron and Pichon (1980).
Identification guides: Veron (1986), Sheppard and Sheppard (1991), Nishihira and Veron (1995), Coles (1996).

1 The typical appearance of a large colony in the Red Sea. SINAI PENINSULA, EGYPT *Photograph: author*

2, 3, 5 Variation in corallite structure and colour. **2** BOLINAO, PHILIPPINES **3** GREAT BARRIER REEF, AUSTRALIA **5** MALDIVE ISLANDS *Photographs: 2 author 3 Ed Lovell 5 Neville Coleman*

4, 6 Surface detail. **4** GREAT BARRIER REEF, AUSTRALIA **6** SINAI PENINSULA, EGYPT *Photograph: 4 Len Zell 6 Mary Stafford-Smith*

Blastomussa wellsi
Wijsman-Best, 1973

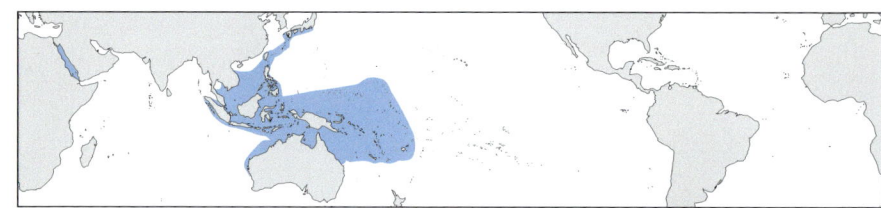

Characters: Colonies are phaceloid, rarely subplocoid. Corallites are 9-14 millimetres diameter. Septa are not arranged in cycles and are numerous. They have small blunt teeth. Mantles, but not tentacles, are extended during the day and may form a continuous surface obscuring the underlying growth-form. **Colour:** Mantles are usually dark grey, but may be red or green. Oral discs are usually green but may be red or dark grey. **Similar species:** *Blastomussa merleti*. See also the euphyllid *Nemenzophyllia turbida* and the faviid *Montastrea multipunctata*, both of which have fleshy mantles. **Habitat:** Lower reef slopes protected from wave action, and turbid environments. **Abundance:** Uncommon, rare in the Red Sea.

x2

Taxonomic references: Wijsman-Best (1973), Chevalier (1975), Veron and Pichon (1980).
Identification guides: Veron (1986), Nishihira and Veron (1995).

Family Mussidae | Genus *Blastomussa*

1-3, 6 Variation in corallite structure and colour. **1** Great Barrier Reef, Australia **2** Sinai Peninsula, Egypt **3** Papua New Guinea **6** Vietnam *Photographs: 1 Ed Lovell 2, 3, 6 author*

4, 5 Surface detail of corallites. **4** Great Barrier Reef, Australia **5** Calamian Islands, Philippines *Photographs: author*

GENUS *MICROMUSSA*
Veron, this publication

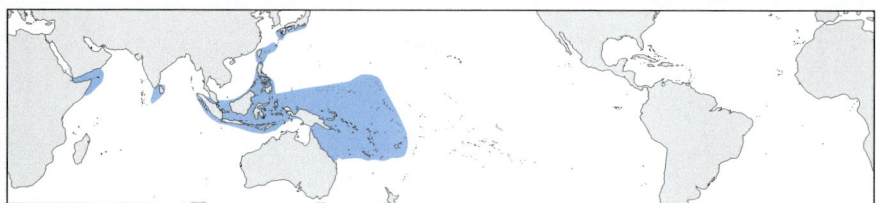

Characters: Colonies are submassive or encrusting and usually flat. Corallites are cerioid or subplocoid, either circular or angular in shape and up to 8 millimetres diameter. Septa are thickened at the corallite wall, and have conspicuous teeth. Colonies may have fleshy tissue over the skeleton, but skeletal structures remain visible. Tentacles are extended only at night. **Similar genera:** *Acanthastrea*, which has larger polyps. Underwater this genus is readily confused with the faviid genera *Favia* and *Favites* which have corallites of similar size. **Fossil record:** None.

Taxonomic note: This genus was created because of the very wide range of corallite size and structure of species formally included in genus *Acanthastrea*. **Type species:** *Acanthastrea amakusensis* Veron, 1990.

Micromussa minuta
(Moll and Borel-Best, 1984)

Characters: Colonies are massive, cerioid, with circular corallites 5-6 millimetres diameter. Walls are thick. Septa are beaded and columellae are poorly developed. Colonies do not have thick fleshy tissue over the skeleton. Colonies may have 'groove and tubercle' structures separating corallites. **Colour:** Pale grey. **Similar species:** *Micromussa diminuta*. See also *M. amakusensis*, which has larger, fleshy, angular corallites with thin walls and no costae. Underwater this species can readily be mistaken for a faviid with small corallites. **Habitat:** Shallow reef environments. **Abundance:** Rare.

x3

Taxonomic reference: Moll and Borel-Best (1984, as *Acanthastrea minuta*).

1

Family Faviidae | Genus *Micromussa*

Micromussa diminuta
Veron, this publication

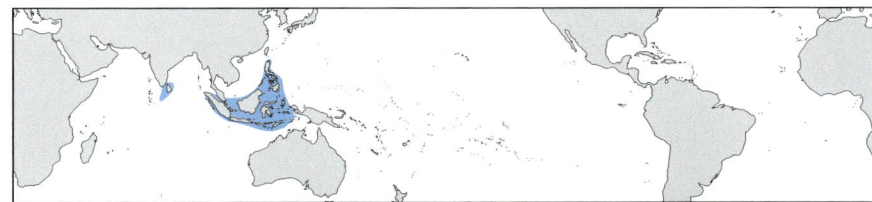

Characters: Colonies are massive and cerioid, with pentagonal or hexagonal corallites 3-4 millimetres diameter. Walls are thin. Septa are beaded and columellae are poorly developed. Colonies do not have thick fleshy tissue over the skeleton. **Colour:** Brown, with pale grey centres. The surface is usually speckled white. **Similar species:** *Micromussa minuta*, which has larger, more rounded corallites with fewer, thicker septa. Underwater this species can readily be mistaken for a faviid with small corallites. **Habitat:** Shallow reef environments. **Abundance:** Uncommon.

Taxonomic note: See 'New species described in *Corals of the World*' (Veron, in preparation) for further information.

1 *Micromussa minuta*. This species looks more like a faviid than a mussid underwater. PAPUA NEW GUINEA *Photograph: author*

2 *Micromussa diminuta*. Surface of a large colony. SRI LANKA *Photograph: author*

3 *Micromussa diminuta*. Corallite detail. CALAMIAN ISLANDS, PHILIPPINES *Photograph: author*

4 *Micromussa diminuta*. Showing angular corallites. SRI LANKA *Photograph: author*

Micromussa amakusensis
(Veron, 1990)

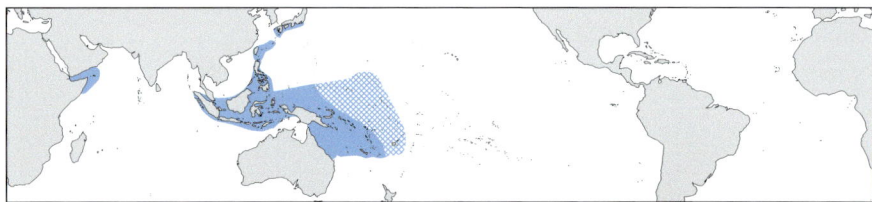

Characters: Colonies are cerioid with neat angular corallites up to 8 millimetres diameter. Septa have 1-3 large teeth. Colonies have a thick fleshy mantle usually covered with fine papillae. **Colour:** Brick-red or green in northern Japan, grey elsewhere. **Similar species:** *Acanthastrea lordhowensis*, which has larger, less uniform corallites containing more septa. See also *Micromussa minuta*. **Habitat:** Protected reef environments and rocky foreshores. **Abundance:** Uncommon in mainland Japan, rare elsewhere.

Distribution note: Records from the central Pacific are doubtful. **Taxonomic reference:** Veron (1990a). **Identification guide:** Nishihira and Veron (1995, as *Acanthastrea amakusensis*).

Family Mussidae   Genus *Micromussa*

1 An encrusting colony. The teeth of septa can just be seen. HONSHU, JAPAN *Photograph: author*

2 Two adjacent colonies with partly extended tentacles. HONSHU, JAPAN *Photograph: author*

3 Corallite detail. HONSHU, JAPAN *Photograph: author*

4, 5 With polyp mantles retracted (4) and extended (5). CALAMIAN ISLANDS, PHILIPPINES *Photographs: author*

GENUS *ACANTHASTREA*

Milne Edwards and Haime, 1848

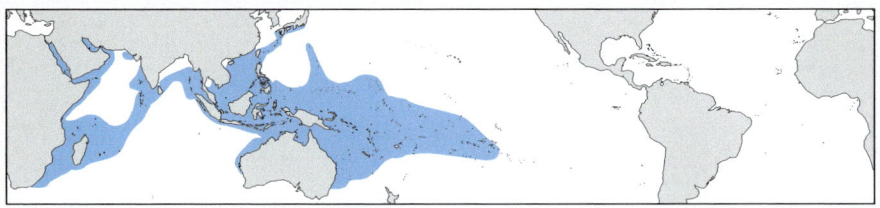

Characters: Colonies are massive or encrusting and are usually flat. Corallites are cerioid or subplocoid, monocentric, and either circular or angular in shape. Septa are thickened at the corallite wall, and have tall teeth. Colonies have thick fleshy tissue over the skeleton. Tentacles are extended only at night. **Similar genera:** *Acanthastrea* with large polyps may be *Lobophyllia*-like and may also resemble cerioid *Mussismilia*. Species with small polyps may be confused with faviids, especially *Favites*. *Micromussa* is distinguished by having smaller corallites (less than 8 mm diameter). **Earliest fossil record:** Miocene of the Pacific (doubtful record).

Taxonomic references: Chevalier (1975), Veron and Pichon (1980).
Summary key: See pp455-6.

Identification of *Acanthastrea* species

With *Micromussa* now separated from *Acanthastrea*, this genus is divisible into two groups, distinguished primarily by corallite size.

Group 1: Species with small corallites (less than 15 mm diameter), opposite.

Group 2: Species with large corallites (over 15 mm diameter), page 27.

Group 1: Species with small corallites (less than 15 mm diameter)

Acanthastrea subechinata
Veron, this publication

Characters: Colonies are encrusting to massive and are usually small. Corallites are subplocoid and circular, with thick walls. Septa have rounded evenly spaced teeth. Colonies have thick fleshy tissue over the skeleton, but this tissue does not form concentric folds. Septal dentations appear as evenly spaced rows of bead-like tissue down the septal margins. **Colour:** Uniform grey or green, sometimes mottled. **Similar species:** *Acanthastrea echinata*, which has more fleshy corallites with tissue forming concentric folds, and larger less regular septal teeth. These species are not easily separated unless they occur together. **Habitat:** Shallow reef environments. **Abundance:** Uncommon.

Taxonomic note: See 'New species described in *Corals of the World*' (Veron, in preparation) for further information.

x2

1 *Acanthastrea* (including this *A. lordhowensis*) closely resemble *Favia* at night when polyps are extended. NORFOLK ISLAND, WESTERN PACIFIC *Photograph: Neville Coleman*

2 Differences in the size of corallites helps in the identification of *Acanthastrea* species. Here, a species with large corallites (*A. hillae*) is adjacent to one with small corallites (*A. lordhowensis*). NORFOLK ISLAND, WESTERN PACIFIC *Photograph: author*

3-5 *Acanthastrea subechinata*. Variation in structure and colour of colonies on upper reef slopes. **3, 5** CALAMIAN ISLANDS, PHILIPPINES **4** FLORES, INDONESIA *Photographs: author*

Acanthastrea lordhowensis
Veron and Pichon, 1982

×2

Characters: Colonies are massive and cerioid, with laterally compressed corallites of uneven height. Walls are acute: septa are thick, with large teeth. Columellae are barely developed. Colonies have a thick fleshy mantle which is covered by fine papillae. **Colour:** Very colourful: red, purple and green are the most common colours, with corallites and walls almost always of contrasting colours. **Similar species:** *Micromussa amakusensis*. **Habitat:** Shallow reef environments especially of subtropical localities. **Abundance:** Sometimes common.

Taxonomic reference: Veron and Pichon (1982). **Identification guides:** Veron (1986), Nishihira and Veron (1995).

1 Colonies vary in appearance according to colour, also the degree of extension of the corallite mantles as seen here (central colonies) with the faviid *Goniastrea australensis* (top left, top right and bottom left). NORFOLK ISLAND, WESTERN PACIFIC *Photograph: author*

2-5 Common colours. **2** GREAT BARRIER REEF, AUSTRALIA **3** SOLITARY ISLANDS, SOUTH-EAST AUSTRALIA **4, 5** NORFOLK ISLAND, WESTERN PACIFIC *Photographs: 2-4 author 5 Neville Coleman*

6 A large colourful colony. GREAT BARRIER REEF, AUSTRALIA *Photograph: author*

Family Mussidae   Genus *Acanthastrea*

Acanthastrea regularis
Veron, this publication

Characters: Colonies are massive and subplocoid. Septa are uniformly spaced with 8-10 evenly spaced, rounded teeth. Teeth on adjacent septa are often aligned, forming concentric circles. Some septa are more prominent than others. Columellae are weakly developed. Colonies do not have thick tissue over the skeleton. **Colour:** Variable brown and yellow-brown, usually with contrasting corallite walls and centres. **Similar species:** May resemble the faviids *Montastrea* and *Favia* species more than other *Acanthastrea* underwater. See also *A. faviaformis*. **Habitat:** Shallow reef environments. **Abundance:** Uncommon.

Taxonomic note: See 'New species described in *Corals of the World*' (Veron, in preparation) for further information.

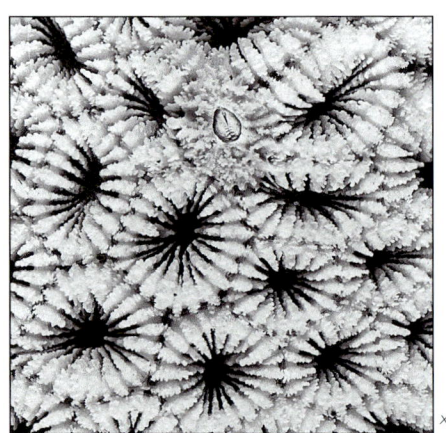

Acanthastrea brevis
Milne Edwards and Haime, 1849

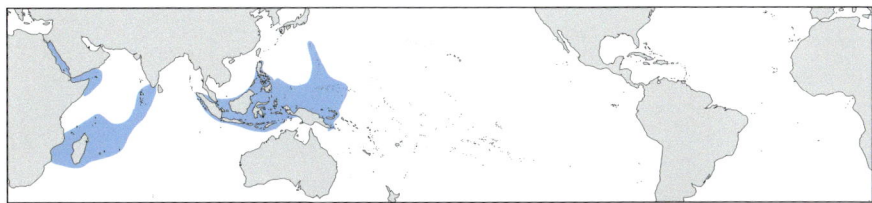

Characters: Colonies are mostly submassive. Corallites are cerioid to subplocoid with moderately thin walls. Septa are thin and widely spaced. Larger septa have very long upwardly projecting teeth giving colonies a spiny appearance. Colonies are usually not fleshy. **Colour:** Uniform or mottled brown, yellow or green. **Similar species:** *Acanthastrea echinata*, which has relatively fleshy corallites with thicker walls and less elongate septal teeth. These two species may be difficult to distinguish unless they occur together. **Habitat:** Shallow reef environments. **Abundance:** Uncommon.

Taxonomic reference: Milne Edwards and Haime's original description/specimens.

x2

1-4 *Acanthastrea regularis*. Variation in corallite structure and colour. **1** CALAMIAN ISLANDS, PHILIPPINES **2** PAPUA NEW GUINEA **3** GREAT BARRIER REEF, AUSTRALIA **4** GUAM *Photographs: 1-3 author 4 Gustav Paulay*

5, 7 *Acanthastrea brevis*. Surface of large colonies. **5** SEYCHELLES **7** GUAM *Photographs: 5 author 7 Gustav Paulay*

6, 8 *Acanthastrea brevis*. Corallite detail. **6** MALDIVE ISLANDS **8** SINAI PENINSULA, EGYPT *Photographs: 6 Neville Coleman 8 author*

Family Mussidae   Genus *Acanthastrea*

Acanthastrea echinata
(Dana, 1846)

Characters: Colonies are encrusting to massive and are rarely over one metre across. Corallites are cerioid or subplocoid, circular and have thick walls. Septa have long pointed teeth. Colonies have thick fleshy tissue over the skeleton which usually forms concentric folds. **Colour:** Uniform or mottled dull brown, grey or green, but sometimes brightly coloured. **Similar species:** *Acanthastrea subechinata* and *A. brevis*. See also *A. hemprichii*. Skeletons may be confused with those of the faviids *Favites abdita* or *F. flexuosa*, but underwater they do not resemble faviids because of the fleshy polyps. **Habitat:** Most reef environments. **Abundance:** Usually uncommon, but by far the most common *Acanthastrea* on tropical reefs.

Taxonomic references: Chevalier (1975), Veron and Pichon (1980). **Identification guides:** Randall and Myers (1983), Veron (1986), Sheppard and Sheppard (1991), Nishihira and Veron (1995), Coles (1996), Carpenter *et al.* (1997).

Family Mussidae Genus *Acanthastrea*

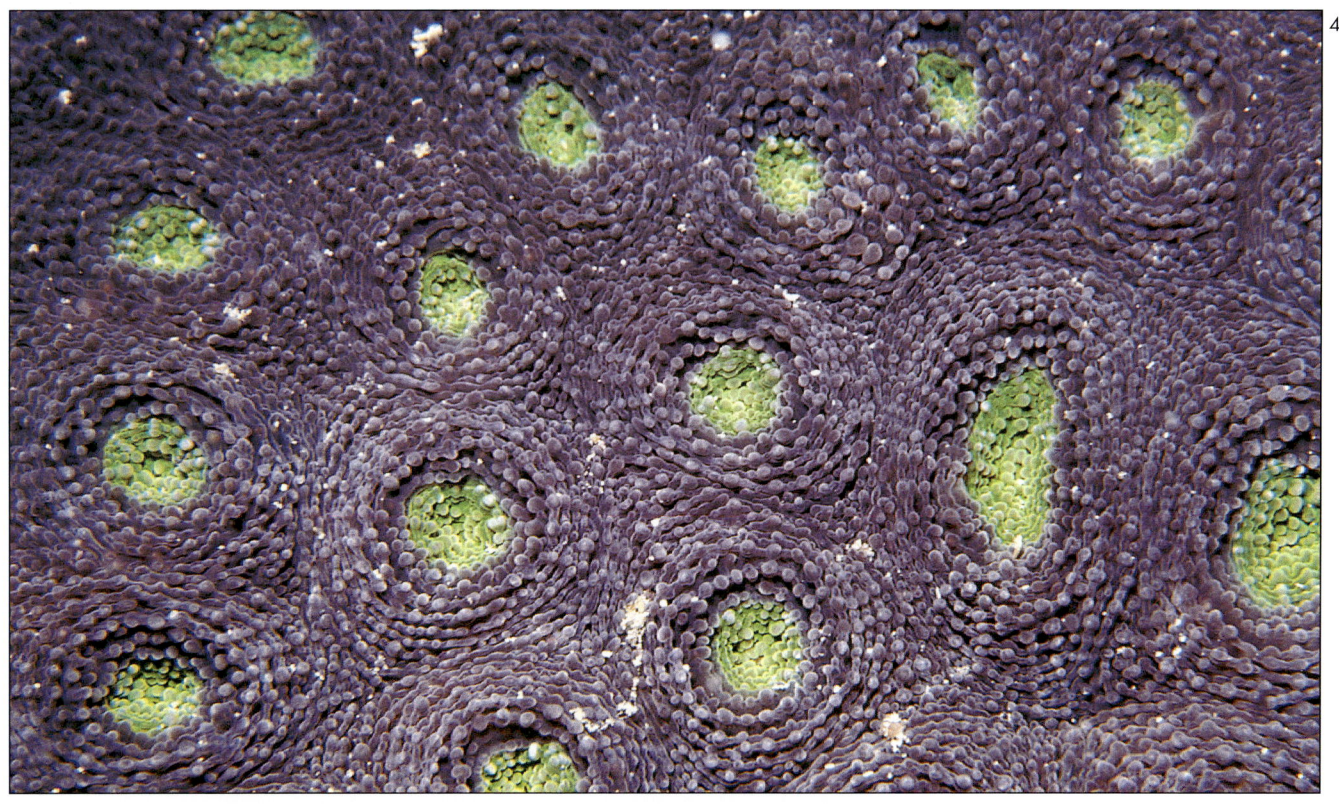

1 Colonies are usually mound-shaped or flat. PEMBA ISLAND, TANZANIA *Photograph: author*

2-6 Colonies have a very wide range of colours and surface texture. Textural differences are primarily due to the thickness of the polyp flesh rather than to underlying skeletal structure. **2** PUERTO GALERA, PHILIPPINES **3** ASHMORE REEF, WESTERN AUSTRALIA **4** GREAT BARRIER REEF, AUSTRALIA **5** SOLITARY ISLANDS, SOUTH-EAST AUSTRALIA **6** GUAM *Photographs: 2, 3 author 4 Valerie Taylor 5 Neville Coleman 6 Gustav Paulay*

Acanthastrea rotundoflora
Chevalier, 1975

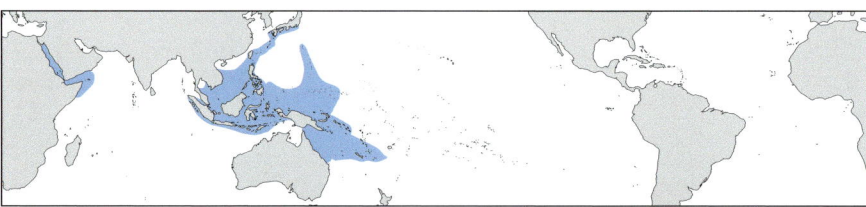

Characters: Colonies are encrusting, sometimes submassive. Small colonies have a conspicuous central corallite. Corallites are plocoid, becoming widely spaced and *Echinophyllia*-like (Pectiniidae) at colony edges. Septa have long pointed teeth. Colonies have fleshy tissue over the skeleton. **Colour:** Dark brown, rust-red or green. **Similar species:** *Acanthastrea echinata*, which has similar but smaller corallites which are less widely spaced, without an *Echinophyllia*-like appearance towards the colony periphery. **Habitat:** Protected reef environments. **Abundance:** Usually uncommon.

Taxonomic reference: Chevalier (1975). **Identification guide:** Nishihira and Veron (1995).

Family Mussidae   Genus *Acanthastrea*

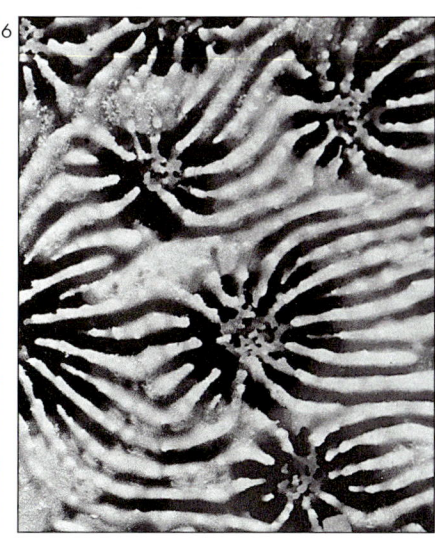

1 A large encrusting colony. CALAMIAN ISLANDS, PHILIPPINES *Photograph: author*

2, 3 Colony surface. **2** BOLINAO, PHILIPPINES **3** GUAM *Photographs: 2 author 3 Gustav Paulay*

4 Thick tissue covering the coenosteum and corallite wall. CALAMIAN ISLANDS, PHILIPPINES *Photograph: author*

5 The *Echinophyllia*-like edge of a colony. RYUKYU ISLANDS, JAPAN *Photograph: author*

6 Showing the central corallite, widely spaced peripheral corallites and toothed costae. VIETNAM *Photograph: author*

Family Mussidae | Genus *Acanthastrea*

Acanthastrea hemprichii
(Ehrenberg, 1834)

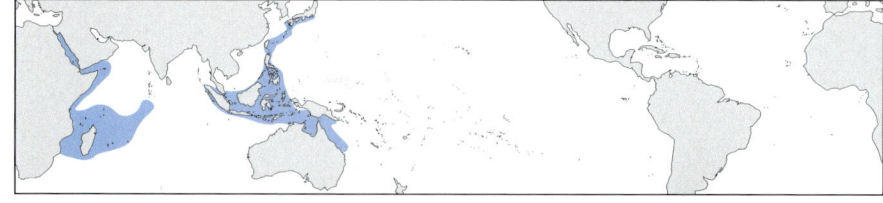

Characters: Colonies are encrusting to massive and frequently over one metre across. Corallites are cerioid. Septa have exsert teeth. Colonies have fleshy tissue over the skeleton, but this is not thick enough to mask underlying skeletal structures. **Colour:** Mottled browns and greens, commonly with brown walls and green oral discs. **Similar species:** *Acanthastrea echinata*, which has more widely spaced, fleshy, less cerioid corallites. See also *A. bowerbanki*, which usually has a central corallite and more angular peripheral corallites and *A. hillae*, which has larger and more fleshy corallites. **Habitat:** Most reef environments. **Abundance:** Uncommon.

Taxonomic reference: Ehrenberg's original description/specimens. **Identification guide:** Nishihira and Veron (1995).

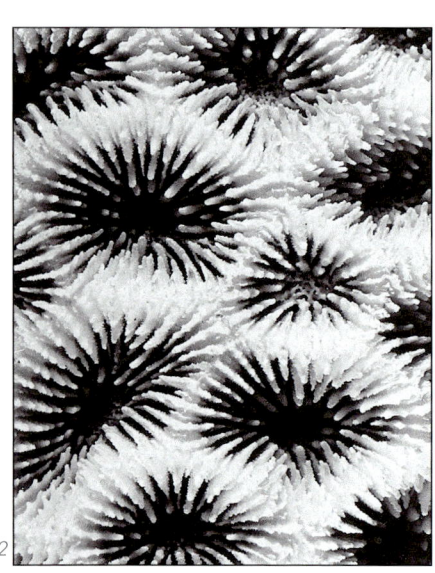

1 A submassive colony. FLORES, INDONESIA
Photograph: author

2-5 Variation in corallite structure and colour. **2** BOLINAO, PHILIPPINES **3** GREAT BARRIER REEF, AUSTRALIA **4** PEMBA ISLAND, TANZANIA **5** NORTHERN TERRITORY, AUSTRALIA *Photographs: 2, 4 author 3, 5 Neville Coleman*

Family Mussidae | Genus *Acanthastrea*

Acanthastrea faviaformis
Veron, this publication

Characters: Colonies are encrusting to massive, less than 0.2 metres across. Most consist of a small number of encrusting corallites. Corallites are plocoid. Septo-costae are prominent and have thick teeth which have ornamented edges. Columellae are deep seated. Colonies lack the fleshy tissue of most *Acanthastrea*. **Colour:** Mottled browns. **Similar species:** *Acanthastrea faviaformis* resembles the faviid *Favia* more than any other *Acanthastrea* and has been called *Favia matthaii* in other publications. See also *A. regularis*. **Habitat:** Shallow reef environments. **Abundance:** Uncommon.

Taxonomic note: See 'New species described in *Corals of the World*' (Veron, in preparation) for further information.

1 A small hemispherical colony. SINAI PENINSULA, EGYPT *Photograph: author*

2, 4 Colonies are commonly encrusting, consisting of small groups of corallites. SINAI PENINSULA, EGYPT *Photographs: author*

3 Corallite detail. SINAI PENINSULA, EGYPT *Photograph: Mary Stafford-Smith*

5 A large colony with corallite characters well developed. CALAMIAN ISLANDS, PHILIPPINES *Photograph: author*

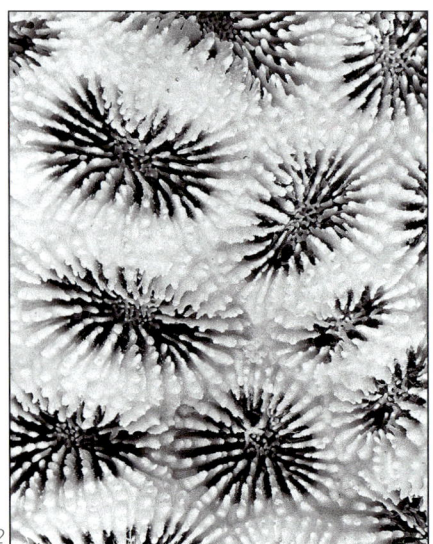

Family Mussidae   Genus *Acanthastrea*

Acanthastrea bowerbanki
Milne Edwards and Haime, 1851

Characters: Colonies are encrusting and usually small. Corallites are cerioid, with irregular angular shapes. A central corallite is usually conspicuous. Septa are compact and columellae are small. Colonies are not fleshy. **Colour:** Usually pale grey, brown or rust coloured, often mottled. **Similar species:** Superficially resembles small colonies of the faviid *Moseleya latistellata*. If a central corallite is inconspicuous it resembles the faviid *Favites*. See also *Acanthastrea hemprichii*. **Habitat:** Lower reef slopes protected from wave action. **Abundance:** Rare except in subtropical localities of eastern Australia.

Taxonomic reference: Veron and Pichon (1980). **Identification guide:** Veron (1986).

1 *Acanthastrea bowerbanki*. Edge of a small colony. LORD HOWE ISLAND, SOUTH-EAST AUSTRALIA *Photograph: Len Zell*

2 *Acanthastrea bowerbanki*. A small colony with a clearly defined central corallite. LORD HOWE ISLAND, SOUTH-EAST AUSTRALIA *Photograph: Neville Coleman*

3 *Acanthastrea bowerbanki*. Detail of corallites. NORFOLK ISLAND, WESTERN PACIFIC *Photograph: Neville Coleman*

Family Mussidae Genus *Acanthastrea*

Group 2: Species with large corallites (over 15 mm diameter)

Acanthastrea maxima
Sheppard and Salm, 1988

Characters: Colonies are cerioid with corallites up to 50 millimetres diameter. Septa are coarsely toothed. Polyps have a fleshy mantle up to 100 millimetres diameter and with a *Lobophyllia*–like texture. **Colour:** Green, grey or brown. **Similar species:** *Symphyllia wilsoni*. The fleshy mantles are more extensive than those of any other *Acanthastrea*. *Acanthastrea ishigakiensis* has slightly smaller polyps. **Habitat:** Known only from moderately deep, turbid water. **Abundance:** Rare.

Taxonomic reference: Sheppard and Salm (1988). **Identification guides:** Sheppard and Sheppard (1991), Coles (1996), Carpenter *et al.* (1997).

4 *Acanthastrea maxima*. Polyp detail. MUSCAT, OMAN *Photograph: Richard Keech*

5 *Acanthastrea maxima*. Polyps of a small colony. MUSCAT, OMAN *Photograph: Richard Keech*

Acanthastrea hillae
Wells, 1955

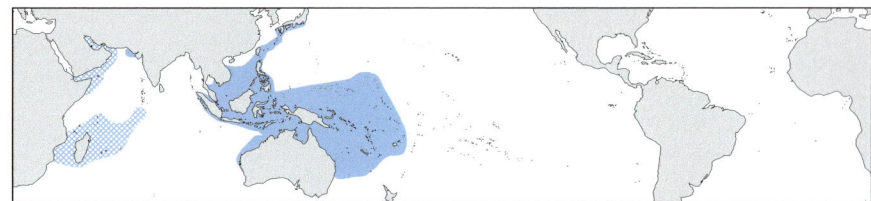

Characters: Colonies are cerioid and usually small but sometimes over 1.5 metres across. Corallites have irregular shapes and sometimes form short valleys with several centres. Colonies have moderately fleshy tissue over the skeleton. **Colour:** Red, cream and brown, with walls and oral discs of contrasting colours; sometimes mottled. **Similar species:** *Acanthastrea ishigakiensis* and *A. maxima*, both of which have relatively fleshy polyps. **Habitat:** Shallow reef environments. **Abundance:** Common only in high latitude locations.

Distribution note: Records from the western Indian Ocean are doubtful. **Taxonomic references:** Chevalier (1975), Veron and Pichon (1980). **Identification guides:** Veron (1986), Nishihira and Veron (1995).

Family Mussidae — Genus *Acanthastrea*

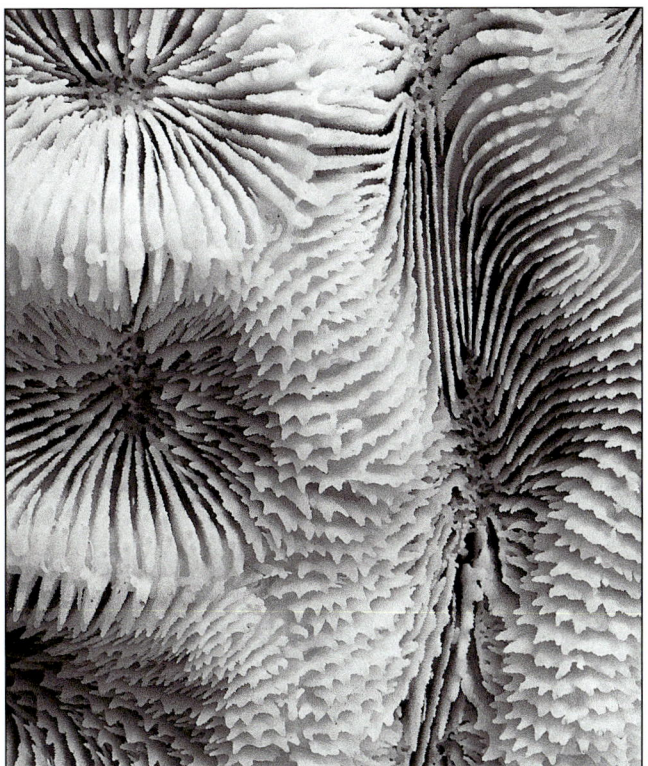

1 A large massive colony. NORFOLK ISLAND, WESTERN PACIFIC *Photograph: author*

2, 3, 5 Variation in corallite structure and colour. NORFOLK ISLAND, WESTERN PACIFIC *Photographs: author*

4 Colony surface. GREAT BARRIER REEF, AUSTRALIA *Photograph: author*

Acanthastrea ishigakiensis
Veron, 1990

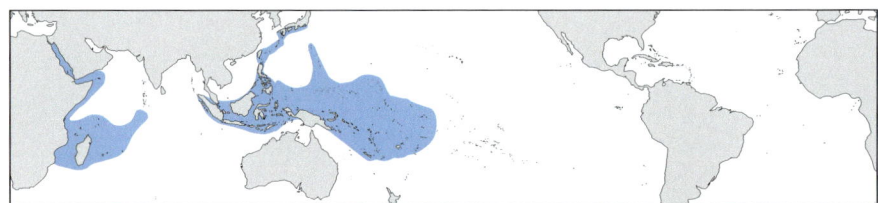

Characters: Colonies are massive, usually hemispherical and often over 0.5 metres across. Corallites are up to 25 millimetres diameter and cerioid, becoming plocoid on the colony sides. Septa are mostly uniform, with large teeth. Colonies have thick fleshy tissue over the skeleton. **Colour:** Uniform blue-grey or mixtures of grey, brown, cream and green, usually with mouth, oral disc and walls of contrasting colours. **Similar species:** *Acanthastrea hillae*, which has smaller corallites with a tendency to form valleys. Resembles *Symphyllia erythraea* underwater. **Habitat:** Shallow, partly protected reef environments. **Abundance:** Uncommon but conspicuous.

Taxonomic reference: Veron (1990a). **Identification guide:** Nishihira and Veron (1995).

1 *Acanthastrea ishigakiensis* (left) adjacent to *Lobophyllia hemprichii* (right). RYUKYU ISLANDS, JAPAN *Photograph: author*

2, 7 The common appearance of whole colonies. **2** RYUKYU ISLANDS, JAPAN **7** MADAGASCAR *Photographs: author*

3-6 Variation in corallite structure and colour. **3** SINAI PENINSULA, EGYPT **4** CALAMIAN ISLANDS, PHILIPPINES **5** RYUKYU ISLANDS, JAPAN **6** PEMBA ISLAND, TANZANIA *Photographs: author*

8 Edge of a colony. MADAGASCAR *Photograph: author*

GENUS MUSSISMILIA
Ortmann, 1890

Characters: Colonies are large and conspicuous, either cerioid or phaceloid. Cerioid colonies are massive and usually flattened. Skeletal detail of all species of the genus is similar, with septa of living colonies having uniform bead-like dentations. Columellae are sponge-like or consist of twisted septal teeth. Colonies have thick fleshy tissue over the skeleton. Tentacles are extended only at night. **Similar genus:** Except for *Mussismilia harttii*, which is phaceloid, this genus is only separable from Indo-Pacific *Acanthastrea* by having bead-like, rather than pointed, septal teeth. **Earliest fossil record:** Miocene of the Tethys.

Taxonomic reference: Laborel (1969).

Mussismilia hispida
(Verrill, 1902)

Characters: Colonies are massive, less than 0.5 metres across and usually flattened. Corallites are rounded, with thick walls, and are 10-15 millimetres diameter. Septa of living colonies have rounded, bead-like dentations. **Colour:** Browns and greys, usually with differently coloured corallite walls and centres, commonly with radial stripes. **Similar species:** *Mussismilia braziliensis*. **Habitat:** Shallow water, tolerant of turbid environments. **Abundance:** Common.

Taxonomic reference: Laborel (1969). **Identification guide:** Hetzel and Castro (1994).

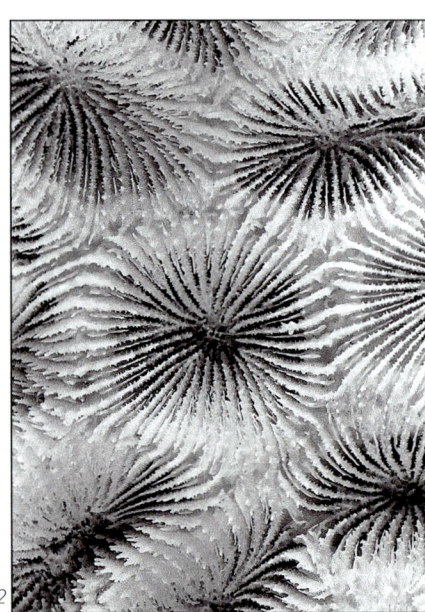

x2

Family Mussidae   Genus *Mussismilia*

1-5 Variation in corallite structure and colour. ABROLHOS ISLANDS, BRAZIL *Photographs: author*

Mussismilia braziliensis
(Verrill, 1867)

Characters: Colonies are massive, usually forming large domes on reef tops. Corallites are cerioid, tightly compacted, irregular in shape and 8-10 millimetres diameter. Septa of living colonies have rounded, bead-like dentations. **Colour:** Various combinations of blue-grey, green and yellow. **Similar species:** *Mussismilia hispida*, which has larger, less irregularly shaped corallites; otherwise these species are similar. **Habitat:** Shallow or subtidal reef environments. **Abundance:** Common.

Taxonomic reference: Laborel (1969). **Identification guide:** Hetzel and Castro (1994).

1 *Mussismilia braziliensis*. Surface detail of a colony. ABROLHOS ISLANDS, BRAZIL *Photograph: author*

2 *Mussismilia braziliensis*. The common appearance of a massive colony. ABROLHOS ISLANDS, BRAZIL *Photograph: author*

3 *Mussismilia braziliensis*. Corallite detail. ABROLHOS ISLANDS, BRAZIL *Photograph: author*

4 *Mussismilia harttii*. Showing the long tubular corallites of a damaged colony. TAMANDARE, BRAZIL *Photograph: author*

5 *Mussismilia harttii*. This species forms very distinctive phaceloid colonies. ABROLHOS ISLANDS, BRAZIL *Photograph: author*

6 *Mussismilia harttii*. Corallite detail. ABROLHOS ISLANDS, BRAZIL *Photograph: author*

Mussismilia harttii
(Verrill, 1868)

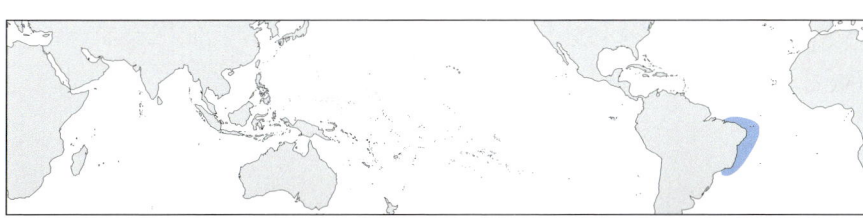

Characters: Colonies have tubular (phaceloid) corallites and are often over one metre across. Individual corallites or small groups of corallites are usually not connected or have dead basal connections and are readily broken apart and scattered. Corallites are rounded, averaging 15-20 millimetres diameter and vary in shape as illustrated. Septa of living colonies have rounded, bead-like, dentations. **Colour:** Mostly bluish-grey. **Similar species:** None. **Habitat:** Shallow turbid reef environments. **Abundance:** Common.

Taxonomic reference: Laborel (1969). **Identification guide:** Hetzel and Castro (1994).

×2

Family Mussidae   Genus *Isophyllia*

GENUS *ISOPHYLLIA*
Milne Edwards and Haime, 1851

Characters: Colonies are small (usually less than 200 mm across) and flat or dome-shaped. Valleys have thin sharp walls. Septo-costae are fine. Columellae are poorly developed. Tentacles are extended only at night. **Similar genera:** *Symphyllia* and *Acanthastrea*. **Earliest fossil record:** Oligocene of the Tethys, Miocene of the Caribbean.

Taxonomic note: Previous publications have placed *Isophyllia rigida* in a separate genus, *Isophyllastrea* Matthai, 1928. There is no adequate basis for doing so as the two species concerned differ only in minor detail. **Taxonomic reference:** Matthai (1928).

Isophyllia sinuosa
(Ellis and Solander, 1786)

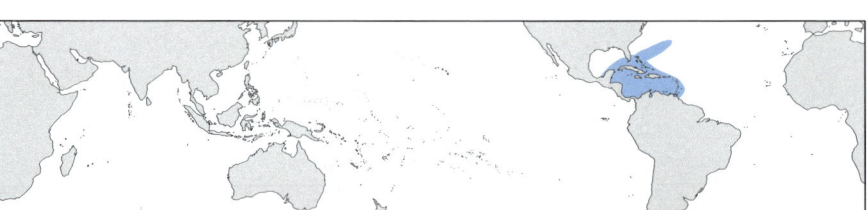

Characters: Colonies are massive and meandroid, with short, sinuous valleys. Septa are thin, with pointed fine teeth. Columella centres are hardly distinguishable. **Colour:** Usually light green, lavender or yellow, with valleys and walls of contrasting colours. **Similar species:** *Isophyllia rigida*, which is monocentric or has short valleys and more distinctive columellae. **Habitat:** Shallow protected reef environments. **Abundance:** Uncommon.

Taxonomic references: Matthai (1928), Zlatarski and Estalella (1982), Cairns (1982), Fenner (1993). **Identification guides:** Colin (1978), Wood (1983), Humann (1993).

Family Mussidae   Genus *Isophyllia*

Isophyllia rigida
(Dana, 1848)

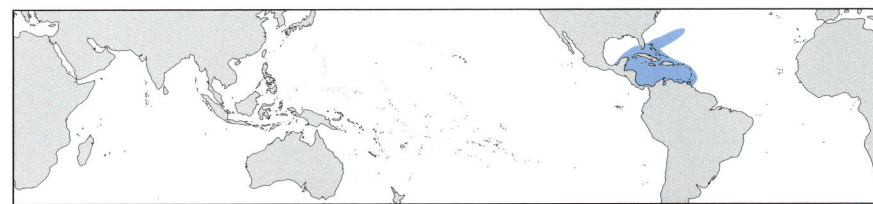

Characters: Colonies are massive and cerioid or have several centres in a short series. Columellae are rudimentary. Septa are thin, with pointed fine teeth. **Colour:** Usually mottled grey-green with valleys and walls of contrasting colours. **Similar species:** *Isophyllia sinuosa*. **Habitat:** Most reef environments, but especially surge channels. **Abundance:** Usually uncommon.

Taxonomic note: This species was previously placed in the genus *Isophyllastrea*. **Taxonomic references:** Matthai (1928), Cairns (1982), Fenner (1993) (all as *Isophyllastrea rigida*). **Identification guides:** Colin (1978), Humann (1993) (both as *Isophyllastrea rigida*).

x2

1, 2 *Isophyllia sinuosa*. This species forms small hemispherical or encrusting colonies with short meandering valleys. COZUMEL, MEXICO *Photographs: Doug Fenner*

3 *Isophyllia rigida*. This species forms small hemispherical colonies which have little variation in growth-form or colour. BELIZE *Photograph: author*

4 *Isophyllia rigida*. Surface detail. BELIZE *Photograph: author*

GENUS *LOBOPHYLLIA*
Blainville, 1830

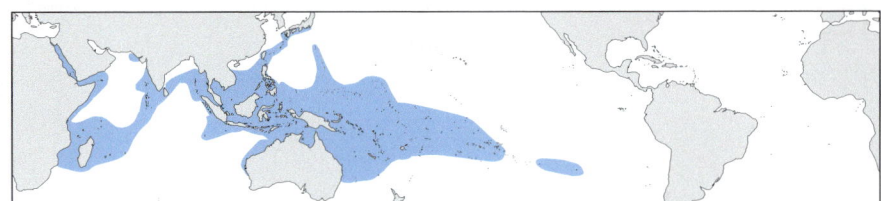

Characters: Colonies are phaceloid to flabello-meandroid, either flat-topped or dome-shaped. Corallites and/or valleys are large. Septa are large with long teeth. Columella centres are broad and compact. Tentacles are extended only at night; they usually have white tips. **Similar genera:** *Symphyllia* in the Indo-Pacific and *Mussa* in the Atlantic have coarse skeletal structures comparable to *Lobophyllia*. Both *L. hataii* (which is partly meandroid) and *L. flabelliformis* (which has a *Symphyllia*-like mantle) can be confused with *Symphyllia* underwater. **Earliest fossil record:** Miocene of the Tethys.

Taxonomic references: Chevalier (1975), Veron and Pichon (1980). **Summary key:** See pp455-6.

Lobophyllia diminuta
Veron, 1985

Characters: Colonies are small, flat or dome-shaped. Corallites are phaceloid and usually monocentric, but may have up to three centres. Corallites average 16 millimetres diameter. Septal teeth are few in number but are long and conspicuous. Columellae are circular or oval and are well developed.
Colour: Mottled orange, green or white, usually with pale mouths.
Similar species: *Lobophyllia dentatus*, which has closely compacted corallites.
Habitat: Upper reef slopes and lagoons. **Abundance:** Uncommon.

Taxonomic reference: Veron and Pichon (1980). **Identification guide:** Veron (1986).

1 Polyps with partly extended tentacles. All *Lobophyllia* have long tentacles and are voracious feeders. Note the distinction between tentacles and the filamentous structures that cover the polyp wall in many species. GREAT BARRIER REEF, AUSTRALIA *Photograph: Neville Coleman*

2 *Lobophyllia* (left) may sometimes be mistaken for *Symphyllia* (right), especially when the fleshy mantle is extended enough to mask the underlying structure. ESSINGTON PENINSULA, NORTHERN AUSTRALIA *Photograph: author*

3 *Lobophyllia diminuta*. A few corallites of a large colony. ASHMORE REEF, WESTERN AUSTRALIA *Photograph: author*

4 *Lobophyllia diminuta*. Corallite detail. The large teeth on the few primary septa are clearly visible. GREAT BARRIER REEF, AUSTRALIA *Photograph: author*

Lobophyllia pachysepta
Chevalier, 1975

Characters: Colonies are flat or hemispherical, phaceloid or partly flabello-meandroid and are seldom more than 0.5 metres across. Corallites are 40–50 millimetres diameter and are circular to irregular in outline. Primary septa are thick, with three to five long lobed teeth. Columellae are large and usually diffuse. Costae are poorly developed. **Colour:** Uniformly dark green or grey with yellowish primary septa. **Similar species:** *Lobophyllia corymbosa*. **Habitat:** Protected upper reef slopes and lagoons. **Abundance:** Usually uncommon.

Taxonomic references: Chevalier (1975), Veron and Pichon (1980). **Identification guides:** Veron (1986), Nishihira and Veron (1995).

Family Mussidae | Genus *Lobophyllia*

Lobophyllia serratus
Veron, this publication

Characters: Well developed colonies are hemispherical and commonly over 2 metres across. Corallites are large (averaging 50 mm diameter), becoming flabello-meandroid, with one to three centres. Septa have tall sharp teeth. Polyps are thick and fleshy, with a rough surface. The margins of polyps are extended to form a distinct serrated rim. **Colour:** Blue-grey to mustard with white polyp margins. **Similar species:** *Lobophyllia flabelliformis*, which has corallites of similar size and fleshy appearance, but these are always flabello-meandroid and have an extensive covering of mantle papillae. **Habitat:** Protected reef environments. **Abundance:** Uncommon.

Taxonomic note: See 'New species described in *Corals of the World*' (Veron, in preparation) for further information.

1, 3 *Lobophyllia pachysepta*. This species commonly occurs as single polyps. The large septal teeth are prominent. GREAT BARRIER REEF, AUSTRALIA *Photographs: 1 Valerie Taylor 3 author*

2 *Lobophyllia pachysepta*. Corallite detail. GREAT BARRIER REEF, AUSTRALIA *Photograph: author*

4 *Lobophyllia pachysepta*. The appearance and characteristic colour of a large colony. CAROLINE ISLANDS, MICRONESIA *Photograph: Pat Colin*

5 *Lobophyllia serratus*. A large hemispherical colony with compact corallites. CALAMIAN ISLANDS, PHILIPPINES *Photograph: author*

6 *Lobophyllia serratus*. A small colony with widely spaced corallites. BALI, INDONESIA *Photograph: Roger Steene*

Lobophyllia corymbosa
(Forskål, 1775)

Characters: Colonies are flat or hemispherical and mostly phaceloid with one to three centres per branch. They seldom exceed 0.5 metres across except in the Red Sea where they commonly exceed 2 metres. Calices are deep, with well defined walls. Septa are thick near the walls and thin within the calice. Septal teeth are tall and blunt, decreasing in size towards the columella. **Colour:** Greenish brown, grey or mustard, usually with pale centres. **Similar species:** *Lobophyllia dentatus*, which is distinguished from phaceloid *L. corymbosa* by the shape of the calices and septa. *Lobophyllia pachysepta* has thickened primary septa with large lobed dentations. These species are distinct underwater. **Habitat:** Upper reef slopes. **Abundance:** Sometimes common.

Taxonomic references: Chevalier (1975), Veron and Pichon (1980). **Identification guides:** Veron (1986), Sheppard and Sheppard (1991), Nishihira and Veron (1995).

1 Small colonies have compact corallites identical in size and shape to those of large colonies. GUAM *Photograph: Jim Maragos*

2 A large spherical colony. APO ISLAND, PHILIPPINES *Photograph: author*

3 *Lobophyllia corymbosa* (right) is easily distinguished from *L. hemprichii* (left). GREAT BARRIER REEF, AUSTRALIA *Photograph: Len Zell*

4 A large irregular colony. SINAI PENINSULA, EGYPT *Photograph: author*

5 Surface detail. SINAI PENINSULA, EGYPT *Photograph: Mary Stafford-Smith*

Lobophyllia hemprichii
(Ehrenberg, 1834)

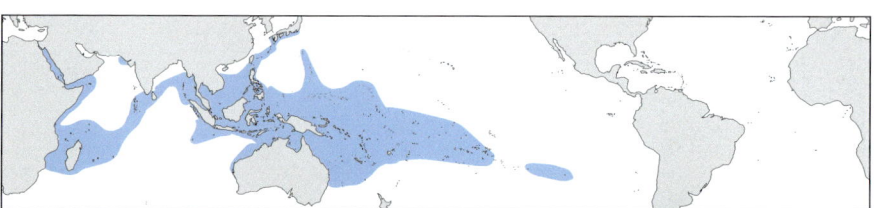

Characters: Colonies are flat to hemispherical and may be over 5 metres across. Several colonies (which may be different colours and have polyp mantles of different texture) may grow together to form a single composite stand. Colonies are phaceloid to flabello-meandroid, the latter with valleys dividing irregularly as growing space permits. Septa taper in thickness from the wall to the columella and have tall sharp teeth. Retracted polyps are thick and fleshy, with either smooth or rough mantles.
Colour: Uniform in colour or with two or more colours concentric to mouths or valley walls. All corallites of the same colony have the same colours.
Similar species: *Lobophyllia dentatus*, which is always monocentric and has more exsert septa with relatively prominent teeth. See also *L. corymbosa*, which is also monocentric and *L. robusta*, which has larger, more fleshy corallites.
Habitat: Upper reef slopes.
Abundance: Frequently a dominant species and may form extensive single species stands.

Taxonomic references: Chevalier (1975), Veron and Pichon (1980). **Identification guides:** Randall and Myers (1983), Veron (1986), Sheppard and Sheppard (1991), Nishihira and Veron (1995).

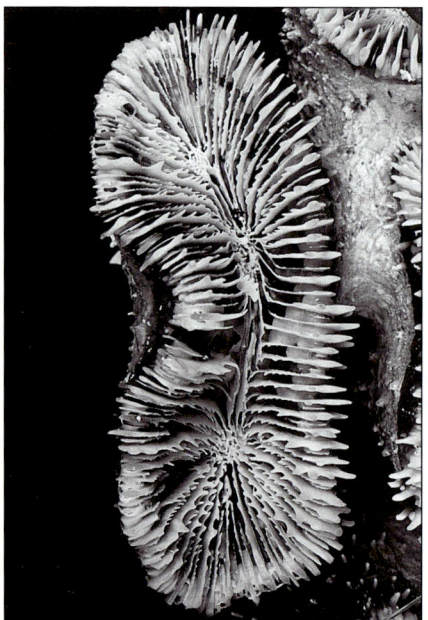

x1

Family Mussidae — Genus *Lobophyllia*

1-3, 5 Variation in the surface appearance of colonies. This variation is due to polyp shape, texture and colour. **1** PAPUA NEW GUINEA, **2** VIETNAM **3** MIDDLETON REEF, SOUTH-EAST AUSTRALIA **5** BALI, INDONESIA *Photographs: 1 Neville Coleman 2, 3 author 5 Valerie Taylor*

4 An extensive monospecific stand. SEYCHELLES *Photograph: author*

6, 7 Variation in corallite detail. **6** GREAT BARRIER REEF, AUSTRALIA **7** SUNDA ISLANDS, INDONESIA *Photographs: 6 Neville Coleman 7 Valerie Taylor*

8 A solitary polyp. These are readily mistaken for *Scolymia*. CALAMIAN ISLANDS, PHILIPPINES *Photograph: author*

‡ See also surface detail, page 2.

Lobophyllia dentatus
Veron, this publication

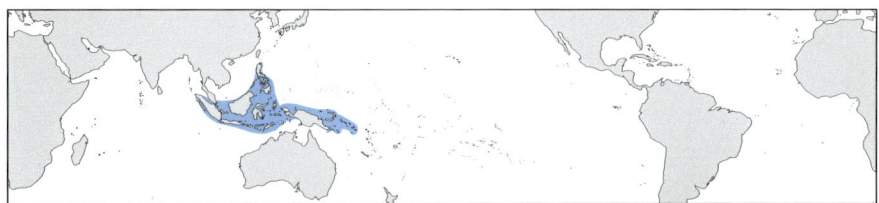

Characters: Colonies are flat to hemispherical and up to 2 metres across. They have tusk-like, elongate, closely compacted monocentric corallites. Individual corallites unite only at the base of the colony. Primary septa are very exsert, with long teeth. The exsert primary septa have a spoke-like appearance underwater. **Colour:** Uniform grey. **Similar species:** *Lobophyllia corymbosa*, which lacks the spoke-like appearance of septa and has a wide range of colours. See also *L. hemprichii*. **Habitat:** Upper reef slopes and lagoons. **Abundance:** Uncommon.

Taxonomic note: See 'New species described in *Corals of the World*' (Veron, in preparation) for further information.

Lobophyllia hataii
Yabe and Sugiyama, 1936

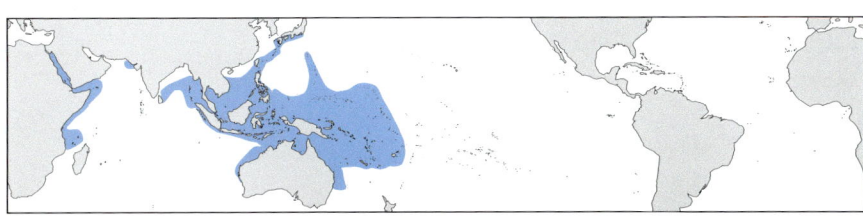

Characters: Colonies are flabello-meandroid at the periphery, and submeandroid at the centre. Valleys are shallow, with flat floors. Columellae are usually in two rows except at valley ends, or are distributed evenly on flat areas. **Colour:** Usually brown or green. Valley floors and walls are usually of contrasting colours. **Similar species:** *Symphyllia valenciennesi*. See also *Lobophyllia flabelliformis*. **Habitat:** Upper reef slopes protected from wave action. **Abundance:** Uncommon.

Taxonomic references: Chevalier (1975), Veron and Pichon (1980). **Identification guides:** Veron (1986), Sheppard and Sheppard (1991), Nishihira and Veron (1995).

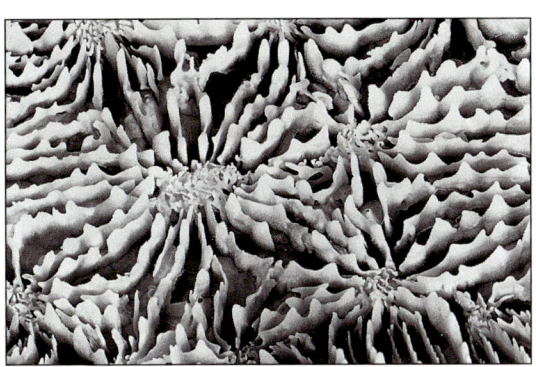

1 *Lobophyllia dentatus*. Colonies are commonly large and are usually hemispherical. PAPUA NEW GUINEA *Photograph: author*

2-4 *Lobophyllia dentatus*. Corallites have a spoke-like appearance due to long primary septa which have long teeth. PAPUA NEW GUINEA *Photographs: author*

5 *Lobophyllia hataii*. Detail of a central corallite. GREAT BARRIER REEF, AUSTRALIA *Photograph: author*

6 *Lobophyllia hataii*. The central corallite and the characteristically radiating valleys. FIJI *Photograph: Valerie Taylor*

Lobophyllia flabelliformis
Veron, this publication

Characters: Colonies are large, usually dome-shaped. They are flabello-meandroid with closely compacted elongate valleys. Despite a robust appearance, large colonies readily break apart. Polyps have a thick fleshy mantle which obscures the underlying skeletal structure and thus this species appears to be a *Symphyllia* underwater. If the mantle is touched it retracts revealing the underlying growth-form, where valleys have no walls in common. The mantle is covered with elongate papillae that may resemble tentacles. **Colour:** Uniform dark grey-brown. **Similar species:** *Lobophyllia robusta*, which does not have such a completely flabello-meandroid growth-form. **Habitat:** Most shallow reef environments. **Abundance:** Usually uncommon.

Taxonomic note: See 'New species described in *Corals of the World*' (Veron, in preparation) for further information.

Family Mussidae | Genus *Lobophyllia*

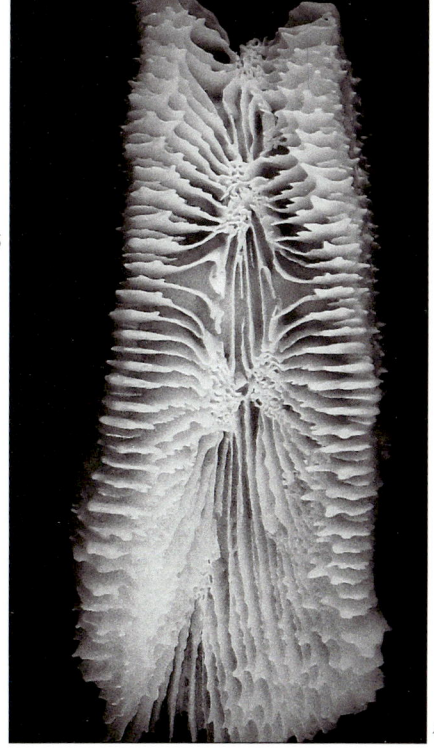

1 Usual appearance of a small colony. PAPUA NEW GUINEA *Photograph: author*

2 Large colonies are superficially *Symphyllia*-like. PAPUA NEW GUINEA *Photograph: author*

3 Surface of a colony which has long mantle papillae. These resemble tentacles. BALI, INDONESIA *Photograph: Gerry Allen*

4 Colony with fully expanded polyps. RYUKYU ISLANDS, JAPAN *Photograph: author*

5 If the colony surface is touched the mantle contracts, showing that adjacent corallites do not have common walls. PAPUA NEW

Lobophyllia robusta
Yabe and Sugiyama, 1936

Characters: Colonies usually consist of a few corallites but may become large and hemispherical. Corallites are large, phaceloid and mostly monocentric. Septa have tall sharp teeth. Polyps are thick and fleshy, with a rough surface. **Colour:** Blue-grey, sometimes with pale valley floors or septo-costae. All corallites of the same colony have the same colour. **Similar species:** *Lobophyllia hemprichii*, which has smaller, less fleshy polyps. *Lobophyllia flabelliformis* has corallites of similar size and fleshy appearance, but these are always flabello-meandroid. **Habitat:** Most reef environments. **Abundance:** Uncommon.

Taxonomic reference: Yabe and Sugiyama's original description/specimens. **Identification guide:** Nishihira and Veron (1995).

Family Mussidae Genus *Lobophyllia*

1 Adjacent colonies of *Lobophyllia robusta* (right) and *L. hemprichii* (left). These species are sometimes difficult to distinguish. RYUKYU ISLANDS, JAPAN *Photograph: author*

2 Colonies commonly consist of just a few corallites. PAPUA NEW GUINEA *Photograph: author*

3 A large solitary corallite. PAPUA NEW GUINEA *Photograph: Jim Maragos*

4 *Lobophyllia robusta* (paler colony) has larger, more fleshy corallites than *L. hemprichii* (darker colony). PAPUA NEW GUINEA *Photograph: Valerie Taylor*

GENUS SYMPHYLLIA
Milne Edwards and Haime, 1848

Characters: Colonies are meandroid and either flat-topped or dome-shaped. Valleys are wide. A groove usually runs along the top of the walls. Septa are large, with long teeth. Columellae are broad and compact. Tentacles are extended only at night. **Similar genera:** Only *Lobophyllia* in the Indo-Pacific. Resembles the Atlantic *Isophyllia*. *Symphyllia erythraea* is *Acanthastrea*-like. **Earliest fossil record:** Pliocene of the Pacific.

Taxonomic references: Chevalier (1975), Veron and Pichon (1980). **Summary key:** See pp455-6.

Symphyllia hassi
Pillai and Scheer, 1976

Characters: Colonies are hemispherical to flat and are seldom more than 0.2 metres across. Valleys are short and usually less than 25 millimetres wide. Walls are thick and in extreme cases can have an ambulacral groove which is wider than the valleys (similar to the faviid *Diploria labyrinthiformis*). Costae and septa are both thick. Septa have large blunt spines. Columellae are large and centred in valleys. **Colour:** Cream. **Similar species:** *Symphyllia agaricia*, which forms large colonies and usually has columellae in two rows. **Habitat:** Exposed upper reef slopes and reef flats. **Abundance:** Uncommon.

Taxonomic reference: Pillai and Scheer (1976).

Symphyllia wilsoni
Veron, 1985

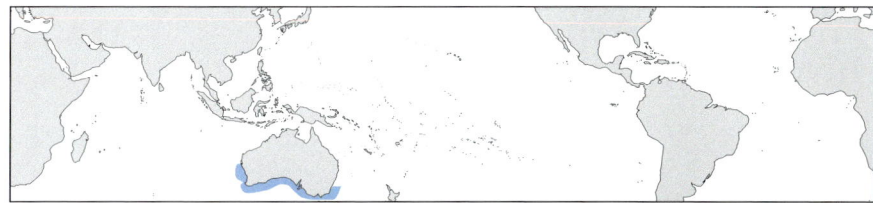

Characters: Colonies are massive or submassive and flattened. Valleys are irregular in length and shape. All skeletal structures are small for a *Symphyllia*, with septa being generally like those of *Acanthastrea*. Polyps are fleshy, with a groove along the tops of valley walls. **Colour:** A wide range of mottled, green, grey, purple and brown. Valley walls and floors have contrasting colours. **Similar species:** Colonies superficially resemble the faviid *Oulophyllia* underwater. **Habitat:** Subtidal rocky foreshores of temperate localities. **Abundance:** Usually uncommon but conspicuous.

Taxonomic reference: Veron (1985). **Identification guide:** Veron (1986).

1 *Symphyllia hassi.* Colonies are seldom larger than this. Walls are thick with prominent costae. NEGROS, PHILIPPINES
Photograph: Doug Fenner

2 *Symphyllia wilsoni.* A large colony. ROTTNEST ISLAND, SOUTH-WEST AUSTRALIA
Photograph: Barry Hutchins

3 *Symphyllia wilsoni.* Encrusting a rock in shallow water. DONGARA, SOUTH-WEST AUSTRALIA *Photograph: Clay Bryce*

4 *Symphyllia wilsoni.* The usual size of hemispherical colonies. ROTTNEST ISLAND, SOUTH-WEST AUSTRALIA *Photograph: Barry Hutchins*

Family Mussidae   Genus *Symphyllia*

Symphyllia erythraea
(Klunzinger, 1879)

Characters: Colonies are massive, becoming hemispherical. Valleys are short, usually with less than three centres. Septa are in two orders, with tall teeth. Columellae are well formed.
Colour: Mottled or uniform brown, grey, green or cream.
Similar species: *Acanthastrea hillae* also has a tendency to form valleys with multiple centres. See also *A. ishigakiensis*.
Habitat: Most reef environments. **Abundance:** Sometimes common.

Taxonomic note: The generic assignment of this distinct species is almost arbitrarily apportioned between *Symphyllia*, *Acanthastrea* and *Isophyllia*.
Taxonomic reference: Scheer and Pillai (1983, as *Acanthastrea erythraea*).
Identification guide: Sheppard and Sheppard (1991).

×1

Family Mussidae | Genus *Symphyllia*

1 The typical appearance of a colony. SINAI PENINSULA, EGYPT *Photograph: author*

2 A small colony in shallow water with short meandering valleys. SINAI PENINSULA, EGYPT *Photograph: author*

3-6 Variation in the shape and colour of valleys. **3, 5, 6** SINAI PENINSULA, EGYPT **4** SAUDI ARABIA, RED SEA *Photographs: 3, 5, 6 author 4 Emre Turak*

Family Mussidae | Genus *Symphyllia*

Symphyllia recta
(Dana, 1846)

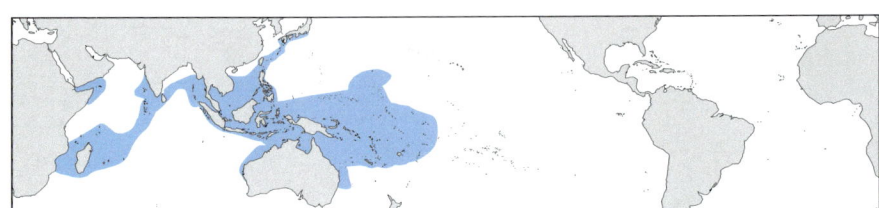

Characters: Colonies are hemispherical to flat. Valleys are 12-15 millimetres wide, and are highly sinuous. Polyp walls have a moderately thick fleshy appearance and usually have a groove along the top. **Colour:** Brown, grey or green, either mottled or with walls and valleys of contrasting colours. **Similar species:** *Symphyllia radians*, which has larger and less sinuous valleys. **Habitat:** Upper reef slopes and fringing reefs. **Abundance:** Common.

Taxonomic reference: Veron and Pichon (1980). **Identification guide:** Veron (1986).

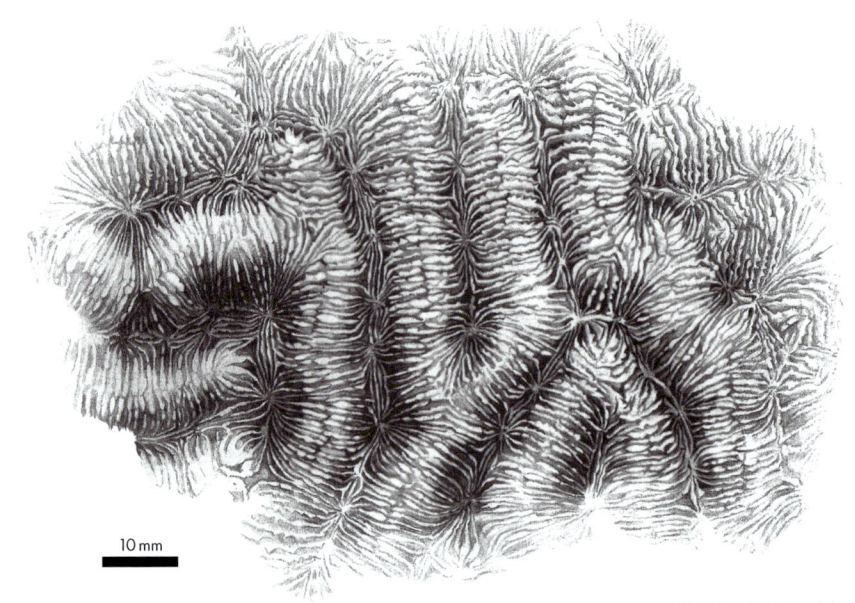

10 mm

Drawing: Doris Englehardt

1 A large hemispherical colony. PAPUA NEW GUINEA *Photograph: author*

2 Surface of a large flat colony. ASHMORE REEF, WESTERN AUSTRALIA *Photograph: author*

3-5 Variation in valley colour and structure. **3** PAPUA NEW GUINEA **4** GREAT BARRIER REEF, AUSTRALIA **5** NORTHERN TERRITORY, AUSTRALIA *Photographs: Neville Coleman*

6 Surface texture of valleys. CALAMIAN ISLANDS, PHILIPPINES *Photograph: author*

Family Mussidae — Genus *Symphyllia*

Symphyllia radians
Milne Edwards and Haime, 1849

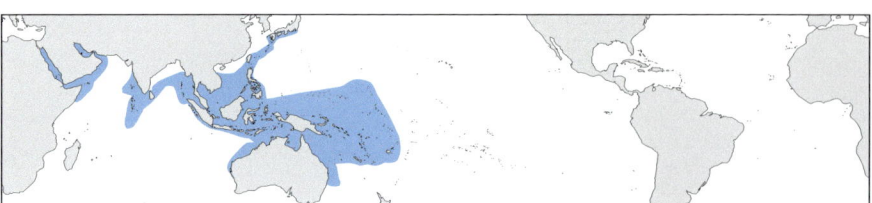

Characters: Colonies are hemispherical to flat. Valleys average 20-25 millimetres wide, and are irregularly meandroid, becoming straight in flat colonies. Walls have a moderately thick fleshy appearance and usually have a groove along the top. **Colour:** A wide range of red, grey and green, with valleys and walls usually of contrasting colours. Red Sea colonies are usually cream. **Similar species:** *Symphyllia radians* has valleys and septa intermediate in size between the smaller *S. recta* and the larger *S. agaricia*. *Symphyllia recta* has more sinuous valleys, *S. agaricia* has a double row of columellae. **Habitat:** Upper reef slopes and fringing reefs. **Abundance:** Common.

Taxonomic references: Chevalier (1975), Veron and Pichon (1980). **Identification guides:** Veron (1986), Sheppard and Sheppard (1991), Nishihira and Veron (1995), Coles (1996).

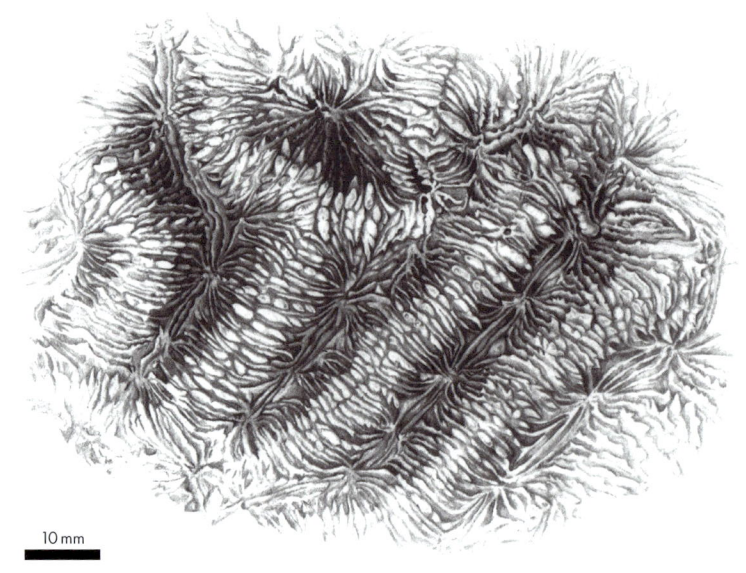

10 mm

Drawing: Doris Englehardt

1, 3, 4 This species forms hemispherical or flat colonies with relatively straight valleys. **1** FLORES SEA, INDONESIA **3** CALAMIAN ISLANDS, PHILIPPINES **4** GREAT BARRIER REEF, AUSTRALIA *Photographs: 1 Roger Steene 3 author 4 Ed Lovell*

2, 5, 6 Variation in valley structure and colour. **2, 5** GREAT BARRIER REEF, AUSTRALIA **6** PAPUA NEW GUINEA *Photographs: 2 Mary Stafford-Smith 5, 6 Neville Coleman*

Family Mussidae	Genus *Symphyllia*

Symphyllia agaricia
Milne Edwards and Haime, 1849

Characters: Colonies are hemispherical to flat. Valleys are sinuous or straight, averaging 35 millimetres wide and are usually separated by a narrow groove. Walls have a thick fleshy appearance. Septa are thick and have large teeth. Columellae are usually in two rows. **Colour:** Brown, green or red, usually with contrasting valleys and walls. **Similar species:** *Symphyllia radians*, which has smaller, straighter valleys. See also *S. hassi* and *Lobophyllia flabelliformis*. **Habitat:** Exposed upper reef slopes. **Abundance:** Uncommon.

Taxonomic reference: Veron and Pichon (1980). **Identification guides:** Veron (1986), Nishihira and Veron (1995).

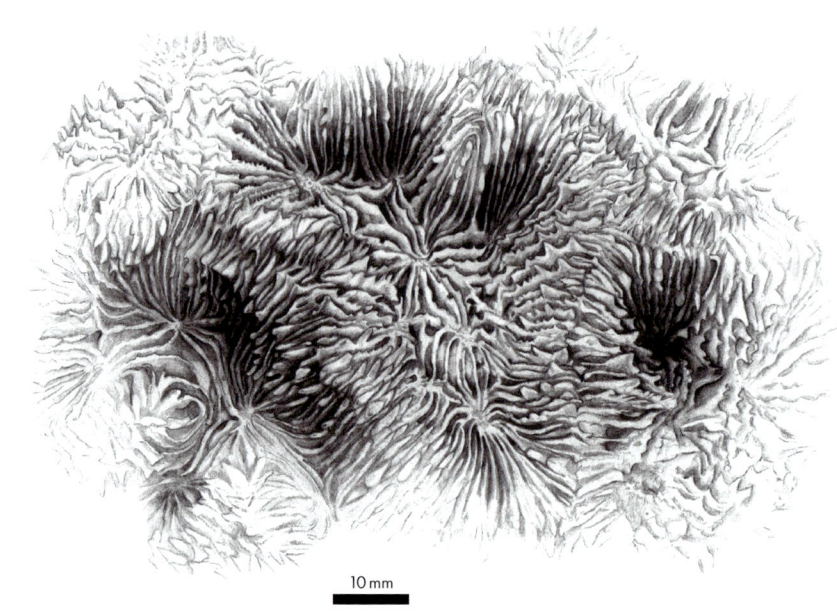

10 mm

Drawing: Doris Englehardt

1, 2 This species forms massive, hemispherical or flat colonies with large valleys. **1** PAPUA NEW GUINEA **2** GREAT BARRIER REEF, AUSTRALIA
Photographs: 1 Neville Coleman 2 Ed Lovell

3 Showing the thick tough fleshy surface of polyps. DAMPIER ARCHIPELAGO, WESTERN AUSTRALIA *Photograph: author*

4 Small colonies have the same surface characters as larger ones. GREAT BARRIER REEF, AUSTRALIA *Photograph: Roger Steene*

5 Mouths are just discernible. DAMPIER ARCHIPELAGO, WESTERN AUSTRALIA *Photograph: author*

Family Mussidae | Genus *Symphyllia*

Symphyllia valenciennesii
Milne Edwards and Haime, 1849

Characters: Colonies are usually flat. Valleys radiate from a flat central area and have steep sides and flat floors. Walls have a moderately fleshy appearance and usually have a groove along the top. Septa are thick, with large teeth. Polyps may be fleshy. **Colour:** Usually grey, brown or mottled, with valley floors and walls of contrasting colours. **Similar species:** *Symphyllia agaricia*, which may have valleys of similar size but these are not flat. See also *Lobophyllia hataii*, which is composed primarily of valleys rather than a flat central area. *Lobophyllia flabelliformis* also has large fleshy polyps. **Habitat:** Lower reef slopes protected from wave action and rocky foreshores of subtropical locations. **Abundance:** Uncommon.

Taxonomic references: Chevalier (1975), Veron and Pichon (1980). **Identification guides:** Veron (1986), Nishihira and Veron (1995).

x1

1

1-4 Colonies are flat plates with a broad central area from which walls radiate. Mouths are clearly visible only if they have a distinctive colour. **1, 4** BALI, INDONESIA **2** HALMAHERA SEA, INDONESIA **3** PAPUA NEW GUINEA *Photographs:* **1** Neville Coleman **2** Gerry Allen **3** author **4** Roger Steene

5 Small colony, showing fluted margins of valleys. BALI, INDONESIA *Photograph:* Neville Coleman

Family Mussidae Genus *Symphyllia*

GENUS *MUSSA*
Oken, 1815

Characters, abundance and distribution: This genus has only one species, see *Mussa angulosa*. **Earliest fossil record:** Miocene of the Caribbean.

Mussa angulosa
(Pallas, 1766)

Characters: Colonies are flat to hemispherical, phaceloid to flabello-meandroid, the latter with short valleys of up to 5 centres. Septa have tall sharp teeth. Columellae are well developed. Retracted polyps are thick and fleshy. **Colour:** Usually grey, purple or green. **Similar species:** *Scolymia cubensis*, which is solitary, less fleshy, and has less excavated corallites with smaller, more even teeth on the septa. **Habitat:** Most reef environments. **Abundance:** Uncommon.

Taxonomic note: Solitary polyps of this species have commonly been called *Scolymia lacera*. **Taxonomic references:** Zlatarski and Estalella (1982), Fenner (1993). **Identification guides:** Colin (1978), Humann (1993).

1 A large colony. BAHAMAS *Photograph: Nancy Sefton*

2 The usual appearance of a small colony composed of a few polyps. BAHAMAS *Photograph: author*

3 Polyps have a rough texture, like those of most *Lobophyllia* species. COZUMEL, MEXICO *Photograph: Doug Fenner*

4, 5 Solitary polyps. These are commonly identified as *Scolymia lacera*. **4** CAYMAN ISLANDS **5** HONDURAS *Photographs: Paul Humann*

6 Polyp tentacles. CAYMAN ISLANDS *Photograph: Nancy Sefton*

Family Mussidae   Genus *Mussa*

GENUS *SCOLYMIA*
Haime, 1852

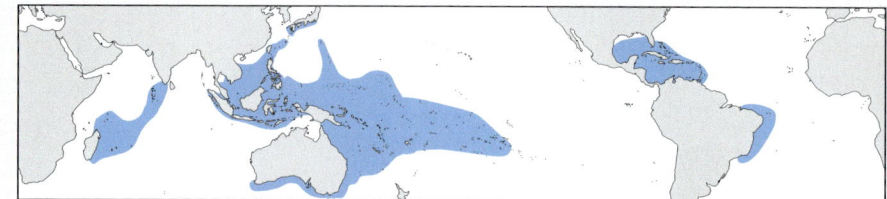

Characters: Corals are usually monocentric, rarely polycentric. Secondary centres may occur inside or outside the original calice and calices may divide. Walls are indistinct beneath the septo-costae. Septa slope evenly, with little fusion. Primary septa have large, regular, blunt teeth. Columellae are broad and compact. Tentacles are extended only at night. **Similar genera:** The skeletal structure of *Scolymia* resembles that of *Australomussa*. *Scolymia* does not have large mantles that are extended with water, like *Cynarina*. **Earliest fossil record:** Oligocene of the Tethys, Miocene of the Caribbean.

Taxonomic note: Vaughan (1901) mistakenly selected *Madrepora lacera* Pallas, 1766 for the type species of *Scolymia*. As *Scolymia* is a well established name, *Madrepora cubensis* Milne Edwards and Haime, 1849 is redesignated type species, in accordance with Haime's (1852) original intention. **Taxonomic references:** Wells (1971b), Veron and Pichon (1980), Fenner (1993).

Scolymia cubensis
(Milne Edwards and Haime, 1849)

Characters: Usually attached but may be free-living, with a tapered base. Polyps have one, rarely two or more centres and are sometimes over 100 millimetres across. They may be only a few millimetres thick, with successive regrowths forming tiers. Sometimes corallites have several mouths. Septa are in two or three orders of slightly different size. A paliform crown may be distinguishable. Polyps are fleshy. **Colour:** Uniform or variegated brown, red, tan or green. Sometimes bright orange (which may photograph brown). **Similar species:** Solitary polyps of *Mussa angulosa*. **Habitat:** Lower reef slopes and soft substrates. **Abundance:** Uncommon.

Taxonomic note: It is commonly believed that there are three, not one, species of *Scolymia* in the Atlantic. The names *Scolymia lacera* (Pallas, 1766), *S. cubensis* (Milne Edwards & Haime, 1849) and *S. wellsi* (Laborel, 1967) have been variously used. In most publications (best depicted by Zlatarski and Estalella, 1982), *S. lacera* is actually *Mussa angulosa* with solitary polyps. The name *S. wellsi* is usually given to small polyps of the present species and to those from Brazil. **Taxonomic reference:** Fenner (1993).

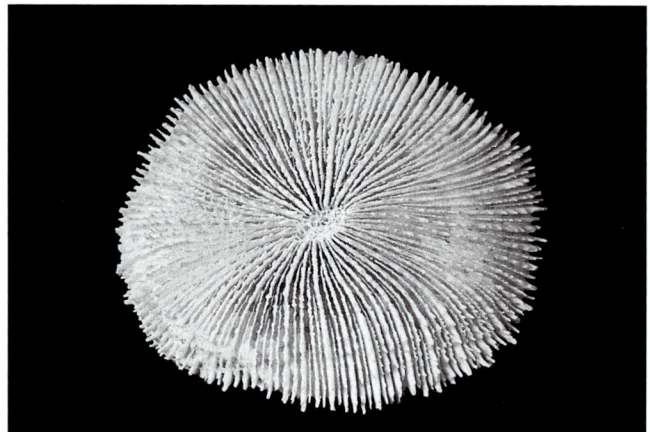

Family Mussidae | Genus *Scolymia*

1 *Scolymia vitiensis* (right) next to *Lobophyllia robusta* (left). SOLOMON ISLANDS *Photograph: Jim Maragos*

2-7 *Scolymia cubensis*. Showing some of the species' wide colour variations. **2, 5** CAYMAN ISLANDS **3, 6** BAHAMAS **4** FLORIDA **7** ABROLHOS ISLANDS, BRAZIL *Photographs: 2-4, 6 Paul Humann 5 Nancy Sefton 7 author*

Scolymia vitiensis
Brüggemann, 1877

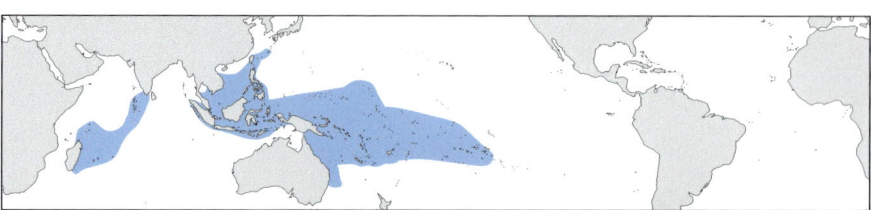

Characters: There is wide latitudinal variation in this species. In subtropical localities it is usually solitary, flat and less than 60 millimetres diameter. In the tropics it is larger and sometimes colonial. Septa slope up from the columella to an indistinct wall then costae slope down to the periphery. This gives the fleshy mantle of the polyps a distinctive concentric texture. Secondary centres occur near the colony centre and also around the periphery. Septo-costae are sturdy, with large blunt teeth. **Colour:** Usually dark green or tan. **Similar species:** *Scolymia australis*, which is much smaller but juveniles may be indistinguishable. **Habitat:** Most reef environments. **Abundance:** Usually uncommon, rare in the south-west Indian Ocean.

Taxonomic reference: Chevalier (1975, as *Parascolymia vitiensis)*, Veron and Pichon (1980). **Identification guides:** Veron (1986), Nishihira and Veron (1995).

Family Mussidae | Genus *Scolymia*

1 Large specimens may have one or several mouths. PAPUA NEW GUINEA *Photograph: Gerry Allen*

2 Groups of corallites are uncommon. These may resemble *Lobophyllia*. BALI, INDONESIA *Photograph: Valerie Taylor*

3-5 Common colour variations. **3** GREAT BARRIER REEF, AUSTRALIA **4** PAPUA NEW GUINEA **5** VANUATU *Photographs: 3 Valerie Taylor 4 Neville Coleman 5 author*

6 With peripheral corallites. These colonies resemble *Australomussa rowleyensis*, but they do not get much larger than this. PAPUA NEW GUINEA *Photograph: Jim Maragos*

7 Surface detail. PAPUA NEW GUINEA *Photograph: author*

Family Mussidae | Genus *Scolymia*

Scolymia australis
(Milne Edwards and Haime, 1849)

Characters: Usually solitary but sometimes two to four centres occur in one corallite, or occasionally in separate corallites. Corallites are saucer-shaped and less than 60 millimetres diameter. Septa are sturdy with blunt saw-like teeth. **Colour:** Colourful, usually mixtures of cream, red, blue and green. **Similar species:** *Scolymia cubensis*. *Scolymia vitiensis* is mostly larger and not cup-shaped. **Habitat:** Reef environments or on rocky headlands in high latitudes. **Abundance:** Relatively common in subtropical localities, uncommon elsewhere.

Distribution note: Records from the central-western Pacific are doubtful. **Taxonomic reference:** Veron and Pichon (1980). **Identification guide:** Veron (1986).

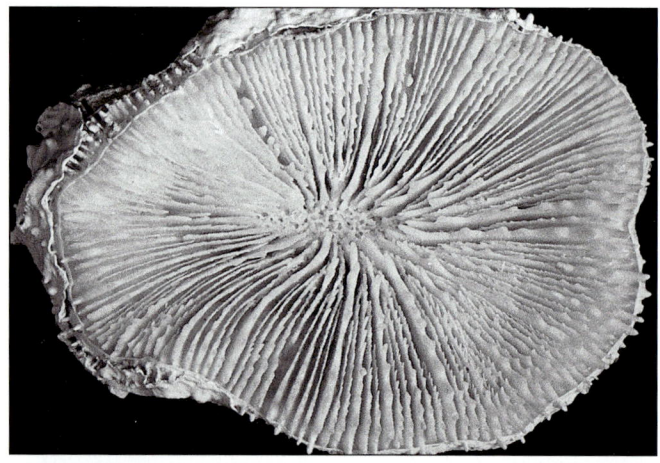
x1

1

2

3

Family Mussidae   Genus *Scolymia*

1, 2, 5, 6 Variation in the structure and colour of solitary polyps. **1** Lord Howe Island, south-east Australia **2** Great Barrier Reef, Australia **5** Norfolk Island, western Pacific **6** Papua New Guinea *Photographs: 1 Neville Coleman 2 Valerie Taylor 5, 6 author*

3 This species sometimes occurs as two or more polyps. Norfolk Island, western Pacific *Photograph: Neville Coleman*

4 Colour variation between two adjacent polyps. Ambon, Indonesia *Photograph: Valerie Taylor*

7 With tentacles extended. Great Barrier Reef, Australia *Photograph: Neville Coleman*

GENUS MYCETOPHYLLIA

Milne Edwards and Haime, 1848

Characters: Colonies are flat plates with radiating valleys. Septo-costae are exsert, columellae are poorly developed. Tentacles are usually small or apparently absent except at the margins of colonies. **Similar genera:** *Symphyllia* in the Indo-Pacific is similar to the larger *Mycetophyllia*. **Earliest fossil record:** Oligocene of the Tethys, Miocene of the Caribbean.

Taxonomic reference: Wells (1973b).

Mycetophyllia lamarckiana
Milne Edwards and Haime, 1848

Characters: Colonies are solid, rounded, often circular plates. Continuous broad shallow valleys radiate from the original point of growth. Corallite centres are vaguely concentric to plate margins. Valleys have one row of mouths. Columellae are rudimentary. **Colour:** Mostly mottled grey or brown. **Similar species:** *Mycetophyllia danaana*, which has well formed valleys. See also *M. aliciae*. **Habitat:** Most reef environments. **Abundance:** Uncommon.

Taxonomic references: Wells (1973b), Zlatarski and Estalella (1982), Cairns (1982). **Identification guides:** Colin (1978), Humann (1993).

x2

1 *Mycetophyllia danaana* with tentacles extended at night. PUERTO RICO *Photograph: Pat Colin*

2 Two common corals of the Caribbean, *Mycetophyllia danaana* (left) and *Colpophyllia natans* (right). JAMAICA *Photograph: author*

3 *Mycetophyllia lamarckiana*. Showing the usual shape of valleys. PUERTO RICO *Photograph: Pat Colin*

Mycetophyllia ferox
Wells, 1973

Characters: Colonies are thin, weakly attached plates with interconnecting, slightly sinuous narrow valleys. Corallite centres are usually in single rows. Columellae are rudimentary or absent. **Colour:** Greys and browns are the most common with valleys and walls of contrasting colours. **Similar species:** *Mycetophyllia danaana*, which has longer, wider and more widely spaced valleys. See also *M. aliciae*. **Habitat:** Shallow reef environments. **Abundance:** Usually uncommon.

Taxonomic references: Wells (1973b), Cairns (1982). **Identification guide:** Humann (1993).

x2

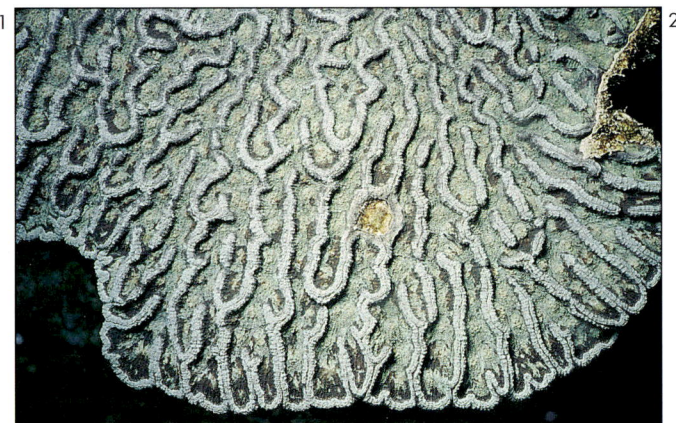

1 *Mycetophyllia ferox*. Forming a thin plate. BELIZE *Photograph: author*

2 *Mycetophyllia ferox*. Showing the characteristically irregular, narrow valleys of a thin plate. BELIZE *Photograph: author*

3 *Mycetophyllia ferox*. Corallite detail at the centre of a large plate. PUERTO RICO *Photograph: Pat Colin*

4 *Mycetophyllia ferox*. Corallite detail showing weak valley formation near the margin of a plate. BELIZE *Photograph: Mary Stafford-Smith*

5 *Mycetophyllia reesi*. A side-attached sheet overgrowing the substrate. CAYMAN ISLANDS *Photograph: Paul Humann*

6 *Mycetophyllia reesi*. A centrally attached plate. BELIZE *Photograph: Paul Humann*

7 *Mycetophyllia reesi*. Variation in structure and colour of adjacent colonies. BELIZE *Photograph: Paul Humann*

Family Mussidae — Genus *Mycetophyllia*

Mycetophyllia reesi
Wells, 1973

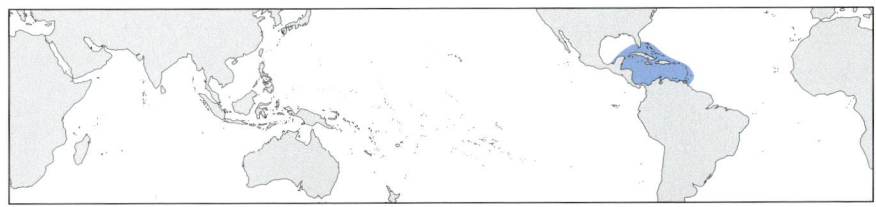

Characters: Colonies are thin laminae, sometimes conforming to the shape of the substrate. They are attached centrally or at a side. There are no radiating ridges or valleys. Corallite centres are parallel to plate margins. **Colour:** Dark grey-brown or bluish-brown, sometimes mottled. **Similar species:** *Mycetophyllia lamarckiana*, which has some formation of radiating ridges. **Habitat:** Lower reef slopes protected from wave action. **Abundance:** Uncommon to rare.

Taxonomic reference: Wells (1973b). **Identification guides:** Colin (1978), Humann (1993).

x2

Mycetophyllia danaana
Milne Edwards and Haime, 1849

Characters: Colonies are solid rounded plates. Long sinuous valleys may radiate from the original point of growth and are sometimes deep. Valley walls may be so short that they form *Hydnophora*-like monticules. Septo-costae are thinner than intervening spaces. Columellae are rudimentary or absent. There is one row of mouths in the valleys. **Colour:** Usually combinations of pink, green and grey, with valleys and walls of contrasting colours. **Similar species:** *Mycetophyllia lamarckiana*. **Habitat:** Most reef environments. **Abundance:** Common.

Taxonomic references: Wells (1973b), Cairns (1982). **Identification guides:** Colin (1978), Humann (1993).

1 Showing deep valleys meandering from the original point of growth. JAMAICA *Photograph: Mary Stafford-Smith*

2-4 Variation in structure and colour. With some colonies, valley walls become *Hydnophora*-like. JAMAICA *Photographs: author*

5 Surface detail. BELIZE *Photograph: author*

Family Mussidae | Genus *Mycetophyllia*

Mycetophyllia aliciae
Wells, 1973

Characters: Colonies are weakly attached plates which are often circular. There is little valley formation, especially towards the colony centre and such valleys that are formed may have two or more rows of mouths. Septo-costae are thick, with large dentations. Columellae are absent. **Colour:** Brown or green, usually with white centres. **Similar species:** *Mycetophyllia lamarckiana*, which has more distinct valleys, thinner septo-costae, one row of mouths in valleys, and some columella formation. See also *M. ferox*. **Habitat:** Most reef environments. **Abundance:** Sometimes common.

Taxonomic references: Wells (1973b), Cairns (1982).
Identification guides: Colin (1978), Humann (1993).

Family Mussidae | Genus *Mycetophyllia*

1 An encrusting colony with a hemispherical shape. BELIZE *Photograph: author*

2 A thin encrusting plate. BELIZE *Photograph: author*

3, 5 Showing the weakly formed valley walls and thick septo-costae. BELIZE *Photographs: author*

4 Surface detail. BELIZE *Photograph: author*

GENUS *AUSTRALOMUSSA*
Veron, 1985

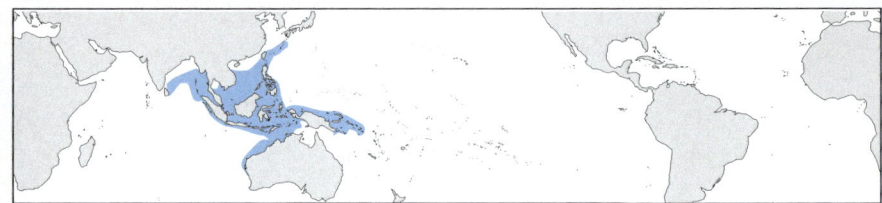

Characters, abundance and distribution: This genus has only one species, see *Australomussa rowleyensis*.
Earliest fossil record: Pliocene of the Pacific.

Australomussa rowleyensis
Veron, 1985

Characters: Colonies are flattened, helmet- or dome-shaped. Corallites are subcerioid or have short, shallow valleys 8-20 millimetres wide, separated by thick walls. Tissue over the septa is usually distinct in colour and/or texture from tissue over the costae. Septa and costae are sturdy, with large blunt teeth. **Colour:** In north-west Australia, colonies are a uniform blue-grey or valleys may have concentric cream and green colours. In south-east Asia colonies have a much wider range of colours including bright red, yellow and green. **Similar species:** Corallite details are like those of *Scolymia vitiensis*. **Habitat:** Lower reef slopes protected from wave action. **Abundance:** Sometimes locally common.

Taxonomic reference: Veron (1985). Identification guides: Veron (1986), Nishihira and Veron (1995).

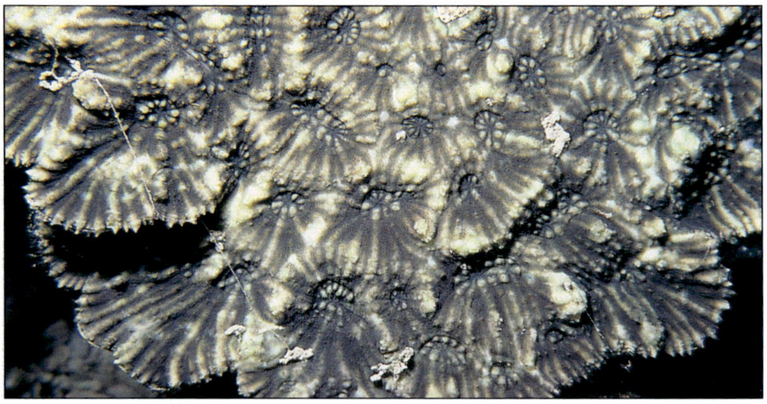

GENUS *INDOPHYLLIA*
Gerth, 1921

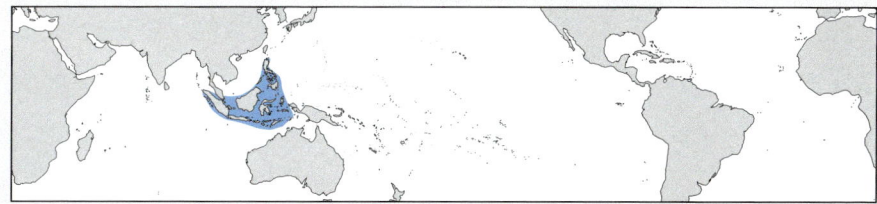

Characters, abundance and distribution: This genus has only one species, see *Indophyllia macassarensis*.
Earliest fossil record: Oligocene, locality unrecorded (doubtful record).

Indophyllia macassarensis
Best and Hoeksema, 1987

Characters: Solitary, free-living when mature, up to 45 millimetres diameter. Walls are indistinct beneath the septo-costae, which are in three orders. Teeth are large and lobed. Columellae are broad and compact. A paliform crown may be distinguishable. Tentacles are fleshy and transparent, making it possible to see the underlying skeleton. **Colour:** Brown. **Similar species:** Closest to *Cynarina lacrymalis*. **Habitat:** Deep sandy substrates where other free-living corals are found. **Abundance:** Rare.

Taxonomic reference: Borel-Best and Hoeksema (1987).

x2.5

x2.5

1 *Australomussa rowleyensis*. A large dome-shaped colony. ROWLEY SHOALS, WESTERN AUSTRALIA Photograph: Ed Lovell
2 *Australomussa rowleyensis*. The edge of a large colony. DAMPIER ARCHIPELAGO, WESTERN AUSTRALIA Photograph: author
3 *Australomussa rowleyensis*. Typical appearance of a small colony. SOLOMON ISLANDS Photograph: Jim Maragos
4 *Australomussa rowleyensis*. An encrusting colony. MERGUI ARCHIPELAGO, BURMA Photograph: author
5 *Australomussa rowleyensis*. Colonies with one corallite resemble *Scolymia vitiensis*. CALAMIAN ISLANDS, PHILIPPINES Photograph: author

GENUS CYNARINA
Brüggemann, 1877

Characters, abundance and distribution: This genus has only one species, see *Cynarina lacrymalis*.
Earliest fossil record: Oligocene, locality unrecorded (doubtful record), Pliocene of the Pacific.

Cynarina lacrymalis
(Milne Edwards and Haime, 1848)

Characters: Corals are monocentric, oval or circular, and are cylindrical with a base for attachment, or have a pointed base when free-living. Primary septa are thick and have extremely large, rounded or lobed teeth. Paliform lobes are usually well developed. Columellae are broad and compact. Tentacles are extended only at night. During the day the mantle is inflated with water and is translucent so that the toothed primary septo-costae are clearly seen. In conditions of low light the mantle may be over twice the diameter of the skeleton. **Colour:** Usually mixtures of green or brown, but may be pink and sometimes other colours. **Similar species:** Closest to *Indophyllia macassarensis*. **Habitat:** Protected reef environments and deep sandy substrates. **Abundance:** Seldom common but always conspicuous.

Taxonomic note: *Acanthophyllia deshayesiana* (Michelin, 1850) (the only species of *Acanthophyllia* Wells, 1937) is a synonym of *Cynarina lacrymalis*. **Taxonomic references:** Chevalier (1975), Veron and Pichon (1980). **Identification guides:** Veron (1986), Sheppard and Sheppard (1991), Nishihira and Veron (1995).

1 This is a very distinctive species. The large white septo-costae can be seen through the polyp walls. GREAT BARRIER REEF, AUSTRALIA *Photograph: Ed Lovell*

2, 4-6 Common colour patterns. **2, 4** GREAT BARRIER REEF, AUSTRALIA **5** DARWIN, NORTHERN AUSTRALIA **6** PAPUA NEW GUINEA *Photographs: 2 Len Zell 4-6 Neville Coleman*

3 With tentacles extended at night. GREAT BARRIER REEF, AUSTRALIA *Photograph: Ed Lovell*

FAMILY
FAVIIDAE
Gregory, 1900

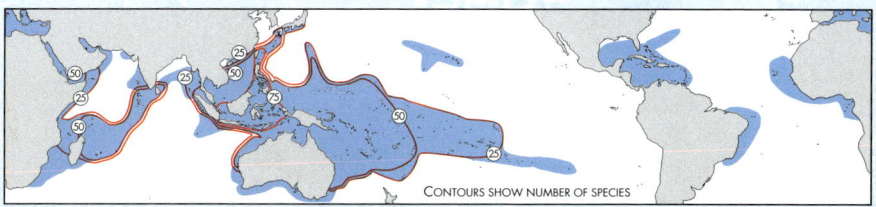

Characters: All species are zooxanthellate and colonial. Septa, paliform lobes, columellae and wall structures, when present, all appear to be structurally similar. Septal structures are simple, columellae are a simple tangle of elongate septal teeth. Walls are composed of thickened septa and cross linkages. **Similar families:** Merulinidae, Trachyphylliidae and Mussidae. **Genera:** Twenty-four genera, more than any other family. The majority of faviid genera are easily recognised because they are composed of a small number of species which have a number of distinctive characters in common. Four genera, *Favia*, *Barabattoia*, *Favites* and *Montastrea* may be confused. *Barabattoia* is really a group of species of doubtful affinity grouped together primarily because they have elongate corallites and predominantly extratentacular budding. *Favia* and *Favites* are usually easily separated because they are plocoid and cerioid respectively, but some species, especially *Favia rotumana*, *Favia rotundata* and *Favia veroni* can be almost completely cerioid (*Favites*-like) in shallow wave washed habitats. Similarly *Montastrea*, which is distinguished from *Favia* by having extratentacular rather than intratentacular budding, may be difficult to recognise in colonies which are not actively budding. Even when they are actively budding, some species, especially *M. valenciennesi*, often display both types of budding in different proportions according to environmental conditions. In the case of Indo-Pacific *Montastrea*, individual species may be more readily recognised than the genus itself.

Taxonomic references: Wijsman-Best (1972), Veron, Pichon and Wijsman-Best (1977).
Summary key: See p452.

Opposite: The Faviidae, represented here by *Diploastrea heliopora*, has more genera than any other zooxanthellate family. GREAT BARRIER REEF, AUSTRALIA *Photograph:* Valerie Taylor

Family Faviidae

Types of colony formation

More than in any other family, faviids display variation in the way colonies are formed through the asexual budding of polyps. These types of colony formation are found in many corals but the terms used to describe them are used more for faviids than for other families, partly because there are so many genera involved. The terms used in this book are: phaceloid, plocoid, cerioid, meandroid and flabello-meandroid, see 'Structure' (v1, p47) and Glossary (p463) for explanations.

Family Faviidae

Fossil record and distribution

Because of their solid construction and wide geographic distribution, most faviid genera are readily preserved as fossils and have a good fossil record. The Faviidae is the only family to have been a dominant reef builder in both the Mesozoic and Cenozoic eras and faviids survived the end-Cretaceous extinctions better than any other scleractinian family.

Most species are widely and relatively uniformly distributed across the Indo-west Pacific. Some species are restricted to intertidal habitats and upper reef slopes, but most occur over a wide range of environments. Most also have a wide latitudinal range. Colonies from subtropical localities are usually distinct, being relatively heavily calcified and often having polyp tentacles extended during the day.

1 An unmistakably Caribbean scene with faviids dominating a shallow reef community. This photograph shows *Montastrea cavernosa* (bottom right), *M. annularis* (top right background), *Diploria labyrinthiformis* (top centre) and two non-faviids, *Meandrina meandrites* (lower centre) and *Agaricia agaricites* (lower left). JAMAICA *Photograph: author*

2 *Cyphastrea serailia* at night, just prior to spawning. GREAT BARRIER REEF, AUSTRALIA *Photograph: Bette Willis*

3 Most faviids, as with this *Barabattoia*, are voracious planktivores. This is best seen by torchlight at night. PALAU *Photograph: Gustav Paulay*

4 Most massive faviids are nearly impossible to identify at night when tentacles are extended. GREAT BARRIER REEF, AUSTRALIA *Photograph: author*

5 *Platygyra pini* releasing egg and sperm bundles at night. GREAT BARRIER REEF, AUSTRALIA *Photograph: Roger Steene*

Family Faviidae   Genus *Cladocora*

GENUS *CLADOCORA*
Ehrenberg, 1834

ZOOXANTHELLATE SPECIES ONLY

Characters: Colonies are phaceloid with long tubular corallites which are closely compacted or branch irregularly. There are no well developed basal structures of attachment. Tentacles are usually extended day and night. Most *Cladocora* species do not have zooxanthellae and are not included in this book. **Similar genera:** Most similar to *Caulastrea*, but is substantially unlike other faviids. **Earliest fossil record:** Eocene of the Caribbean and Tethys.

Taxonomic reference: Zibrowius (1980).

Cladocora caespitosa
(Linnaeus, 1767)

Characters: Colonies are phaceloid, up to 0.5 metres across. Corallites are tubular, 4-5 millimetres diameter, and are compacted together forming clumps. Septo-costae are in two alternating orders, only the first of which usually reaches the columella. Columellae and paliform lobes may both be well developed. Only colonies from water shallow enough for light penetration are zooxanthellate. **Colour:** Cream or brown with brown tentacles. **Similar species:** *Cladocora arbuscula*. **Habitat:** Shallow rocky substrates, to deep water where colonies are azooxanthellate. **Abundance:** Uncommon.

Taxonomic reference: Zibrowius (1980). **Identification guide:** Mojetta and Ghisotti (1994).

Family Faviidae | Genus *Cladocora*

1 An unidentified *Cladocora* resembling *C. caespitosa*. GULF OF MEXICO *Photograph: Julian Sprung*

2 *Cladocora caespitosa*. Colonies consist of uniformly and closely compacted polyps. MEDITERRANEAN SPAIN *Photograph: Maoz Fine*

3 *Cladocora caespitosa*. A clump of compacted corallites with fine elongate tentacles extended during the day. MEDITERRANEAN FRANCE *Photograph: Julian Sprung*.

4 *Cladocora caespitosa*. Colony surface. MEDITERRANEAN SPAIN *Photograph: Maoz Fine*

Family Faviidae Genus *Cladocora*

Cladocora arbuscula
(Lesueur, 1821)

Characters: Colonies consist of clumps of interlocking tubular corallites which are less than 4 millimetres diameter. Clumps may be compact and attached to hard substrates or loose aggregations on soft substrates which readily fall apart if disturbed. Usually only the terminal parts of corallites are alive. Corallite walls and septo-costae are both thin and delicate. The latter are in two alternating orders and have irregular teeth. **Colour:** Tan or brown, with translucent tentacles. **Similar species:** *Cladocora caespitosa* has larger, more solid corallites which are phaceloid rather than irregularly branched and have well developed paliform lobes. **Habitat:** Seagrass beds and other turbid environments where there is a lot of sediment. **Abundance:** Uncommon, rare on reefs.

Taxonomic references: Zibrowius (1980), Cairns (1982). **Identification guides:** Colin (1978), Wood (1983), Humann (1993).

x5

Family Faviidae Genus *Caulastrea*

GENUS *CAULASTREA*
Dana, 1846

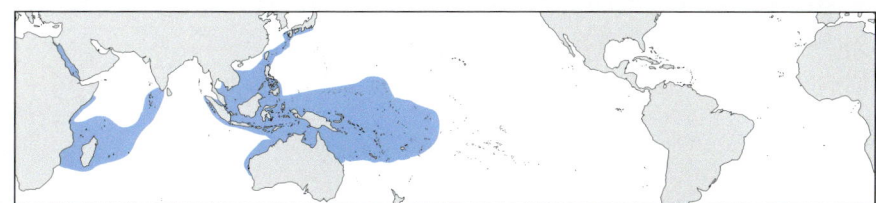

Characters: Colonies are usually phaceloid. Corallites have numerous fine septa and well developed columellae. Paliform lobes are mostly absent. Tentacles are rarely extended during the day. **Similar genera:** *Erythrastrea*. *Caulastrea* also has a superficial resemblance to the mussids *Blastomussa wellsi* and *Mussismilia harttii*. **Earliest fossil record:** Eocene of the Tethys, Oligocene of the Caribbean.

Taxonomic references: Wijsman-Best (1976), Veron, Pichon and Wijsman-Best (1977). **Summary key:** See pp452-3.

Identification of *Caulastrea* species

The genus is easily recognised, except for *Caulastrea tumida*, which can be plocoid rather than phaceloid. Species are also easily recognised except for *C. echinulata* and *C. furcata* which may be difficult to distinguish, mainly because they seldom occur together and because the latter has a wide range of skeletal and colour variation.

Caulastrea connata
(Ortmann, 1892)

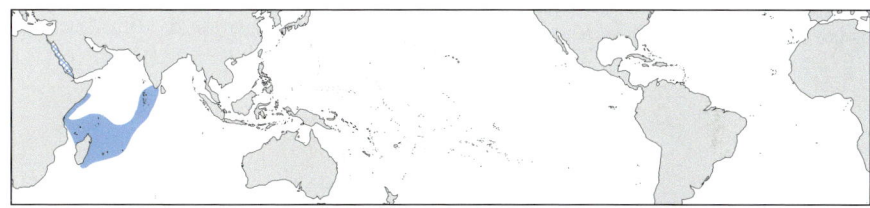

Characters: Colonies are phaceloid. Corallites are prominent and sturdy, 10-25 millimetres diameter, and frequently have more than one mouth. Septa are irregularly spaced, exsert and rough-edged. Costae are poorly developed; paliform lobes are not conspicuous. **Colour:** Grey-brown. **Similar species:** *Caulastrea tumida*, which has smaller, more fleshy corallites and more even septa. **Habitat:** Shallow reef environments. **Abundance:** Rare.

Taxonomic note: This species was formerly placed in *Astreosmilia*, a genus of its own.
Taxonomic reference: Ortmann's original description/specimens.

x2

1 *Cladocora arbuscula*. Corallites are irregular tubes branching perpendicularly. Polyps are extended during the day. FLORIDA *Photograph: Paul Humann*

2 *Cladocora arbuscula*. Irregular tubular corallites attached to dead basal corallites. FLORIDA *Photograph: Paul Humann*

3 *Cladocora arbuscula*. On a sandy substrate. FLORIDA *Photograph: Paul Humann*

4 *Caulastrea connata*. This rare species has large corallites with large irregular septa. ZANZIBAR, TANZANIA *Photograph: author*

Caulastrea furcata
Dana, 1846

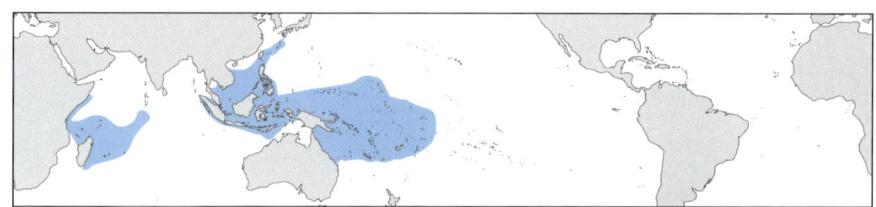

Characters: Colonies are phaceloid. Corallites diverge from the colony base, or are irregular, or crowded if space is restricted. They are less than 10 millimetres diameter. Septa are exsert and irregular, with some septa thicker than others. Polyps are fleshy; thick septa give prominent radiating stripes to the upper corallite surface. **Colour:** Brown or green with green oral discs. **Similar species:** *Caulastrea echinulata* and *C. curvata*. **Habitat:** Protected reef slopes where the substrate is partly sandy. Forms extensive single species stands, sometimes over 5 metres across. **Abundance:** Common.

Taxonomic references: Wijsman-Best (1972), Veron, Pichon and Wijsman-Best (1977).
Identification guides: Veron (1986), Nishihira and Veron (1995).

x2

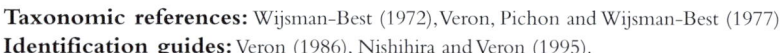
1

Family Faviidae   Genus *Caulastrea*

1 A large colony with crowded corallites. GREAT BARRIER REEF, AUSTRALIA *Photograph: author*

2 Showing the neat appearance of a small hemispherical colony. GREAT BARRIER REEF, AUSTRALIA *Photograph: Ed Lovell*

3 Usual appearance of corallites with some septa forming radiating stripes. GREAT BARRIER REEF, AUSTRALIA *Photograph: Valerie Taylor*

4 Side view of corallites. GREAT BARRIER REEF, AUSTRALIA *Photograph: Valerie Taylor*

5 Thick crowded corallites of a colony in a wave washed environment. PAPUA NEW GUINEA *Photograph: author*

Caulastrea tumida
Matthai, 1928

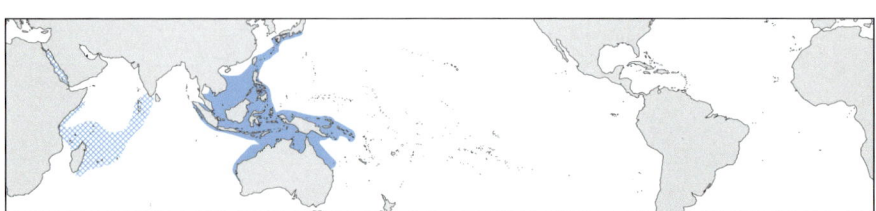

Characters: Colonies are phaceloid or plocoid, the former being predominant in high latitudes and very shallow water. Corallites are short and sturdy, 10-15 millimetres diameter, and frequently have more than one mouth. Costae are poorly developed. **Colour:** Dull cream, grey and green. **Similar species:** *Caulastrea connata*. See also *Favia vietnamensis*. **Habitat:** Shallow reefs, and rocky foreshores of subtropical locations. **Abundance:** Common only in Western Australia and Japan.

Distribution note: Records from the western Indian Ocean and Red Sea are doubtful. **Taxonomic reference:** Veron, Pichon and Wijsman-Best (1977). **Identification guides:** Veron (1986), Sheppard and Sheppard (1991), Nishihira and Veron (1995).

1 With short tubular corallites, the usual appearance of the species. Ryukyu Islands, Japan *Photograph: author*

2-4 This species is more fleshy than other *Caulastrea*, with corallites which tend to fuse. **2** Vietnam **3** Calamian Islands, Philippines **4** Bolinao, Philippines *Photographs: author*

Family Faviidae | Genus *Caulastrea*

Caulastrea curvata
Wijsman–Best, 1972

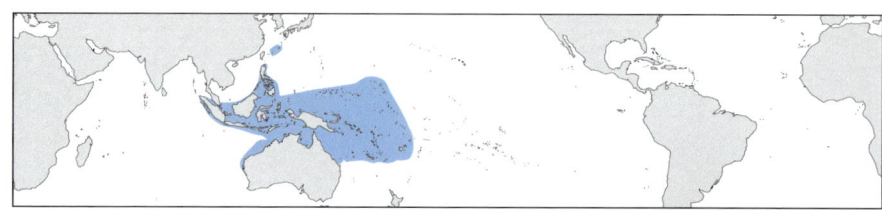

Characters: Colonies are usually small. Corallites are sometimes compact, but usually sprawl irregularly. They are characteristically curved at the colony periphery and average 8 millimetres diameter. **Colour:** Pale brown. **Similar species:** *Caulastrea furcata*, which can be distinguished underwater by its colour; corallites are also more regular, larger, have more numerous septa and better developed costae. **Habitat:** Flat substrates, often found with *C. furcata*. **Abundance:** Uncommon.

Taxonomic references: Wijsman-Best (1972), Veron, Pichon and Wijsman-Best (1977).
Identification guides: Veron (1986), Nishihira and Veron (1995).

x2

Caulastrea echinulata
(Milne Edwards and Haime, 1849)

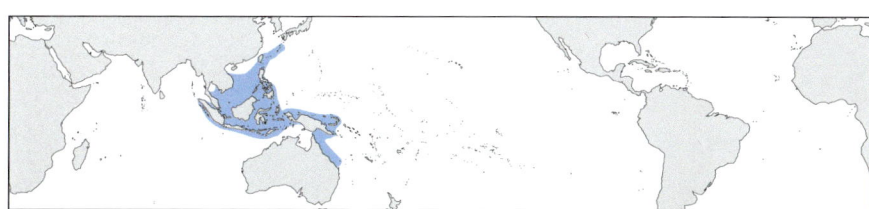

Characters: Colonies are phaceloid. Corallites are usually closely compacted and less than 10 millimetres diameter. Septa are exsert and irregular, but approximately uniform in width, if not height. Polyps are fleshy, concealing septal characters. **Colour:** Tan to dark brown with pale oral discs. **Similar species:** *Caulastrea furcata*, which usually has less compact corallites with some thickened primary septa. Superficially resembles the mussid *Blastomussa wellsi*. **Habitat:** Horizontal substrates protected from wave action and with turbid water. **Abundance:** Uncommon.

Taxonomic references: Wijsman-Best (1972), Veron, Pichon and Wijsman-Best (1977). **Identification guides:** Veron (1986), Nishihira and Veron (1995).

5

1 *Caulastrea curvata*. A small colony with short corallites. PAPUA NEW GUINEA *Photograph: author*

2, 3 *Caulastrea curvata*. Corallites are sometimes closely compacted in shallow water. CALAMIAN ISLANDS, PHILIPPINES *Photographs: author*

4 *Caulastrea curvata*. Showing long thin tubular corallites curving at the colony perimeter, the usual appearance of the species. DAMPIER ARCHIPELAGO, WESTERN AUSTRALIA *Photograph: author*

5 *Caulastrea echinulata*. Forming compact colonies. Corallites are larger than those of *C. furcata* and have ragged rims. BOLINAO, PHILIPPINES *Photograph: author*

6 *Caulastrea echinulata*. Surface detail of compact corallites. Septa form only indistinct radiating stripes. GREAT BARRIER REEF, AUSTRALIA *Photograph: Neville Coleman*

x2

6

Family Faviidae   Genus *Erythrastrea*

GENUS *ERYTHRASTREA*
Scheer and Pillai, 1983

Characters, abundance and distribution: This genus has only one species, see *Erythrastrea flabellata*. **Fossil record:** None.

Erythrastrea flabellata
Scheer and Pillai, 1983

Characters: Colonies are hemispherical, up to one metre across, and are flabello-meandroid. Valleys are 8-11 millimetres wide and have thin walls, only the upper part of which are living. Septa protrude up to 5 millimetres and are equal, with coarse teeth. Columella centres are irregularly spaced and distinct. Costae are fine. **Colour:** Pale cream or green. **Similar species:** *Caulastrea tumida* which has smaller fleshy polyps and forms smaller colonies which are not flabello-meandroid. Superficially resembles the mussid *Lobophyllia*. **Habitat:** Reef slopes protected from wave action. **Abundance:** Uncommon.

Taxonomic reference: Scheer and Pillai (1983). **Identification guide:** Sheppard and Sheppard (1991).

×1

1, 3 *Erythrastrea flabellata.* Colonies typically consist of irregular hemispherical clumps. SINAI PENINSULA, EGYPT *Photographs: author*

2, 4 *Erythrastrea flabellata.* Corallite detail. SINAI PENINSULA, EGYPT *Photographs: author*

5 *Manicina areolata.* A hemispherical colony. This is the most common colouration of colonies in shallow water. BELIZE *Photograph: author*

6 *Manicina areolata.* A free-living colony with an axial valley and short side valleys. NETHERLANDS ANTILLES *Photograph: Paul Humann*

7 *Manicina areolata.* A juvenile colony with mouths forming. HONDURAS *Photograph: Paul Humann*

8 *Manicina areolata.* Surface detail of valleys. BELIZE *Photograph: author*

Family Faviidae

GENUS MANICINA
Ehrenberg, 1834

Characters, abundance and distribution: This genus has only one species, see *Manicina areolata*. **Earliest fossil record:** Oligocene of the Caribbean.

Manicina areolata
(Linnaeus, 1758)

Characters: Colonies are either free-living or attached; the former have a cone-shaped undersurface. Small colonies are oval shaped and consist of a central axial valley with short side valleys. Larger colonies are hemispherical and meandroid. There is a conspicuous ambulacral groove along the tops of valley walls. Septa are thick and even. **Colour**: Usually uniform pale orange-brown but may be a variety of browns, greys and greens with contrasting colours of valleys and walls. **Similar species**: *Colpophyllia natans*. Small colonies resemble *Diploria labyrinthiformis* and the species generally resembles the trachyphyllid *Trachyphyllia geoffroyi*. **Habitat**: Subtidal seagrass beds where colonies are small and free-living, also shallow reef environments where colonies are attached and become hemispherical. **Abundance**: Usually uncommon.

Taxonomic references: Roos (1971), Zlatarski and Estalella (1982), Cairns (1982). **Identification guides**: Colin (1978), Humann (1993).

GENUS *FAVIA*
Oken, 1815

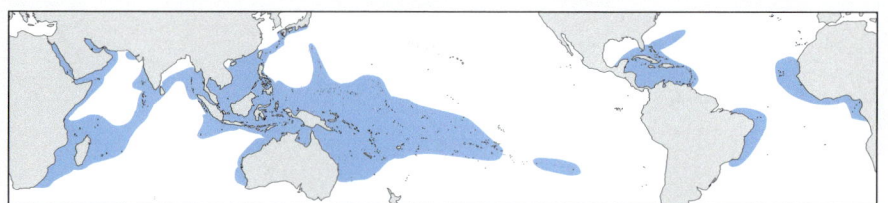

Characters: Colonies are usually massive, either flat or dome-shaped. Corallites are mostly monocentric and plocoid. Daughter corallites are formed by intratentacular division. Tentacles are extended only at night and are tapered, often with pigmented tips. **Similar genera:** *Favia* is similar to *Favites* but the latter has cerioid corallites. This distinction is sometimes arbitrary in which case *Favia* corallites are further characterised by subdividing equally, whereas *Favites* corallites usually subdivide unequally, producing daughter corallites of different sizes. *Favia* is distinguished from *Barabattoia* by having less exsert corallites and intratentacular budding. **Earliest fossil record:** Cretaceous of the Tethys (doubtful record), Eocene of the far eastern Pacific/Caribbean.

Taxonomic references: Wijsman-Best (1972, 1974), Veron, Pichon and Wijsman-Best (1977). **Summary key:** See pp452-3.

1 *Favia* occur in most coral communities of the Indo-Pacific but are seldom a dominant genus. Highest abundance and highest diversity occur in protected upper reef slopes such as this one, which are not dominated by *Acropora*. PAPUA NEW GUINEA *Photograph: author*

2 *Favia* species are most easily identified where several occur in close proximity. Here two similar species, *F. danae* (left) and *F. favus* (right), are readily distinguished. GREAT BARRIER REEF, AUSTRALIA *Photograph: author*

3 Most *Favia* species look similar at night when extended tentacles mask underlying polyp structures. GREAT BARRIER REEF, AUSTRALIA *Photograph: Ed Lovell*

Identification of *Favia* species

Favia is one of the most widely and uniformly distributed of all coral genera. Individual species are also widely distributed in the Indo-west Pacific, and a high proportion occur beyond the latitudinal limits of reefs. *Favia fragum* in the Atlantic is also widespread; it differs substantially from any Indo-Pacific species.

There are often exceptions to characters used in species identification. This is because most species have wide depth ranges on reefs, wide geographic ranges, and are tolerant of a wide range of environmental conditions. As a result, a colony on an intertidal reef flat may be very different from another colony of the same species on the same reef but from deeper water. Both may be very different from other colonies of the same species in similar conditions but from a different country. In many cases *Favia* are also difficult to recognise underwater without substantial practice because of colour variation and also because the soft tissue of living polyps obscures underlying skeletal structures.

The genus can be divided into three groups according to corallite size:

Group 1: Species with small corallites (averaging less than 8 mm diameter), page 102.

Group 2: Species with middle sized corallites (averaging 8-12 mm diameter), page 106.

Group 3: Species with large corallites (averaging more than 12 mm diameter), page 116.

Family Faviidae — Genus *Favia*

> **Group 1: Species with small corallites (averaging less than 8 mm diameter)**
> These species have little resemblance to each other, but may be confused with species of other genera.

Favia stelligera
(Dana, 1846)

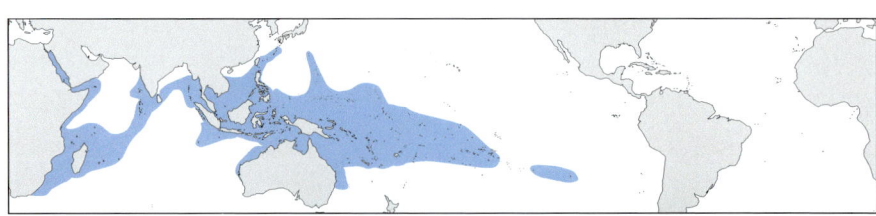

Characters: Colonies are spherical, columnar, hillocky or flat and may be several metres across. Corallites are evenly distributed and conical, with thick walls and small openings. Costae are equal and well developed. Those of adjacent corallites do not join. A crown of paliform lobes is usually clearly visible. **Colour:** Uniform brown or green. **Similar species:** Does not resemble other *Favia* and growth-form alone makes this species distinctive underwater. *Montastrea* species all have larger corallites and extratentacular budding. Of these, *M. salebrosa* is most similar but forms massive colonies and has corallites with numerous closely compacted corallites. See also *Plesiastrea versipora*, which also has extratentacular budding. **Habitat:** Shallow reef environments where water movement is strong. **Abundance:** Common.

Taxonomic references: Wijsman-Best (1972), Veron, Pichon and Wijsman-Best (1977). **Identification guides:** Randall and Myers (1983), Veron (1986), Sheppard and Sheppard (1991), Nishihira and Veron (1995).

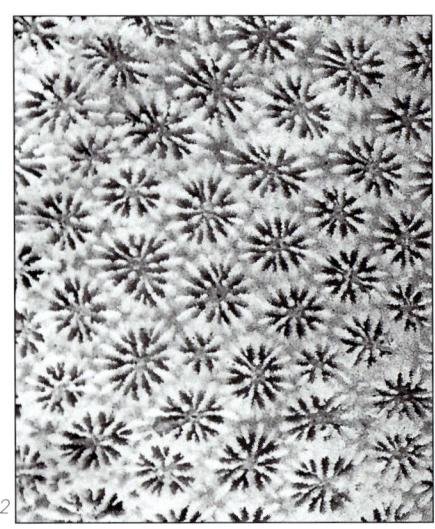

Family Faviidae | Genus *Favia*

1 This species forms tall columns unlike any other *Favia*. GREAT BARRIER REEF, AUSTRALIA *Photograph: Ed Lovell*

2 An encrusting colony. SINAI PENINSULA, EGYPT *Photograph: author*

3 A common colony shape. PHILIPPINES *Photograph: Valerie Taylor*

4 Showing the characteristically uniform appearance of corallites, giving colonies an even surface. GREAT BARRIER REEF, AUSTRALIA *Photograph: author*

5 Corallites on a column tip. GUAM *Photograph: Gustav Paulay*

6 Corallite detail. GREAT BARRIER REEF, AUSTRALIA *Photograph: Valerie Taylor*

Family Faviidae Genus *Favia*

Favia fragum
(Esper, 1797)

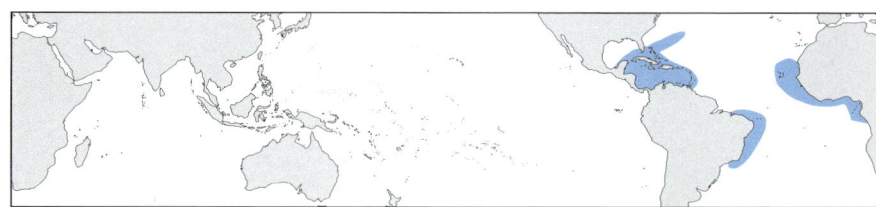

Characters: Colonies are small (usually less than 50 mm across) and hemispherical to encrusting. Corallites have very variable shapes ranging from immersed to conical (plocoid) to tubular (subphaceloid) and may be circular with one mouth, to elongate with many mouths. Encrusting colonies in intertidal habitats may be submeandroid. Spherical colonies with unrestricted growing space commonly develop tubular corallites. Corallites or valleys are seldom more than 5 millimetres across. Whatever the corallite shape, the walls are neatly rounded. Septo-costae are exsert and evenly spaced. **Colour:** Usually tan to light orange-brown with pale green tentacles. Walls and calices may have contrasting colours. **Similar species:** None. **Habitat:** Intertidal rock pools and shallow reef environments. **Abundance:** Common.

Taxonomic note: This species is usually called *Favia gravida* in Brazil after Laborel (1969). Brazilian colonies are usually more meandroid but all characters overlap. **Taxonomic references:** Roos (1971), Zlatarski and Estalella (1982). **Identification guides:** Colin (1978), Humann (1993).

x2

Favia laxa
(Klunzinger, 1879)

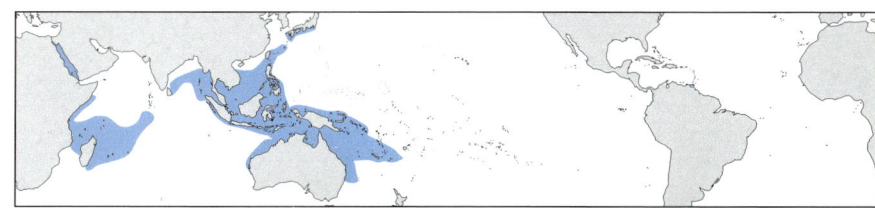

Characters: Colonies are hemispherical. Corallites are conical and uniform in shape. They show both extra- and intratentacular budding. Paliform lobes form a neat crown. Septa are fine and neatly arranged. Costae are also neat; a tiny line of demarcation, visible underwater, separates costae of adjacent corallites. **Colour:** Uniform, usually pale brown or pinkish-brown. **Similar species:** *Favia helianthoides*, which has larger, less irregular corallites. **Habitat:** Shallow reef environments. **Abundance:** Common in the Red Sea, uncommon elsewhere.

Taxonomic references: Wijsman-Best (1972), Veron, Pichon and Wijsman-Best (1977). **Identification guide:** Sheppard and Sheppard (1991).

1, 2 *Favia fragum*. Variation in corallite shape and colour. **1** ABROLHOS ISLANDS, BRAZIL **2** BELIZE *Photographs: author*

3 *Favia fragum*. Corallites are sometimes protuberant and can be almost phaceloid. FLORIDA *Photograph: Julian Sprung*

4 *Favia fragum* (two colonies, left) with other Brazilian corals having small corallites: *Stephanocoenia michelinii* (top right) and *Siderastrea stellata* (bottom right). Note the variation of corallite shape from circular to submeandroid. ABROLHOS ISLANDS, BRAZIL *Photograph: author*

5 *Favia fragum*. In intertidal habitats this species develops submeandroid corallites. ABROLHOS ISLANDS, BRAZIL *Photograph: author*

6 *Favia laxa*. Typical surface appearance. GREAT BARRIER REEF, AUSTRALIA *Photograph: Len Zell*

7 *Favia laxa*. Corallite detail. Note the line of demarcation between costae of adjacent corallites. GREAT BARRIER REEF, AUSTRALIA *Photograph: Ed Lovell*

Family Faviidae | Genus *Favia*

Group 2: Species with middle sized corallites (averaging 8-12 mm diameter)

Favia matthaii
Vaughan, 1918

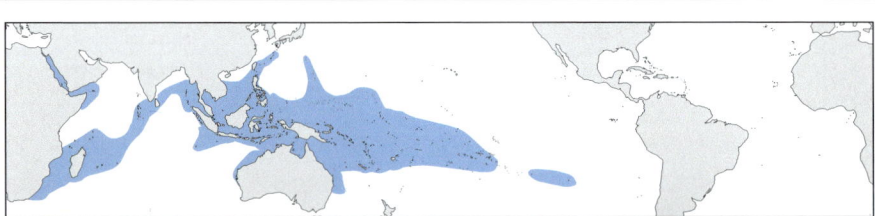

Characters: Colonies are massive and usually small. Corallites are crowded and circular. Septa are thickened, exsert or ragged, with large teeth near the wall. They have well developed paliform lobes forming a crown around the columella. **Colour:** Usually brown or grey or mottled, with walls and calices of contrasting colours. **Similar species:** *Favia albidus*. Readily distinguished from *F. pallida* and *F. speciosa* by the exsert or ragged septa and paliform crown. **Habitat:** Upper reef slopes. **Abundance:** Sometimes common.

Taxonomic references: Wijsman-Best (1972), Veron, Pichon and Wijsman-Best (1977). **Identification guides:** Randall and Myers (1983), Veron (1986), Nishihira and Veron (1995).

x2

Family Faviidae Genus *Favia*

1 Colony surface showing a common colour and the characteristic appearance of corallites with irregularly exsert septa. GREAT BARRIER REEF, AUSTRALIA *Photograph: Valerie Taylor*

2 Corallites of a colony in shallow water with highly irregular septa. PAPUA NEW GUINEA *Photograph: author*

3 A large massive colony. ASHMORE REEF, WESTERN AUSTRALIA *Photograph: author*

4, 5 Colour variation and detail of corallites. GREAT BARRIER REEF, AUSTRALIA *Photographs: Valerie Taylor*

Favia speciosa
Dana, 1846

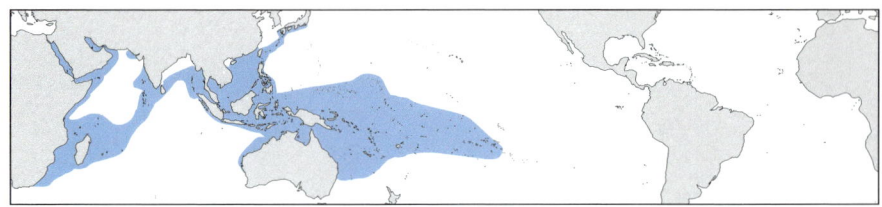

Characters: Colonies are massive. Corallites are circular and closely compacted in shallow water, more widely spaced in deeper water. Septa are fine, numerous and regular. Paliform lobes are usually poorly developed. **Colour:** Pale grey, green or brown, usually with calices of contrasting colours. **Similar species:** *Favia pallida* and *F. truncatus*. See also *F. helianthoides*. **Habitat:** All reef environments. **Abundance:** One of the most common faviids, especially in high latitudes.

Taxonomic reference: Veron, Pichon and Wijsman-Best (1977). **Identification guides:** Veron (1986), Sheppard and Sheppard (1991), Nishihira and Veron (1995), Coles (1996), Carpenter *et al.* (1997).

1 Typical appearance in shallow water with corallites crowded. PAPUA NEW GUINEA *Photograph: Gerry Allen*

2-5 Common variation in corallite shape and colour. **2, 3** GREAT BARRIER REEF, AUSTRALIA **4** FLORES, INDONESIA **5** LORD HOWE ISLAND, SOUTH-EAST AUSTRALIA *Photographs: 2, 3 author 4, 5 Roger Steene*

Family Faviidae   Genus *Favia*

Favia helianthoides
Wells, 1954

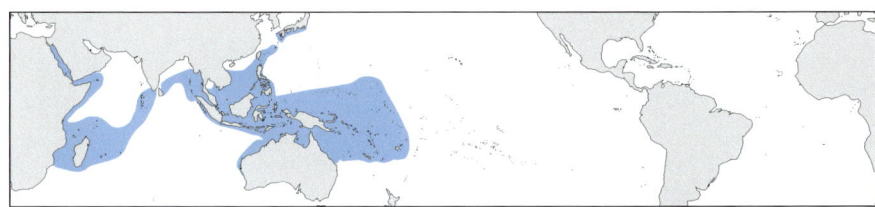

Characters: Colonies are submassive. Corallites are conical and mostly uniform in shape. They show both extra- and intratentacular budding. Paliform lobes are well developed. Septa are neatly arranged and are thickened over the wall. Costae are also neat and those of adjacent corallites adjoin. **Colour:** Brown, tan or blue-grey with cream oral discs. **Similar species:** *Favia laxa*, which has smaller, more widely spaced corallites, and *F. speciosa* which has less conical corallites. See also *Diploastrea heliopora*, which has a similar colour but larger corallites with much thicker septo-costae. **Habitat:** Shallow reef environments. **Abundance:** Sometimes common.

Taxonomic note: This species is distinctive in the central Pacific, but there are unresolved taxonomic issues west of this region. **Taxonomic reference:** Wells (1954). **Identification guide:** Nishihira and Veron (1995).

Favia lacuna
Veron, Turak and DeVantier, this publication

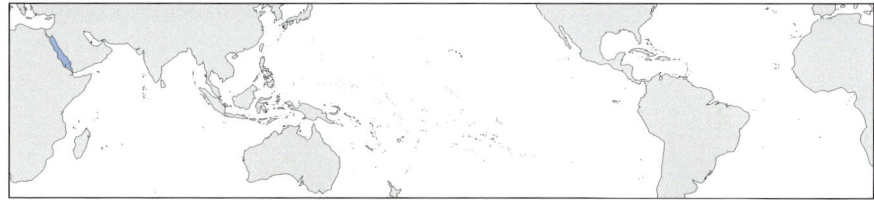

Characters: Colonies are submassive to massive, usually more than one metre across. Corallites are subplocoid, circular to irregular in shape, and crowded. The inner walls of corallites plunge vertically. Septa are thin, uniformly spaced and subequal, most reaching the columella deep within the calice. Costae of adjacent corallites do not meet, leaving a narrow ambulacral groove. There are no paliform lobes. Columellae are small and compact. **Colour:** Tan with white centres. **Similar species:** Looks like a small *Oulophyllia* underwater. **Habitat:** Shallow exposed reef environments. **Abundance:** Common.

Taxonomic note: See 'New species described in *Corals of the World*' (Veron, in preparation) for further information.

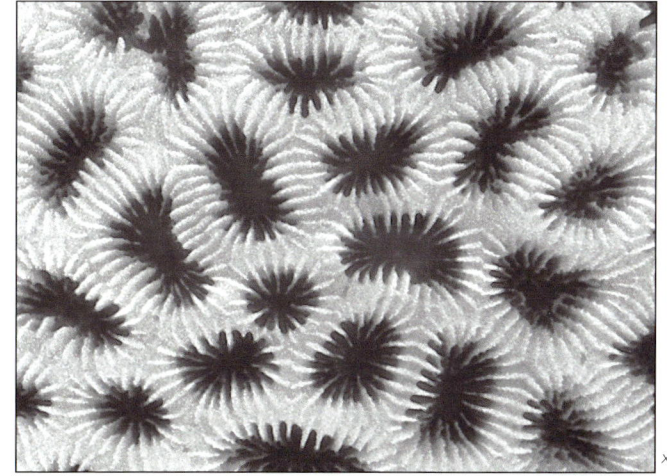

×2

1, 4 *Favia helianthoides*. Colonies usually have evenly spaced corallites. **1** PAPUA NEW GUINEA **4** SEYCHELLES *Photographs: 1 Neville Coleman 4 author*

2, 3 *Favia helianthoides*. Surface of two colonies showing similarity in corallite detail in the Indo-west Pacific. **2** SEYCHELLES **3** CALAMIAN ISLANDS, PHILIPPINES *Photographs: author*

5 *Favia helianthoides*. Surface of a massive colony showing the typically uniform arrangement of corallites. PEMBA ISLAND, TANZANIA *Photograph: author*

6 *Favia lacuna*. A large massive colony. SAUDI ARABIA, RED SEA *Photograph: Emre Turak*

Family Faviidae — Genus *Favia*

Favia albidus
Veron, this publication

Characters: Colonies are massive, usually small. Corallites are crowded, usually circular and monocentric but rarely becoming elongate and polycentric. Septa are thickened over the corallite wall and exsert, with large teeth near the wall. Paliform lobes may form a crown around the columella. Columellae are small. **Colour:** Usually pale brown. **Similar species:** *Favia matthaii*, which has thinner septa, less development of a paliform crown and smaller septal dentations. Septal dentations are sufficiently long to be *Acanthastrea*-like (Mussidae). **Habitat:** Upper reef slopes. **Abundance:** Uncommon.

Taxonomic note: See 'New species described in *Corals of the World*' (Veron, in preparation) for further information.

1 *Favia albidus*. A small massive colony. SAUDI ARABIA, RED SEA *Photograph: Lyndon DeVantier*

2 *Favia albidus*. Corallite detail. SINAI PENINSULA, EGYPT *Photograph: author*

3 *Favia truncatus*. Colony surface. COCOS (KEELING) ATOLL, WESTERN AUSTRALIA *Photograph: author*

4-6 *Favia truncatus*. Variation in corallite structure and colour. **4** GUAM **5** ASHMORE REEF, WESTERN AUSTRALIA **6** GREAT BARRIER REEF, AUSTRALIA *Photographs: 4 Gustav Paulay 5 author 6 Wendy Morris*

Favia truncatus
Veron, this publication

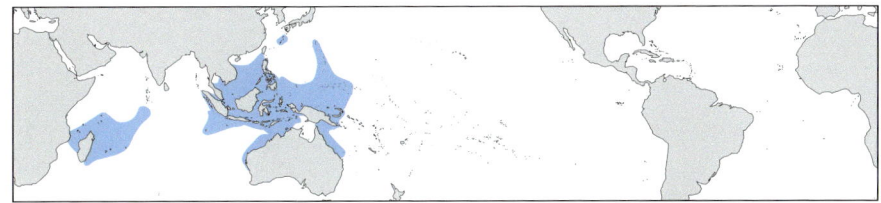

Characters: Colonies are massive, flat or hemispherical. Corallites are typically inclined on the colony surface, facing downwards on hemispherical surfaces. Corallite walls have sharp rims except for colonies from very shallow water. The lower part of the wall of inclined corallites is commonly immersed, giving the upper part a hooded appearance. Septa are widely spaced and irregular in size. Paliform crowns are well developed. **Colour:** Uniform yellowish-green or brown. **Similar species:** *Favia speciosa* and *F. pallida*, both of which have larger corallites with less exsert septo-costae. The inclined corallites, giving a hooded appearance, usually make colonies recognisable underwater. **Habitat:** Most shallow reef environments. **Abundance:** Common in equatorial regions.

Taxonomic note: See 'New species described in *Corals of the World*' (Veron, in preparation) for further information.

x2

Family Faviidae | Genus *Favia*

Favia pallida
(Dana, 1846)

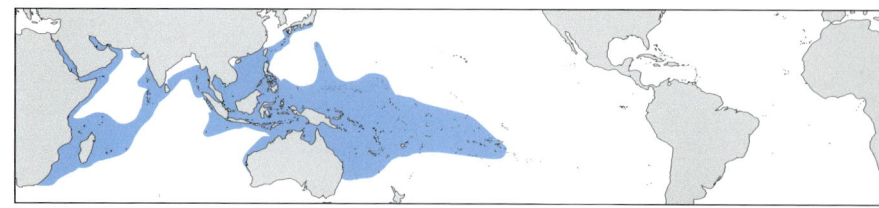

Characters: Colonies are massive. Corallites are circular, closely compacted in shallow water, more widely spaced in deeper water. Septa are widely spaced and characteristically irregular. Paliform lobes are usually poorly developed. **Colour:** Pale yellow, cream or green, with dark brown or green oral discs. **Similar species:** *Favia speciosa*, which has more conical corallites with finer and more numerous septa and seldom has distinctively dark oral discs. *Favia truncatus* has corallites inclined on the colony surface. **Habitat:** All reef environments, often a dominant species of back reef margins. **Abundance:** One of the most common faviids of eastern Australia, but usually much less common elsewhere.

Taxonomic references: Wijsman-Best (1972), Veron, Pichon and Wijsman-Best (1977).
Identification guides: Veron (1986), Sheppard and Sheppard (1991), Nishihira and Veron (1995), Coles (1996), Carpenter *et al.* (1997).

Family Faviidae · Genus *Favia*

1-3 *Favia pallida* usually forms rounded, massive colonies. **1** GREAT BARRIER REEF, AUSTRALIA **2, 3** PAPUA NEW GUINEA *Photographs: 1 author 2, 3 Gerry Allen*

4, 5 Corallite detail. The colour of (4) is unusual. **4** CALAMIAN ISLANDS, PHILIPPINES **5** GREAT BARRIER REEF, AUSTRALIA *Photographs: author*

6 Typical colour of corallites. GREAT BARRIER REEF, AUSTRALIA *Photograph: Mary Stafford-Smith*

Family Faviidae — Genus *Favia*

> **Group 3: Species with large corallites (averaging more than 12 mm diameter)**

Favia favus
(Forskål, 1775)

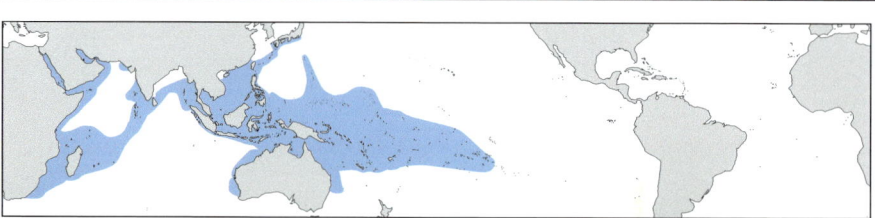

Characters: Colonies are massive, rounded or flat. Corallites are conical. Septa are slightly irregular and widely spaced. Paliform lobes are poorly developed. **Colour:** A wide variety, often mottled, with pale calices. **Similar species:** *Favia speciosa*, which has smaller, usually more compact corallites. *Favia lizardensis* and *F. danae* have corallites of similar size, but the former has more compact corallites with uniform septo-costae and colouration, and the latter has strongly beaded septo-costae. See also *F. rosaria* and *F. maritima*. **Habitat:** May be a dominant species on reef back margins. **Abundance:** Common.

Taxonomic references: Scheer and Pillai (1974), Veron, Pichon and Wijsman-Best (1977). **Identification guides:** Veron (1986), Sheppard and Sheppard (1991), Nishihira and Veron (1995).

Family Faviidae | Genus *Favia*

1 Large colonies are usually hemispherical or flattened. Honshu, Japan *Photograph: author*

2-4 Common variation in corallite shape and colour. **2** Scott Reef, Western Australia **3** Sinai Peninsula, Egypt **4** Flores, Indonesia
Photographs: 2, 3 author 4 Neville Coleman

Family Faviidae | Genus *Favia*

Favia leptophylla
Verrill, 1868

Characters: Colonies are massive, flattened rather than hemispherical and commonly over one metre across. Corallites are compact, neatly rounded and uniform in size. Septo-costae are exsert and are equal and uniformly spaced. **Colour:** Uniform dark grey with pale grey oral discs. **Similar species:** None in the Atlantic Ocean. Similar to *Favia favus* of the Indo-Pacific. See also *F. marshae*. **Habitat:** Shallow reef environments. **Abundance:** Generally uncommon.

Taxonomic references: Laborel (1969), Amaral (1992). **Identification guide:** Hetzel and Castro (1994).

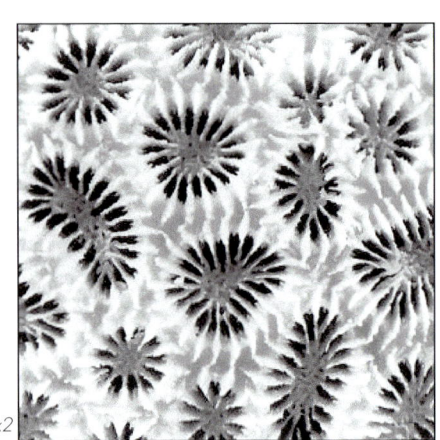

x2

Favia rosaria
Veron, this publication

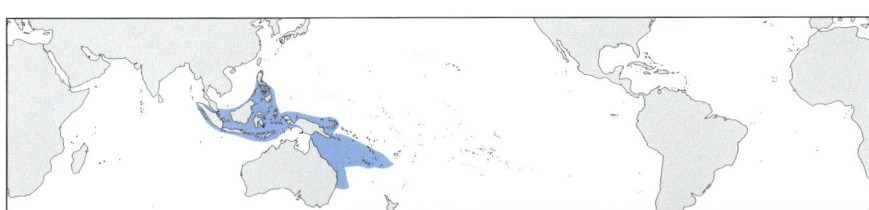

Characters: Colonies are submassive to encrusting and often up to one metre across. Corallites are crowded, up to 20 millimetres diameter, and have low walls. Extratentacular budding is common. Septo-costae are uniform, not exsert. Septa have fine teeth. Paliform lobes are inconspicuous. **Colour:** Distinctive pinkish-brown with darker corallite inner walls and pale oral discs. **Similar species:** The colouration is distinctive underwater. *Favia favus* and *F. danae* have corallites of similar size. The former have coarser septo-costae, the latter have more exsert corallites. **Habitat:** Shallow reef slopes. **Abundance:** Rare.

Taxonomic note: See 'New species described in *Corals of the World*' (Veron, in preparation) for further information.

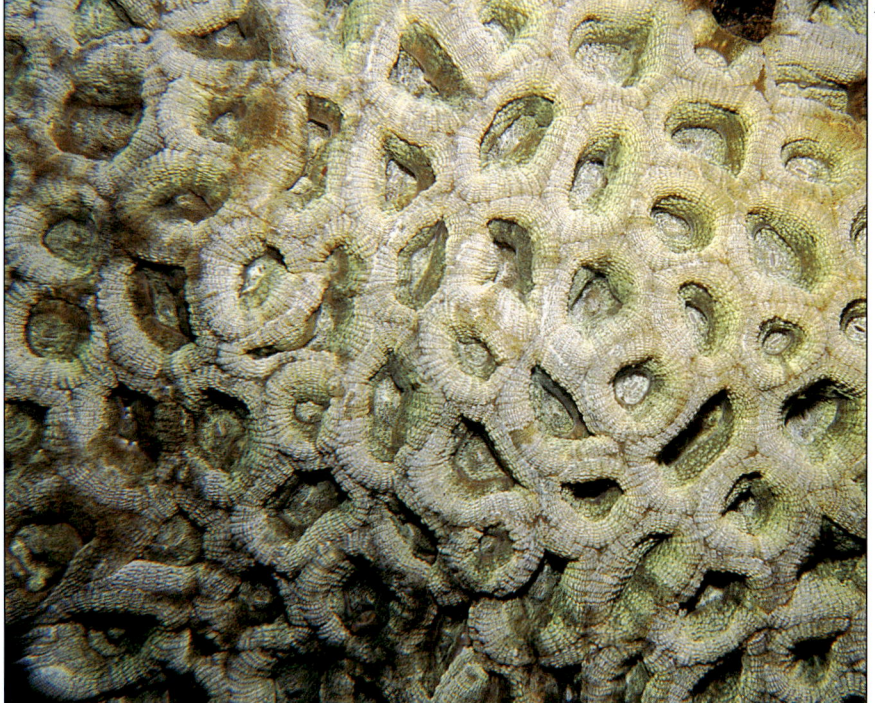

1 *Favia leptophylla.* Showing the typically smooth surface of a large massive colony. ABROLHOS ISLANDS, BRAZIL *Photograph: author*

2 *Favia leptophylla.* Corallite detail. ABROLHOS ISLANDS, BRAZIL *Photograph: author*

3 *Favia rosaria.* Small colony showing the most common colour pattern of the species. PAPUA NEW GUINEA *Photograph: author*

4 *Favia rosaria.* Corallite detail. FLORES, INDONESIA *Photograph: Neville Coleman*

Family Faviidae   Genus *Favia*

Favia lizardensis
Veron and Pichon, 1977

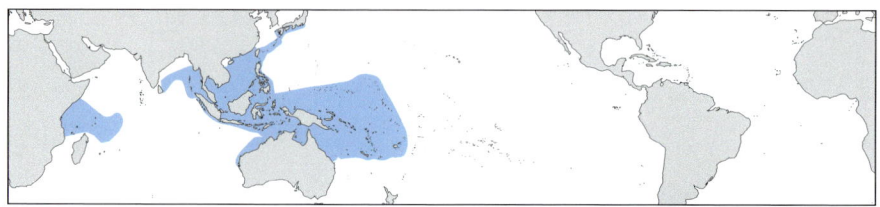

Characters: Colonies are massive and are commonly over one metre across. Corallites are circular, and regularly spaced. Corallite walls are thick but have fine rims. Septa are uniformly thin and widely spaced, without paliform lobes. Costae are well developed. **Colour:** Pinkish brown with cream or greenish oral discs. **Similar species:** A distinct species when seen underwater. Closest to *Favia rosaria*. **Habitat:** Upper reef slopes. **Abundance:** Seldom common.

Taxonomic reference: Veron, Pichon and Wijsman-Best (1977). **Identification guides:** Veron (1986), Nishihira and Veron (1995).

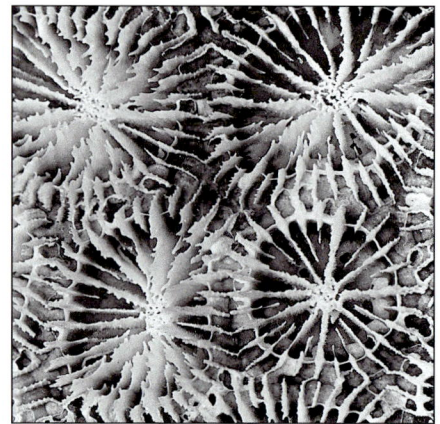
x2

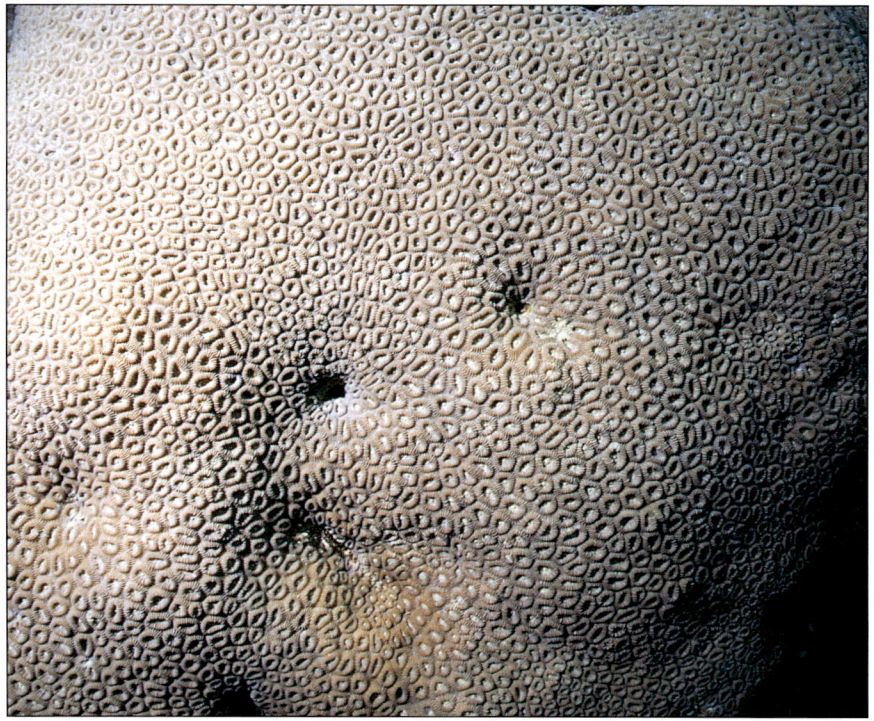

Favia rotumana
(Gardiner, 1899)

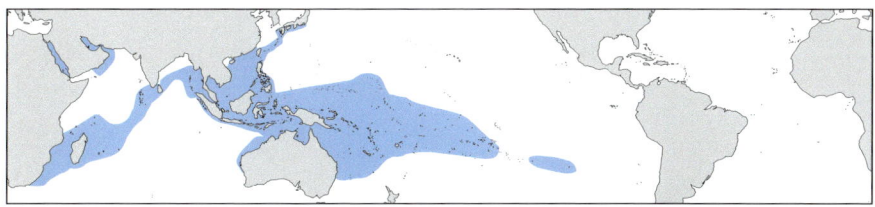

Characters: Colonies are usually flat and subplocoid. Corallites on convex surfaces are clearly subplocoid, those on flat surfaces are crowded, becoming cerioid and irregular in shape, and may have up to three centres. Septa are exsert, thin and irregular; they plunge steeply inside the wall. Paliform lobes are poorly developed or absent. **Colour:** A wide range, usually with contrasting corallite walls and oral discs. **Similar species:** *Favia matthaii*, which also has irregular septa but corallites are circular and smaller. See also *Favites russelli*. **Habitat:** Upper reef slopes. **Abundance:** Uncommon.

Taxonomic references: Wijsman-Best (1972), Veron, Pichon and Wijsman-Best (1977). **Identification guides:** Veron (1986), Sheppard and Sheppard (1991), Nishihira and Veron (1995).

x2

1 *Favia lizardensis*. A large massive colony. SCOTT REEF, WESTERN AUSTRALIA *Photograph: author*

2, 3 *Favia lizardensis*. The characteristic shape and colour of corallites. **2** SEYCHELLES **3** GREAT BARRIER REEF, AUSTRALIA *Photographs: 2 author 3 Wendy Morris*

4 *Favia rotumana*. The typically ragged appearance of cerioid corallites. GREAT BARRIER REEF, AUSTRALIA *Photograph: Ed Lovell*

5 *Favia rotumana*. A small colony with relatively widely separated subplocoid corallites. PAPUA NEW GUINEA *Photograph: author*

Favia marshae
Veron, this publication

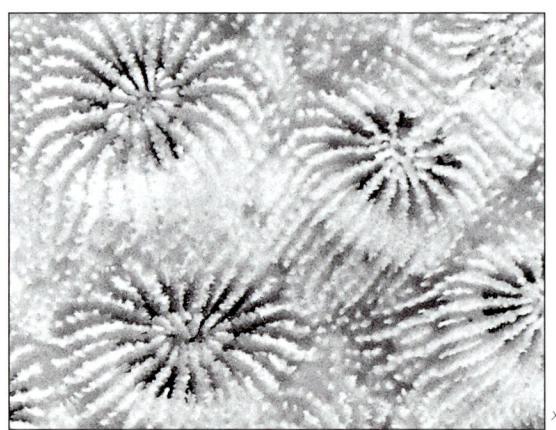
×2

Characters: Colonies are dome-shaped or flat. Corallites are shallow, circular, neatly arranged and 15-20 millimetres diameter. Corallites at the colony margin are frequently arranged in concentric rows. Septa are fine. Paliform lobes are weakly formed or absent. **Colour:** Uniform pale grey with contrasting walls and centres. **Similar species:** *Favia rotundata*, which has larger, more fleshy corallites. Superficially resembles *Favites vasta*. **Habitat:** Shallow reef environments. **Abundance:** Uncommon.

Taxonomic note: See 'New species described in *Corals of the World*' (Veron, in preparation) for further information.

1 *Favia marshae*. Colonies have a smooth surface and neatly rounded corallites. BOLINAO, PHILIPPINES *Photograph: author*

2 *Favia marshae*. Edge of colony. GUAM *Photograph: Gustav Paulay*

3, 4 *Favia danae*. This species has large rough corallites. **3** SINAI PENINSULA, EGYPT **4** GREAT BARRIER REEF, AUSTRALIA *Photographs: author*

5 *Favia danae*. Showing corallite detail, notably the strongly beaded septo-costae. GREAT BARRIER REEF, AUSTRALIA *Photograph: Ed Lovell*

Favia danae
Verrill, 1872

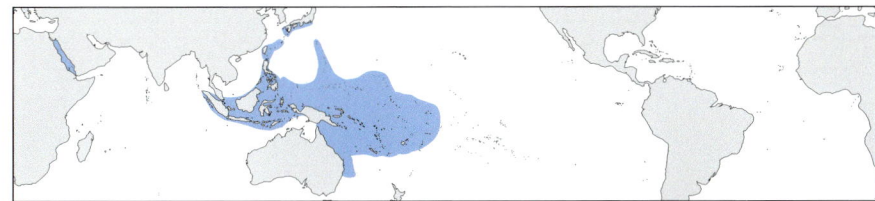

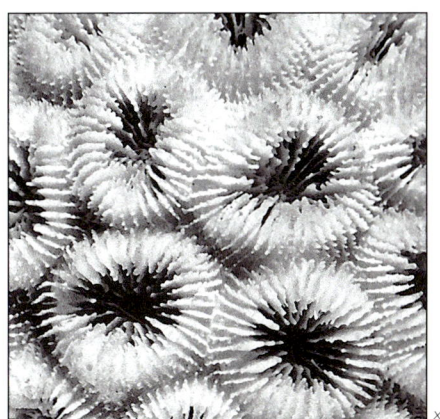

Characters: Colonies are massive and usually small. Corallites are conical, with thick walls. Septo-costae are irregular and thick. Costae are strongly beaded. Paliform lobes are weakly developed. **Colour:** Uniform or mottled brown, green, yellow-green or grey. **Similar species:** *Favia favus*, which has more uniform septo-costae and costae which are less beaded. See also *F. maritima*. **Habitat:** A wide range of reefs and rocky foreshores of subtropical locations. **Abundance:** Usually uncommon.

Taxonomic reference: Verrill's original description/specimens. **Identification guide:** Nishihira and Veron (1995).

Favia rotundata
(Veron and Pichon, 1977)

Characters: Colonies are dome-shaped or flat. Corallites are thick walled and circular, tending to be cerioid and 19-22 millimetres diameter. Polyps are fleshy and circular in outline. **Colour:** Pale grey, yellowish or brown. Corallite rims are usually distinctively coloured. **Similar species:** *Favia maxima* and *F. veroni*, which have corallites of similar size, but these are generally more exsert. **Habitat:** Reef slopes and lagoons. **Abundance:** Sometimes common.

Taxonomic reference: Veron, Pichon and Wijsman-Best (1977). **Identification guides:** Veron (1986), Sheppard and Sheppard (1991), Nishihira and Veron (1995).

Family Faviidae Genus *Favia*

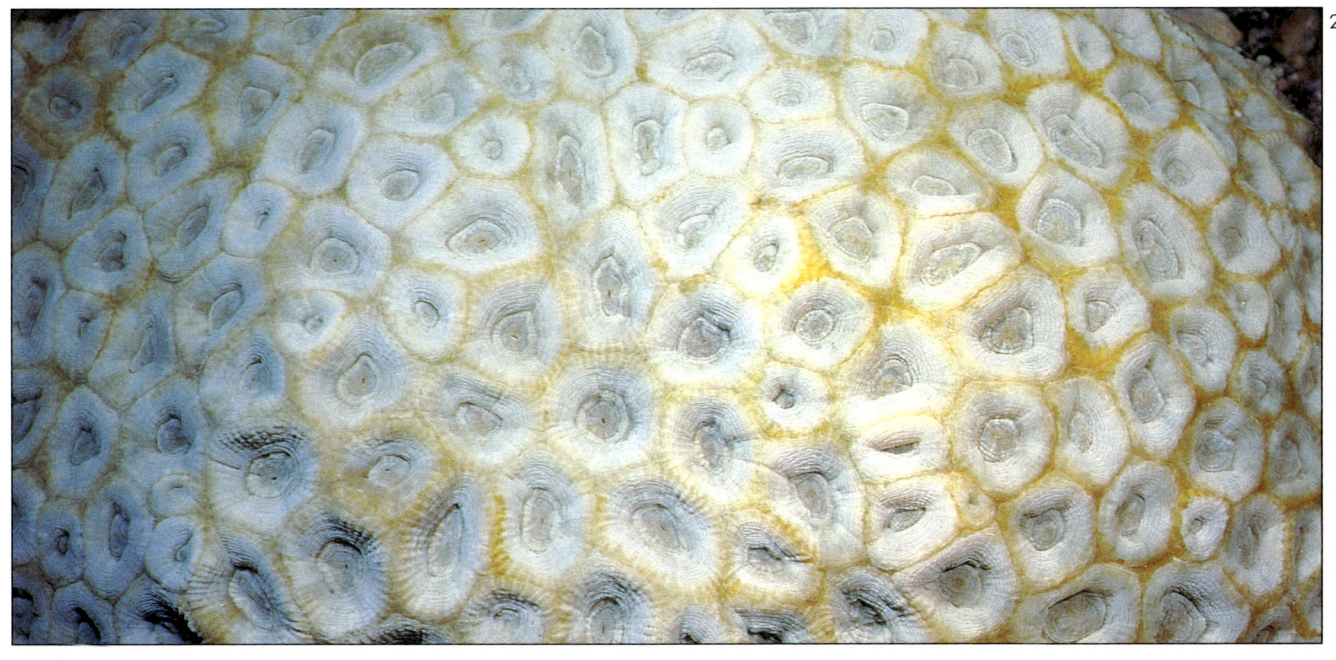

1 Typical appearance of colonies in a protected environment. BALI, INDONESIA *Photograph: Jim Maragos*

2 A large massive colony on a reef flat. GREAT BARRIER REEF, AUSTRALIA *Photograph: Ed Lovell*

3-5 Variation in corallite detail. **3** BALI, INDONESIA **4** GREAT BARRIER REEF, AUSTRALIA **5** HONSHU, JAPAN *Photographs: 3 Neville Coleman 4, 5 author*

Family Faviidae Genus *Favia*

Favia maxima
Veron and Pichon, 1977

Characters: Colonies are massive and usually small. Corallites have well defined walls. Septa are regular, thickened at the wall and with conspicuous paliform lobes forming a crown around the columella. Polyps may be fleshy. **Colour:** Brown or yellow-brown with dull green or white oral discs. **Similar species:** *Favia maritima*, which has more exsert corallites and does not have prominent paliform lobes. See also *F. veroni* and *F. vietnamensis*. **Habitat:** Upper reef slopes. **Abundance:** Rare.

Taxonomic reference: Veron, Pichon and Wijsman-Best (1977).

x2

1

2

Favia vietnamensis
Veron, this publication

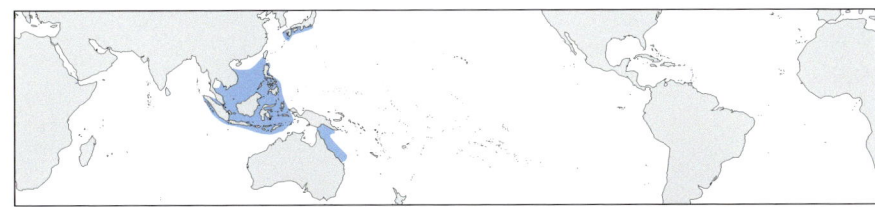

Characters: Colonies are usually small. Corallites are irregularly shaped and deeply excavated. Septa are irregular in height. Paliform lobes are weakly developed. Colonies are very fleshy. **Colour:** Brown or grey, either uniform or with distinctive oral discs. **Similar species:** Unlike other *Favia* but resembles subphaceloid growth-forms of *Caulastrea tumida* underwater. **Habitat:** Protected reef environments. **Abundance:** Rare.

Taxonomic note: See 'New species described in *Corals of the World*' (Veron, in preparation) for further information.

1 *Favia maxima*. Colony surface. GUAM *Photograph: Gustav Paulay*

2 *Favia maxima*. Corallites are large with a conspicuous paliform crown that can be seen underwater. CALAMIAN ISLANDS, PHILIPPINES *Photograph: author*

3 *Favia vietnamensis*. Surface of a colony at the northern extremity of the distribution range. HONSHU, JAPAN *Photograph: author*

4, 5 *Favia vietnamensis*. Variation in corallite shape and colour. **4** GREAT BARRIER REEF, AUSTRALIA **5** VIETNAM *Photographs: 4 Ed Lovell, 5 author*

Favia veroni
Moll and Borel-Best, 1984

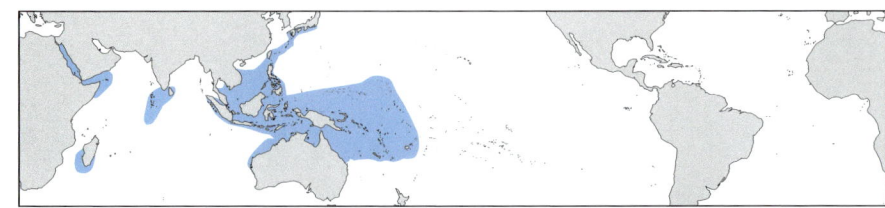

Characters: Colonies are massive. Corallites are compacted together, often irregularly projecting and irregular in outline, with large calices up to 10 millimetres deep. Paliform lobes are absent. **Colour:** Corallite walls are usually rich brown or red with cream oral discs. **Similar species:** *Favia maxima*, which has similar sized corallites with conspicuous paliform lobes. *Favia maritima* has smaller, more exsert, more widely spaced corallites. See also *F. vietnamensis*. **Habitat:** Reef slopes. **Abundance:** Rare.

Taxonomic reference: Moll and Borel-Best (1984). **Identification guide:** Nishihira and Veron (1995).

1 A small massive colony. CALAMIAN ISLANDS, PHILIPPINES *Photograph: author*

2 Large colony with excavated corallites. SINAI PENINSULA, EGYPT *Photograph: Mary Stafford-Smith*

3 Colony surface. BOLINAO, PHILIPPINES *Photograph: author*

4 Corallite detail. BALI, INDONESIA *Photograph: Neville Coleman*

Favia maritima
(Nemenzo, 1971)

Characters: Colonies are massive and usually hemispherical. Corallites are exsert. Septa are uniform, fine and numerous. Paliform lobes are poorly developed or absent. **Colour:** Dark brown or greenish, sometimes with pale oral discs. **Similar species:** *Favia maxima*. See also *F. favus* and *F. danae*, both of which have smaller corallites with less numerous and less regular septa. **Habitat:** Reef slopes. **Abundance:** Uncommon.

Taxonomic reference: Nemenzo (1971).
Identification guides: Veron (1986), Nishihira and Veron (1995).

Family Faviidae Genus *Favia*

1 Colony surface. GREAT BARRIER REEF, AUSTRALIA *Photograph: Ed Lovell*

2 Characteristic appearance of a small colony. GREAT BARRIER REEF, AUSTRALIA *Photograph: Ed Lovell*

3, 4 Variation in corallite shape and colour. **3** GREAT BARRIER REEF, AUSTRALIA **4** BALI, INDONESIA *Photographs: 3 author 4 Valerie Taylor*

Family Faviidae

GENUS *BARABATTOIA*
Yabe and Sugiyama, 1941

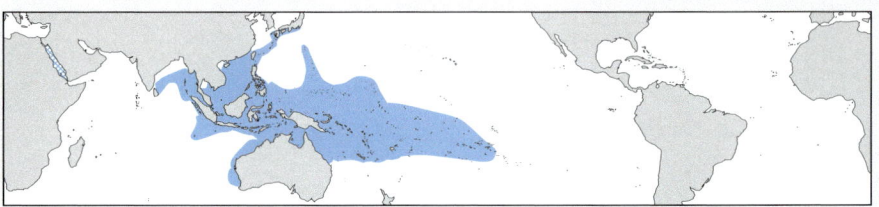

Characters: This is a genus of convenience to accommodate two species that appear to have affinities with each other, but which are outside the boundaries of *Favia*. Colonies have tubular corallites which fuse irregularly. Budding is primarily extratentacular. **Similar genera**: *Favia*, which has plocoid rather than tubular corallites and primarily intratentacular budding. See also *Caulastrea* and *Montastrea*. **Fossil record:** none.

Taxonomic reference: Veron, Pichon and Wijsman-Best (1977).

Barabattoia laddi
(Wells, 1954)

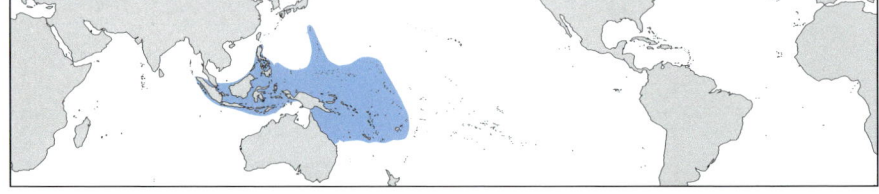

Characters: Colonies are clusters of tubular corallites. Budding is extratentacular. Corallites bifurcate at approximately 10 millimetre intervals and frequently join. Costae are in two alternating orders and are strongly beaded. **Colour**: Pale brown. **Similar species**: *Barabattoia amicorum* and *Montastrea*. **Habitat**: Recorded only from shallow lagoons. **Abundance**: Rare.

Taxonomic reference: Wells (1954).

Barabattoia amicorum
(Milne Edwards and Haime, 1850)

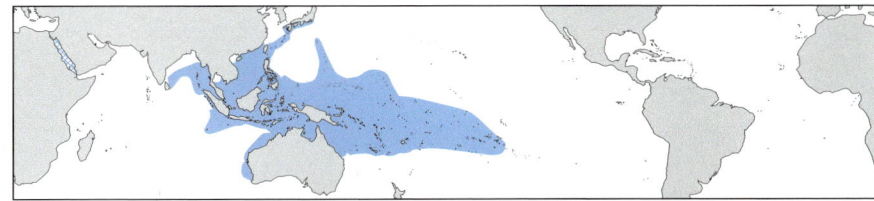

Characters: Colonies are massive and usually small. Corallites are plocoid to tubular. Budding is primarily extratentacular. Costae are equal and well developed. Paliform lobes are usually poorly developed. Columellae are small and compact. Tentacles are extended only at night. **Colour**: Usually brown, cream or green with pale oral discs. **Similar species**: *Barabattoia laddi*, which has longer corallites and alternating costae. **Habitat**: Shallow reef environments, especially reef backs protected from strong wave action. **Abundance**: Uncommon.

Taxonomic reference: Veron, Pichon and Wijsman-Best (1977).
Identification guides: Veron (1986), Nishihira and Veron (1995).

1 *Barabattoia laddi*. A *Montastrea*-like colony with compact corallites. BALI, INDONESIA Photograph: Jim Maragos

2 *Barabattoia laddi*. A small colony showing extreme development of plocoid corallites. GUAM Photograph: Gustav Paulay

3, 5 *Barabattoia amicorum*. Common variation in corallite shape and colour. **3** GREAT BARRIER REEF, AUSTRALIA **5** VIETNAM Photographs: **3** Ed Lovell, **5** author

4 *Barabattoia amicorum*. Characteristic appearance of a medium sized colony. Note the prevalence of extratentacular budding. CALAMIAN ISLANDS, PHILIPPINES Photograph: author

Family Faviidae — Genus *Favites*

GENUS *FAVITES*
Link, 1807

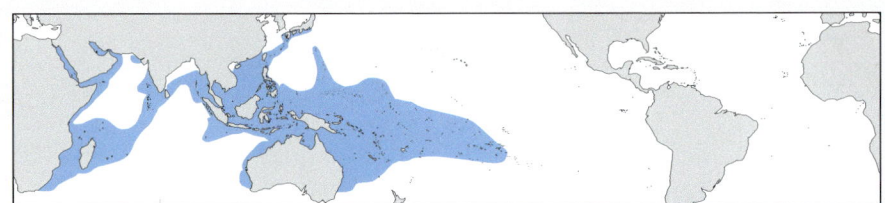

Characters: Colonies are usually massive and either flat or dome-shaped. Corallites are monocentric and cerioid, occasionally subplocoid. Adjacent corallites mostly share common walls. Paliform lobes are seldom well developed. Tentacles are extended only at night and are tapered, like those of *Favia*. **Similar genera:** *Favites* is similar to *Favia* and also to *Goniastrea*. *Goniastrea* may be cerioid like *Favites*, in which case it is distinguished by the presence of exsert paliform lobes, also by having a regular pattern of septa with relatively fine teeth. **Earliest fossil record:** Eocene of the Caribbean, Oligocene of the Tethys.

Taxonomic references: Wijsman-Best (1976), Veron, Pichon and Wijsman-Best (1977). **Summary key:** See pp452–3.

1 *Favites* species are often easier to identify underwater if they occur together. Here, a difficult species pair, *F. abdita* (left) and *F. flexuosa* (right), are easily distinguished. Papua New Guinea Photograph: Gerry Allen

2 *Favites vasta* showing progressive stages in intratentacular budding. Sinai Peninsula, Egypt Photograph: author

3 Tentacles extended at night. Great Barrier Reef, Australia Photograph: author

Identification of *Favites* species

These provide the same sorts of taxonomic problems as *Favia* species because they have similar habitat preferences and geographic ranges. *Favites* is particularly common on subtropical reefs and in non-reef habitats where they have relatively well calcified skeletons and where colonies are generally colourful.

Favites can be divided into four species groups according to corallite size:

Group 1: Species with very small corallites (averaging less than 6 mm diameter), page 136.

Group 2: Species with small corallites (averaging 6-10 mm diameter), page 138.

Group 3: Species with middle sized corallites (10-14 mm diameter), page 144.

Group 4: Species with large corallites (over 14 mm diameter), page 152.

Group 1: Species with very small corallites (averaging less than 6 mm diameter)

Favites stylifera
(Yabe and Sugiyama, 1937)

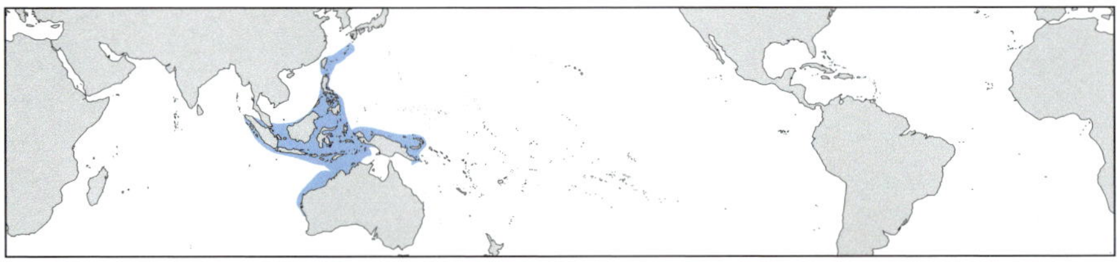

Characters: Colonies are encrusting to submassive. Corallites are irregular in shape, 3-6 millimetres diameter. Septa are few and highly contorted, with irregular teeth. The paliform crown is weakly developed. **Colour**: Pale brown, sometimes with green oral discs. **Similar species**: *Favites micropentagona*, which has corallites of similar size but with a uniform shape. See also *Platygyra yaeyamaensis* which has larger corallites forming valleys. **Habitat**: Upper reef slopes. **Abundance**: Uncommon to rare.

Taxonomic reference: Yabe and Sugiyama's original description/specimens. **Identification guide:** Nishihira and Veron (1995).

Favites micropentagona
Veron, this publication

Characters: Colonies are encrusting to submassive. Corallites are pentagonal in shape and 3-4 millimetres diameter. Septa are in two alternating cycles, with irregular teeth. The paliform crown is clearly developed. **Colour:** Pale brown, sometimes with dark oral discs. **Similar species:** *Favites pentagona*, which has larger but otherwise similar corallites. See also *F. stylifera*. **Habitat:** Upper reef slopes. **Abundance:** Uncommon.

Taxonomic note: See 'New species described in *Corals of the World*' (Veron, in preparation) for further information.

1-5 *Favites stylifera*. Variation in corallite shape and colour. **1, 2** PAPUA NEW GUINEA **3** BOLINAO, PHILIPPINES **4** ASHMORE REEF, WESTERN AUSTRALIA **5** RYUKYU ISLANDS, JAPAN *Photographs: author*

6-9 *Favites micropentagona*. Corallite variation in encrusting colonies. **6-8** CALAMIAN ISLANDS, PHILIPPINES **9** RYUKYU ISLANDS, JAPAN *Photographs: author*

10 *Favites micropentagona*. Corallite detail. CALAMIAN ISLANDS, PHILIPPINES *Photograph: author*

Group 2: Species with small corallites (6–10 mm diameter)

Favites pentagona
(Esper, 1794)

Characters: Colonies are submassive to encrusting, sometimes forming irregular columns. They commonly exceed one metre across. Corallites are thin walled and angular. Septa are few in number. Paliform lobes are well developed, commonly forming a conspicuous crown. **Colour:** Often brightly coloured, brown or red, commonly with green oral discs. **Similar species:** *Favites bestae* and *F. micropentagona*. The paliform crown makes this species *Goniastrea*-like. **Habitat:** Shallow reef environments. **Abundance:** Sometimes common.

Taxonomic note: This species is divisible into several smaller semi-distinct taxonomic units, see 'What are species?' p425.
Taxonomic references: Wijsman-Best (1972), Veron, Pichon and Wijsman-Best (1977).
Identification guides: Veron (1986), Sheppard and Sheppard (1991), Nishihira and Veron (1995), Carpenter *et al.* (1997).

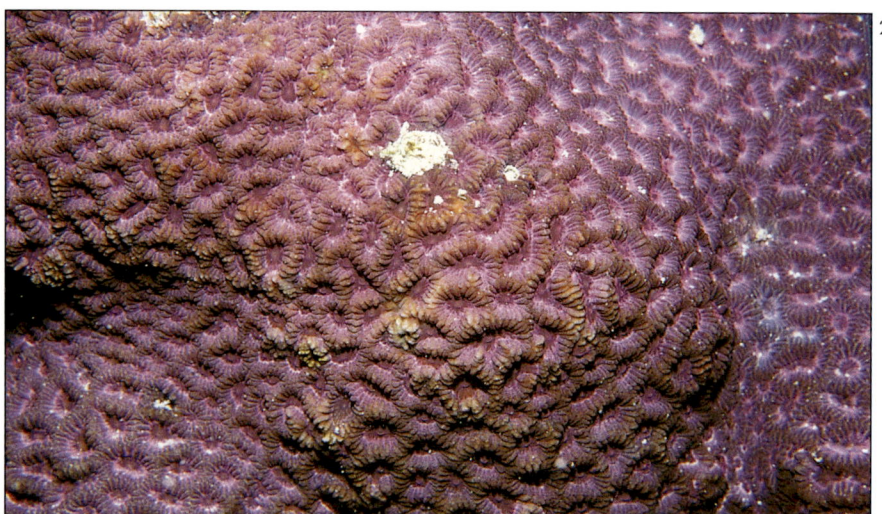

Family Faviidae Genus *Favites*

1 Central parts of colonies in shallow water commonly have irregular upper surfaces. DAMPIER ARCHIPELAGO, WESTERN AUSTRALIA *Photograph: author*

2, 4 Encrusting colonies, exposed to strong wave action. **2** PEMBA ISLAND, TANZANIA **4** SEYCHELLES *Photographs: author*

3, 5, 6 Showing some of the wide range of corallite shapes and colours found in different habitats. **3** PEMBA ISLAND, TANZANIA **5** PORT HEDLAND, WESTERN AUSTRALIA **6** CALAMIAN ISLANDS, PHILIPPINES *Photographs: 3, 6 author 5 Neville Coleman*

Family Faviidae — Genus *Favites*

Favites bestae
Veron, this publication
(new name)

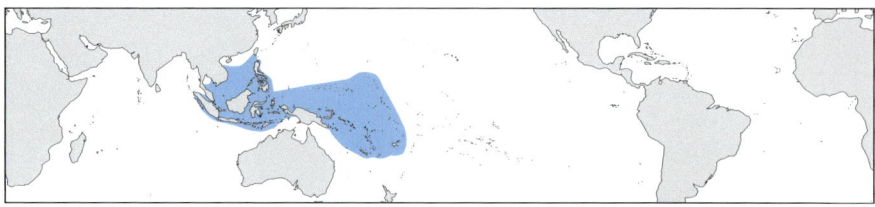

Characters: Colonies are submassive to encrusting. Corallites are thick walled and rounded, becoming subplocoid. Septa are few in number, uniform in height and are usually in two (alternating) orders. Paliform lobes and columellae are well developed. **Colour:** Browns and greens, usually with contrasting walls and centres. **Similar species:** *Favites pentagona*, which has angular corallites and irregular septa and *F. chinensis*, which has no paliform lobes. See also *Montastrea colemani*. **Habitat:** Shallow reef environments. **Abundance:** Rare.

Taxonomic note and references: This is a new name for *Favites melicerum* as discussed by Wijsman-Best (1972) and Veron, Pichon and Wijsman-Best (1977). See 'New species described in *Corals of the World*' (Veron, in preparation) for further information.

x2

1 *Favites bestae*. A small colony. BALI, INDONESIA *Photograph: Jim Maragos*

2 *Favites bestae*. Corallites are similar to those of *F. pentagona*, but are rounded and lack exsert septa. VIETNAM *Photograph: author*

3-5 *Favites acuticollis*. Variation in corallite structure and colour. **3** TAIWAN **4** RYUKYU ISLANDS, JAPAN **5** CALAMIAN ISLANDS, PHILIPPINES *Photographs: author*

Favites acuticollis
(Ortmann, 1889)

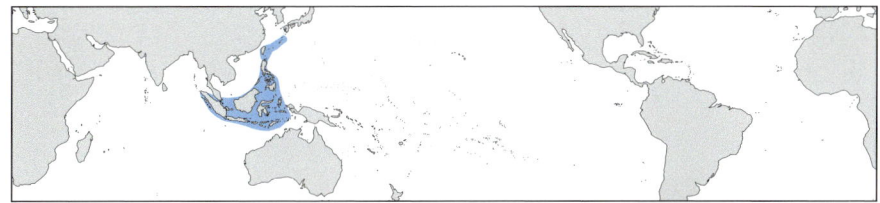

Characters: Colonies are submassive to encrusting. Corallites are deep with very thin angular walls giving colonies a honeycomb appearance. Corallites are usually less than 7 millimetres diameter. Septa are few and widely spaced, paliform lobes are absent. **Colour:** Dark colours, often with white upper margins to walls. **Similar species:** *Favites pentagona*, which has thicker walls, slightly larger corallites and well developed paliform lobes. See also *F. spinosa*, which has very exsert septa and *Goniastrea columella*, which has paliform lobes. **Habitat:** Shallow reef environments. **Abundance:** Rare.

Taxonomic reference: Wijsman-Best (1972).

Favites spinosa
(Klunzinger, 1879)

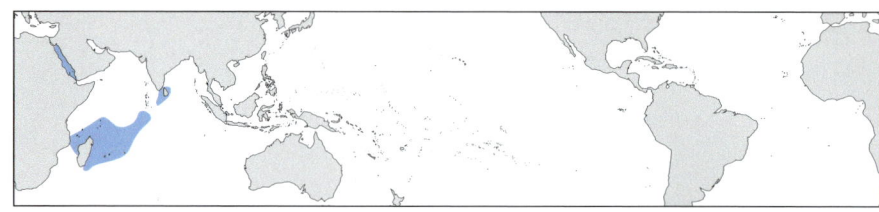

Characters: Colonies are small, massive and rounded. Corallites are deeply excavated, with angular walls. Septa are straight, widely spaced and are usually in two alternating orders. Paliform lobes are weakly developed. Septa have very prominent teeth, which have ragged margins. Columellae are small and compact. **Colour:** Walls are off-white, centres are dark. **Similar species:** *Favites acuticollis*. See also *F. flexuosa*, which has much larger corallites and usually less spiny septa. Can be confused with *Acanthastrea* species (Mussidae), especially *A. subechinata*, underwater. **Habitat:** A wide range of reef environments. **Abundance:** Uncommon.

Taxonomic reference: Klunzinger's original description/specimens.

1 *Favites spinosa*. An encrusting colony. SEYCHELLES *Photograph: author*

2, 4 *Favites spinosa*. Corallite detail. SEYCHELLES *Photographs: author*

3 *Favites spinosa*. Angular corallites with a rough surface. SRI LANKA *Photograph: author*

5-8 *Favites chinensis*. Common variation in corallite shape and colour. **5, 7, 8** GREAT BARRIER REEF, AUSTRALIA **6** CALAMIAN ISLANDS, PHILIPPINES *Photographs:* **5** Len Zell **6-8** *author*

Favites chinensis
(Verrill, 1866)

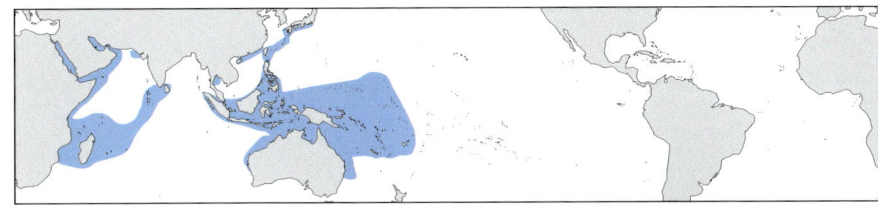

Characters: Colonies are massive and rounded. Corallites are shallow, angular to subplocoid, with thin walls. Septa are straight and even. Those of adjacent corallites are aligned across the wall. There are no paliform lobes. **Colour:** Usually yellow or greenish-brown. **Similar species:** *Favites complanata*, which has larger, more excavated corallites with thicker walls. See also *F. bestae*. **Habitat:** A wide range of reef environments. **Abundance:** Uncommon.

Taxonomic references: Wijsman-Best (1972), Veron, Pichon and Wijsman-Best (1977). **Identification guides:** Veron (1986), Sheppard and Sheppard (1991), Nishihira and Veron (1995).

Group 3: Species with middle sized corallites (10-13 mm diameter)

Favites halicora
(Ehrenberg, 1834)

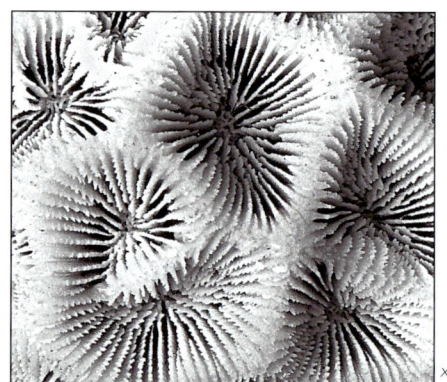
x2

Characters: Colonies are massive, either rounded or hillocky. Corallites have very thick walls and tend to become subplocoid. Paliform lobes may be developed. **Colour:** Usually uniform pale yellowish- or greenish-brown. **Similar species:** *Favites abdita*, which has more angular corallites with thinner walls and no paliform lobes. **Habitat:** Shallow reef environments. **Abundance:** Usually uncommon.

Taxonomic reference: Veron, Pichon and Wijsman-Best (1977). **Identification guides:** Veron (1986), Sheppard and Sheppard (1991), Nishihira and Veron (1995).

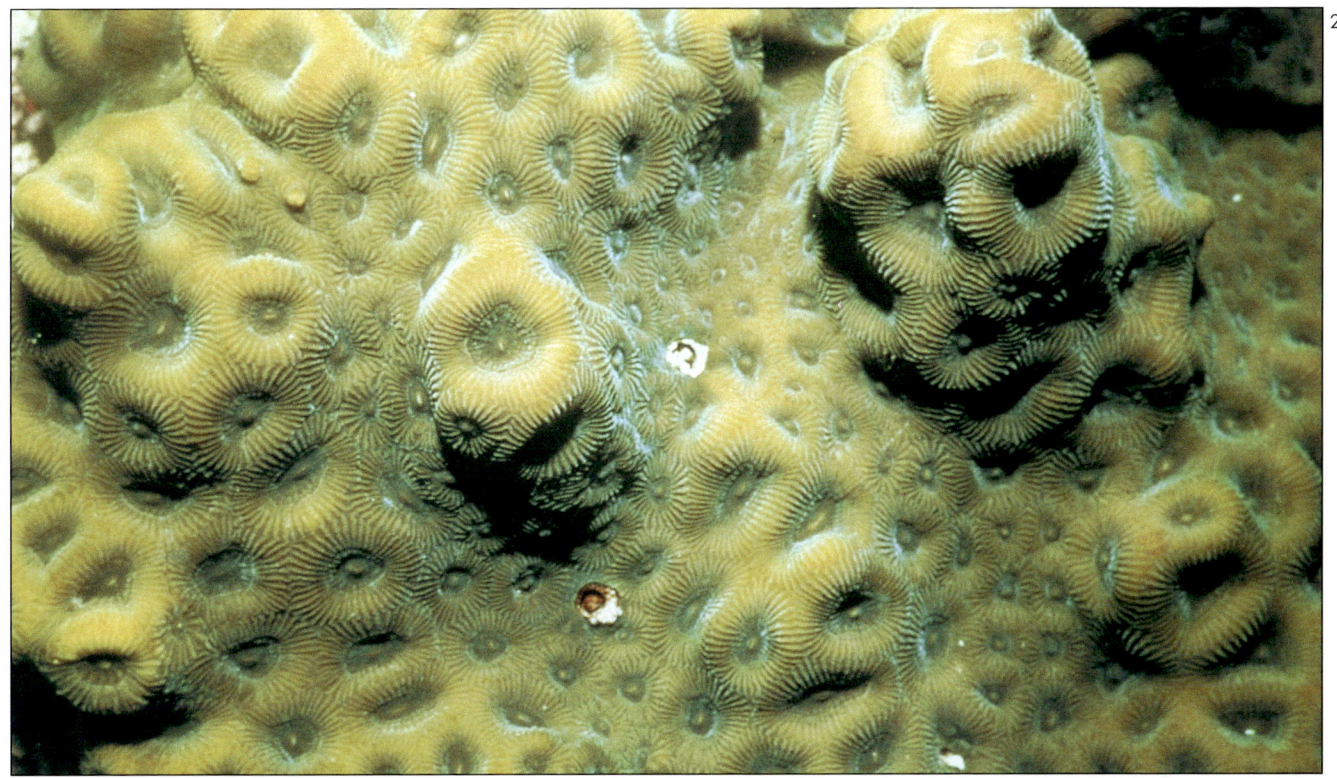

1 Showing the common appearance of colonies in tropical locations. BALI, INDONESIA *Photograph: Gerry Allen*

2 This species has a solid appearance and often has an irregular upper surface. HOUTMAN ABROLHOS ISLANDS, SOUTH-WEST AUSTRALIA *Photograph: Clay Bryce*

3 Corallite detail. GUAM *Photograph: Gustav Paulay*

4 Colonies commonly have a hillocky surface in shallow water. TAIWAN *Photograph: author*

Favites abdita
(Ellis and Solander, 1786)

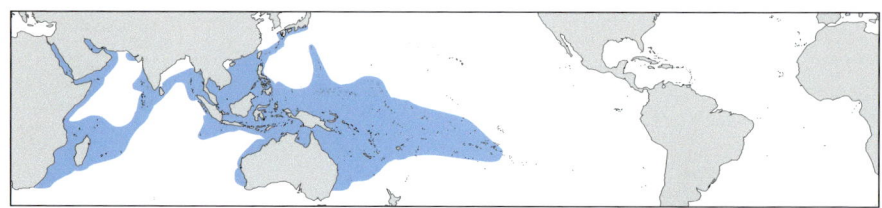

Characters: Colonies are massive, either rounded or hillocky and sometimes over one metre across. Corallites are rounded, with thick walls. Septa are straight, with exsert teeth. **Colour:** Dark in turbid environments, otherwise pale brown with brown or green oral discs. **Similar species:** *Favites halicora* and *F. flexuosa*. **Habitat:** Most reef environments. **Abundance:** Common.

Taxonomic references: Wijsman-Best (1972), Veron, Pichon and Wijsman-Best (1977).
Identification guides: Veron (1986), Sheppard and Sheppard (1991), Nishihira and Veron (1995).

Family Faviidae — Genus *Favites*

1 A massive colony with an encrusting base. PAPUA NEW GUINEA Photograph: Gerry Allen

2-4 Common variation in corallite shape and colour. **2, 3** GREAT BARRIER REEF, AUSTRALIA **4** ASHMORE REEF, WESTERN AUSTRALIA Photographs: 2, 4 author 3 Neville Coleman

5 Corallite detail. GREAT BARRIER REEF, AUSTRALIA Photograph: Valerie Taylor

6 Colony surface. GUAM Photograph: Gustav Paulay

Favites russelli
(Wells, 1954)

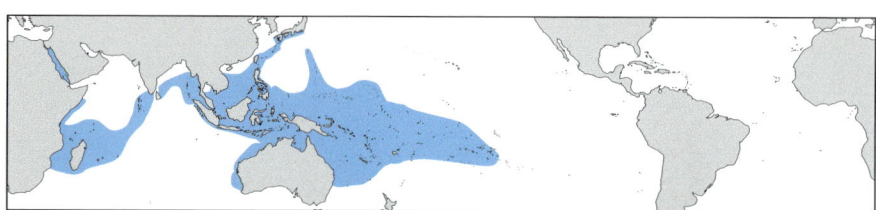

Characters: Colonies are submassive to encrusting. Corallites have thin to thick irregular walls. Paliform lobes are well developed. **Colour:** Usually green, brown or mottled or with green or cream oral discs. **Similar species:** *Favites pentagona*, which has thinner walls and smaller corallites. See also *Favia rotumana*. **Habitat:** Most reef environments. **Abundance:** Usually uncommon.

Taxonomic reference: Veron, Pichon and Wijsman-Best (1977). **Identification guides:** Veron (1986), Nishihira and Veron (1995).

Family Faviidae Genus *Favites*

1, 2, 5 Variation in corallite shape and colour. **1** GREAT BARRIER REEF, AUSTRALIA **2** CALAMIAN ISLANDS, PHILIPPINES **5** RYUKYU ISLANDS, JAPAN *Photographs: 1 Len Zell 2, 5 author*

3, 6 Corallite detail. GREAT BARRIER REEF, AUSTRALIA *Photographs: Ed Lovell*

4 A small encrusting colony. GREAT BARRIER REEF, AUSTRALIA *Photograph: Ed Lovell*

Favites complanata
(Ehrenberg, 1834)

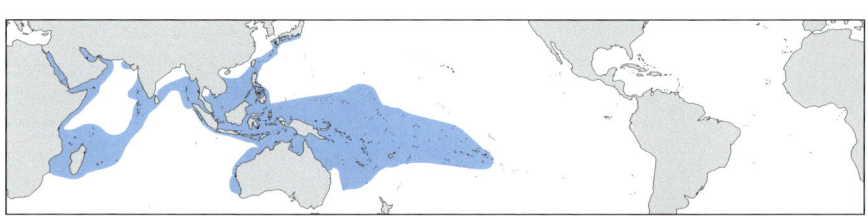

Characters: Colonies are massive with slightly angular corallites. Corallites have thick, rounded walls. Paliform lobes are weakly developed. Columellae are large. Septal spines may be prominent. Costae commonly form a three pointed star where three corallites adjoin. **Colour:** Usually brown, sometimes with green or grey oral discs. **Similar species:** *Favites abdita*, which has more angular corallites and lacks the star-like costal pattern. See also *F. chinensis*. **Habitat:** Most reef environments. **Abundance:** Sometimes common.

Taxonomic reference: Veron, Pichon and Wijsman-Best (1977). **Identification guides:** Veron (1986), Sheppard and Sheppard (1991), Nishihira and Veron (1995).

Family Faviidae　　　　　　　　　　　　　　　　　　　　　　　　　　　　　　　　　Genus *Favites*

1-3 Variation in corallite shape, colour and septal spine development. **1** PEMBA ISLAND, TANZANIA **2** MALDIVE ISLANDS **3** GREAT BARRIER REEF, AUSTRALIA *Photographs: 1, 3 author 2 Neville Coleman*

4 Massive colony. SEYCHELLES *Photograph: author*

5 Polyps are sometimes fleshy, making the species difficult to recognise. GREAT BARRIER REEF, AUSTRALIA *Photograph: author*

Family Faviidae | Genus *Favites*

Group 4: Species with large corallites (over 14 mm diameter)

Favites vasta
(Klunzinger, 1879)

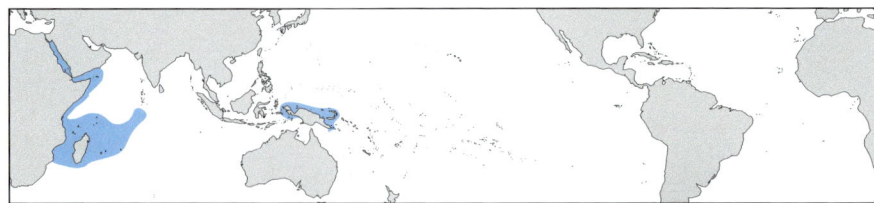

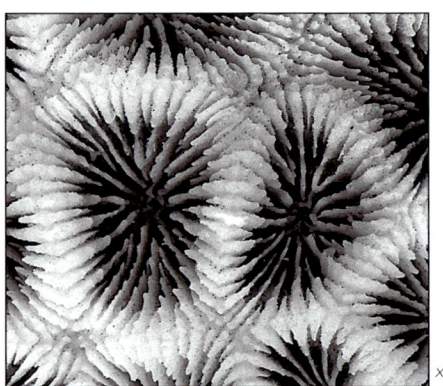

Characters: Colonies are massive and are commonly over one metre across. Corallites are deep and angular and have very thick walls. Septa are equal, uniform, not exsert and finely toothed. Paliform lobes are usually present. **Colour:** Walls are always uniform amber with cream or white oral discs. **Similar species:** *Favites flexuosa*, which has more angular corallites with thinner walls and prominent septal teeth. **Habitat:** Most reef environments. **Abundance:** Uncommon.

Taxonomic note: This species was incorrectly synonymised with *Favites flexuosa* by Veron, Pichon and Wijsman-Best (1977). **Taxonomic reference:** Scheer and Pillai (1983, as *F. flexuosa*). **Identification guide:** Sheppard and Sheppard (1991, as *F. flexuosa*).

x2

1

Family Faviidae Genus *Favites*

1 A hemispherical colony. PAPUA NEW GUINEA *Photograph: author*

2 The usual appearance of a small colony. PEMBA ISLAND, TANZANIA *Photograph: author*

3 Surface of a large colony. ZANZIBAR, TANZANIA *Photograph: author*

4, 5 Corallite detail. PAPUA NEW GUINEA *Photographs: 4 author 5 Neville Coleman*

Favites flexuosa
(Dana, 1846)

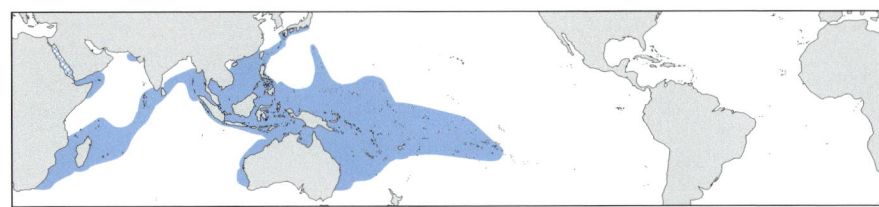

Characters: Colonies are hemispherical or flat. Corallites are angular and deep. Septa are prominent, with large conspicuous teeth. Paliform lobes are weakly developed. **Colour:** A wide range, usually with contrasting walls and oral discs. **Similar species:** *Favites paraflexuosa* and *F. vasta*. *Favites abdita* has smaller, usually less angular corallites. See also the mussid *Acanthastrea echinata*, which has larger septal teeth and thick fleshy polyps. **Habitat:** A wide range of reef environments and rocky foreshores. **Abundance:** Sometimes common, especially in subtropical locations.

Taxonomic references: Wijsman-Best (1972), Veron, Pichon and Wijsman-Best (1977).
Identification guides: Veron (1986), Nishihira and Veron (1995).

x2

Family Faviidae   Genus *Favites*

Favites paraflexuosa
Veron, this publication

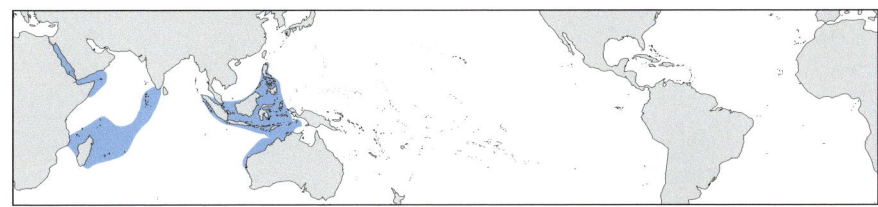

Characters: Colonies are hemispherical or flat. Corallites are angular and deep. Septa are even, with fine teeth. Paliform lobes are weakly developed. **Colour:** Brown with pale oral discs. **Similar species:** *Favites flexuosa*, which has identical corallite shape but septa have conspicuous teeth, a distinction which is clear underwater. *Favites vasta* has larger, less angular corallites and a distinctive colouration. **Habitat:** Shallow reef environments. **Abundance:** Uncommon.

Taxonomic note: See 'New species described in *Corals of the World*' (Veron, in preparation) for further information.

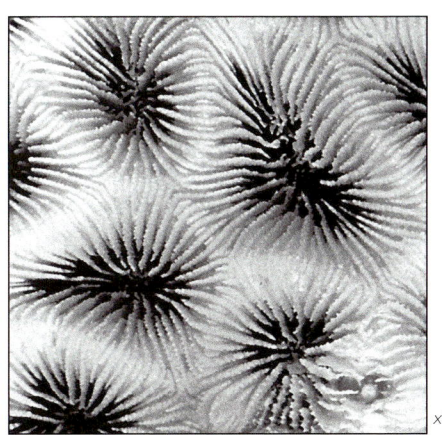
x2

1 *Favites flexuosa*. Surface of an *Acanthastrea*-like massive colony showing angular corallites with prominent teeth. CALAMIAN ISLANDS, PHILIPPINES *Photograph: author*

2, 3 *Favites flexuosa*. Corallite detail. **2** DARWIN, NORTHERN AUSTRALIA **3** LORD HOWE ISLAND, SOUTH-EAST AUSTRALIA *Photographs: Neville Coleman*

4 *Favites paraflexuosa*. Surface of a large colony showing typically angular corallites. SEYCHELLES *Photograph: author*

5 *Favites paraflexuosa*. Corallite detail. CALAMIAN ISLANDS, PHILIPPINES *Photograph: author*

6 *Favites paraflexuosa*. A hemispherical colony. NINGALOO REEFS, WESTERN AUSTRALIA *Photograph: Clay Bryce*

Family Faviidae — Genus *Goniastrea*

GENUS *GONIASTREA*
Milne Edwards and Haime, 1848

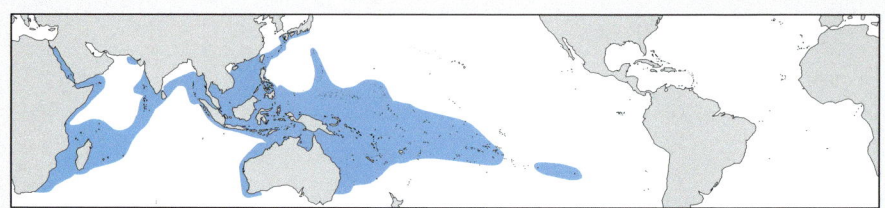

Characters: Colonies are massive and usually spherical or form thick flat plates. Corallites are monocentric and cerioid, to polycentric and meandroid. Paliform lobes are well developed. Meandroid colonies have well defined columella centres. Tentacles are extended only at night. Among the hardiest of corals and often a dominant species of intertidal communities. **Similar genera:** *Goniastrea* has similarities with *Favites*, *Leptoria* and also *Platygyra*. *Platygyra*, like *Goniastrea*, can be cerioid or meandroid but has weakly developed (if any) paliform lobes and columella centres which are seldom distinguishable. **Earliest fossil record:** Eocene of the Caribbean, Oligocene of the Pacific (doubtful record) and Tethys.

Taxonomic references: Wijsman-Best (1976), Veron, Pichon and Wijsman-Best (1977). **Summary key:** See pp452-3.

1 Encrusting *Goniastrea australensis* colonies of different colours compete for space. NORFOLK ISLAND, WESTERN PACIFIC Photograph: author

2 *Goniastrea* colonies commonly encrust rocky substrates in shallow water. SRI LANKA Photograph: author

3 *Goniastrea* species typically have a wide range of growth-forms. Many can form columnar colonies like this *G. pectinata* growing in shallow water exposed to strong currents. BALI, INDONESIA Photograph: author

4 *Goniastrea* are commonly found in intertidal environments where colonies withstand both pounding waves and summer sun. SCOTT REEF, WESTERN AUSTRALIA Photograph: author

Identification of Goniastrea species

For most *Goniastrea* species there are significant differences in colour and/or skeletal detail between colonies from subtropical and tropical central Indo-Pacific locations. *Goniastrea* are often the dominant corals of intertidal mudflats, rock platforms and some outer reef flats. The genus includes some of the most tolerant of all coral species to aerial exposure, the same set of species occurring in intertidal environments throughout much of the Indo-west Pacific. Extreme skeletal aberrations often occur in intertidal habitats where, for example, normally cerioid species may develop meandroid polyps on their upper surface.

Goniastrea can be divided into three groups, based on corallite shape and size:

Group 1: Monocentric species with corallite diameter mostly less than 5 millimetres, page 158.

Group 2: Predominantly monocentric species with corallite diameter over 5 millimetres, page 164.

Group 3: Predominantly meandroid species, page 170.

Family Faviidae — Genus *Goniastrea*

> **Group 1: Monocentric species with corallite diameter mostly less than 5 millimetres**

Goniastrea minuta
Veron, this publication

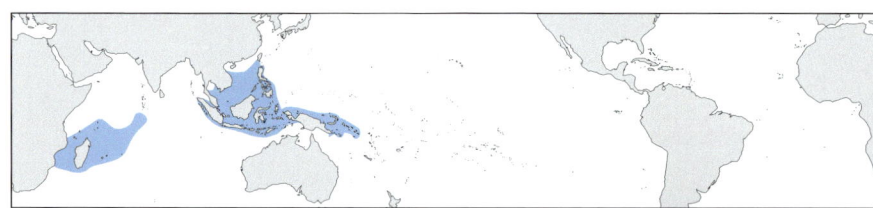

Characters: Colonies are usually encrusting, becoming submassive. Corallites are angular, with a uniform appearance. Walls are usually thin. Long and short septa strongly alternate. Paliform lobes are well developed, forming a neat crown. **Colour:** Uniform pale brown or greenish-brown. Wall tops are pale. **Similar species:** *Goniastrea retiformis*, which has similar corallite characters but corallites are much larger. May be mistaken for the poritids *Porites* and *Poritipora* underwater. **Habitat:** Shallow reef environments. **Abundance:** Uncommon.

Taxonomic note: See 'New species described in *Corals of the World*' (Veron, in preparation) for further information.

x2

1

Family Faviidae | Genus *Goniastrea*

1 A submassive colony. PAPUA NEW GUINEA
Photograph: Gerry Allen

2 The rounded surface of massive colonies.
SEYCHELLES *Photograph: author*

3, 4 Corallite detail. **3** PAPUA NEW GUINEA **4** CALAMIAN ISLANDS, PHILIPPINES *Photographs: author*

5 Corallites are superficially *Porites*-like.
PAPUA NEW GUINEA *Photograph: author*

Goniastrea ramosa
Veron, this publication

Characters: Colonies are small irregular clumps of branches, usually less than 0.3 metres across. Corallites are angular, with thick rounded walls. The ends of branches are ragged, being primarily composed of exsert costae. Septa are in two alternating orders and have large dentations. Columellae are small and compact. Paliform lobes form a distinct crown. **Colour:** Uniform cream or brown. **Similar species:** This is the only branching *Goniastrea*. Corallites are similar to those of *G. retiformis*. *Australogyra zelli* has similar branch shapes but is much larger. **Habitat:** Reef flats sheltered from strong wave action. **Abundance:** Rare.

Taxonomic note: See 'New species described in *Corals of the World*' (Veron, in preparation) for further information.

x2

1 *Goniastrea ramosa*. A small colony on a sheltered reef flat. FLORES, INDONESIA Photograph: author

2 *Goniastrea ramosa*. Colony detail. FLORES, INDONESIA Photograph: author

3 *Goniastrea edwardsi*. A massive colony in shallow water. GREAT BARRIER REEF, AUSTRALIA Photograph: author

4, 5 *Goniastrea edwardsi*. Showing common variation in corallite shape and colour. **4** BOLINAO, PHILIPPINES **5** GREAT BARRIER REEF, AUSTRALIA Photographs: 4 author 5 Ed Lovell

Family Faviidae   Genus *Goniastrea*

Goniastrea edwardsi
Chevalier, 1971

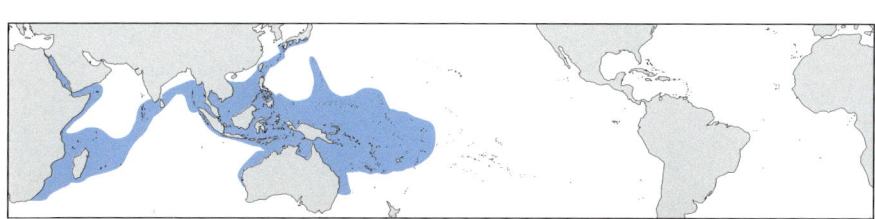

Characters: Colonies are massive, hemispherical or columnar and often over one metre across. Corallites are only slightly angular, with thick rounded walls. Septa are irregular in length and taper from the wall to the columellae, which are small. Paliform lobes are thick. **Colour:** Uniform cream or brown, occasionally with orange centres. **Similar species:** *Goniastrea retiformis*. See also *G. aspera*, which has larger corallites. **Habitat:** Mostly shallow subtidal communities. **Abundance:** Common.

Taxonomic reference: Veron, Pichon and Wijsman-Best (1977). **Identification guides:** Veron (1986), Sheppard and Sheppard (1991), Nishihira and Veron (1995).

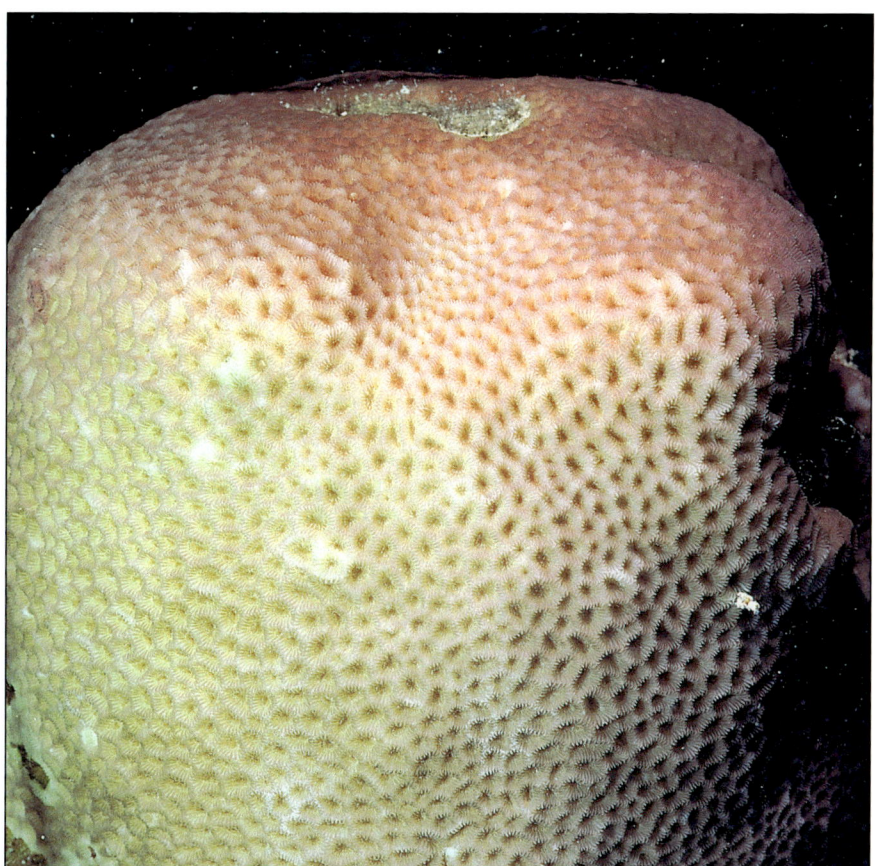

Goniastrea retiformis
(Lamarck, 1816)

Characters: Colonies are massive, hemispherical, flat or columnar, and commonly over one metre across. Corallites are four to six sided. Long and short septa clearly alternate and are thin and straight with well developed thin paliform lobes. **Colour:** Uniform cream or pale brown, occasionally brown, pink or green. **Similar species:** *Goniastrea edwardsi*, which has thicker walls and septa and more irregular corallites. See also *G. minuta*. **Habitat:** Sometimes a dominant species of intertidal habitats. **Abundance:** Common.

Taxonomic references: Wijsman-Best (1972), Veron, Pichon and Wijsman-Best (1977). **Identification guides:** Veron (1986), Sheppard and Sheppard (1991), Nishihira and Veron (1995).

Family Faviidae   Genus *Goniastrea*

1 A small massive colony in shallow water. SEYCHELLES *Photograph: author*

2 Forming flat paving-like colonies on a wave washed reef flat. BALI, INDONESIA *Photograph: Neville Coleman*

3 A flat plate with neatly arranged corallites. SOLOMON ISLANDS *Photograph: Julian Sprung*

4 An encrusting colony. SINAI PENINSULA, EGYPT *Photograph: author*

5, 6 Corallite detail. **5** GUAM **6** GREAT BARRIER REEF, AUSTRALIA *Photographs:* 5 *Gustav Paulay* 6 *Ed Lovell*

163

Family Faviidae | Genus *Goniastrea*

> Group 2: Predominantly monocentric species with corallite diameter over 5 millimetres

Goniastrea palauensis
(Yabe and Sugiyama, 1936)

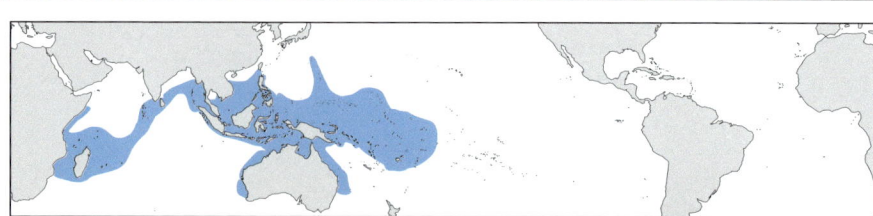

Characters: Colonies are massive and usually flattened or hillocky. Polyps are cerioid, with thick walls. Columellae are small and are surrounded by a crown of tall paliform lobes. Septa are straight, with a neat appearance. **Colour:** Usually brown or dull green becoming dark in deep water. Oral discs are usually cream but may be bright green. **Similar species:** *Goniastrea aspera* and *G. peresi*, which have smaller corallites. See also *Favites russelli*, which is usually subplocoid, has smaller corallites, relatively irregular septa and smaller paliform lobes. **Habitat:** Most reef environments. **Abundance:** Usually uncommon.

Taxonomic reference: Veron, Pichon and Wijsman-Best (1977). **Identification guide:** Veron (1986).

1-3 *Goniastrea palauensis*. Common variation in corallite shape and colour. **1, 2** Great Barrier Reef, Australia **3** Dampier Archipelago, Western Australia *Photographs: 1, 2 author 3 Ed Lovell*

4 *Goniastrea palauensis*. Showing the typical appearance of a small colony. Great Barrier Reef, Australia *Photograph: Ed Lovell*

5 *Goniastrea columella*. An encrusting colony. Socotra, Yemen *Photograph: Lyndon DeVantier*

Goniastrea columella
Crossland, 1948

Characters: Colonies are submassive to encrusting with irregular corallites that are either monocentric up to 10 millimetres diameter or form short valleys. Walls are acute. Septa have moderately large teeth. Paliform lobes are moderately developed. Columellae are large and spongy. **Colour:** Dark green. **Similar species:** *Goniastrea pectinata*, which has rounded walls, more even septa and better development of paliform lobes. See also *Platygyra contorta* which has more irregular corallites. **Habitat:** Recorded only from rocky foreshores. **Abundance:** Common but inconspicuous.

Taxonomic reference: Crossland (1948).

Goniastrea peresi
(Faure and Pichon, 1978)

Characters: Colonies are encrusting and helmet-shaped, with neatly scalloped lower margins. Corallites are angular and characteristically aligned in short shallow radiating valleys at the colony margin. Septa are strongly beaded. A small neat paliform crown is usually present. Budding is both intra- and extratentacular. **Colour:** Pinkish-tan. **Similar species:** *Goniastrea aspera*, which does not have corallites aligned in valleys. See also *G. palauensis* and *Favites abdita*. **Habitat:** Shallow reef environments. **Abundance:** Common.

Taxonomic note: Previous authors have placed this species in *Favites*. **Taxonomic reference:** Scheer and Pillai (1983, as *Favites peresi*). **Identification guide:** Sheppard and Sheppard (1991), Coles (1996), (both as *Favites peresi*).

Family Faviidae   Genus *Goniastrea*

Goniastrea deformis
Veron, 1990

Characters: Colonies are massive with irregular corallites up to 9 millimetres diameter. Septa have large teeth and granulated sides. Paliform lobes are usually well developed. 'Groove and tubercle' formations are sometimes present. **Colour:** Dark colours, usually reddish-brown with paler oral discs. **Similar species:** *Goniastrea aspera*, which has more regular corallites with less granulated, more evenly spaced septa. See also *G. columella* and *Platygyra contorta*. **Habitat:** Rocky foreshores of temperate locations. **Abundance:** Uncommon.

Taxonomic reference: Veron (1990a). **Identification guide:** Nishihira and Veron (1995).

x2

1 *Goniastrea peresi*. A small massive helmet-shaped colony. SINAI PENINSULA, EGYPT *Photograph: author*

2 *Goniastrea peresi*. Surface of a submassive colony. PEMBA ISLAND, TANZANIA *Photograph: author*

3 *Goniastrea peresi*. Showing surface characters of a small colony. SINAI PENINSULA, EGYPT *Photograph: author*

4 *Goniastrea peresi*. Showing the characteristic tendency to form radiating walls at the colony margin. SINAI PENINSULA, EGYPT *Photograph: author*

5 *Goniastrea deformis*. An encrusting colony. KYUSHU, JAPAN *Photograph: author*

6 *Goniastrea deformis*. Corallite detail. HONSHU, JAPAN *Photograph: author*

Goniastrea aspera
Verrill, 1905

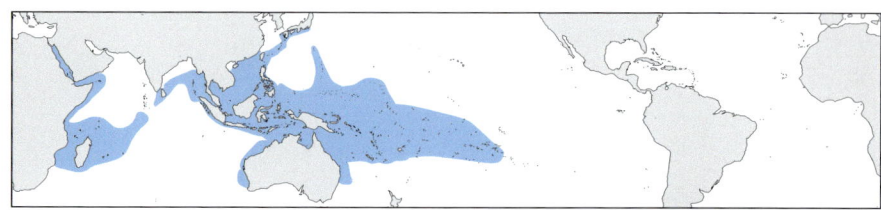

Characters: Colonies are massive to encrusting. Corallites are angular in shape and have thick walls. Long and short septa generally alternate. Paliform lobes are well developed in colonies from turbid water but may be absent in colonies from exposed habitats. **Colour:** Usually pale brown. Corallite centres are often cream. **Similar species:** *Goniastrea edwardsi*, which has similar skeletal structures but is much smaller. See also *Favites pentagona* and *F. halicora*. **Habitat:** Usually intertidal habitats where different colonies may adjoin to form flat expanses frequently over 5 metres across. Also occurs in protected turbid environments. **Abundance:** Sometimes common and may be a dominant species.

Taxonomic reference: Veron, Pichon and Wijsman-Best (1977). **Identification guides:** Veron (1986), Nishihira and Veron (1995).

x2

1 *Goniastrea aspera*. This species commonly covers extensive areas of intertidal reef flat. GREAT BARRIER REEF, AUSTRALIA *Photograph: Len Zell*

2 *Goniastrea aspera*. Typical surface view of a massive colony. GREAT BARRIER REEF, AUSTRALIA *Photograph: Neville Coleman*

3 *Goniastrea aspera*. Corallite detail. GREAT BARRIER REEF, AUSTRALIA *Photograph: Len Zell*

4 *Goniastrea aspera*. Forming a small massive colony in shallow water. GUAM *Photograph: Gustav Paulay*

5 *Goniastrea thecata*. A massive colony in shallow water. SOCOTRA, YEMEN *Photograph: Lyndon DeVantier*

Goniastrea thecata
Veron, DeVantier and Turak, this publication

Characters: Colonies are massive and more than one metre across. Corallites are irregular in shape, with 1-3 centres. Walls are thick. Septa are evenly spaced and strongly alternate with short septa developed only near the corallite rim. Paliform lobes are poorly developed or absent. Columellae form distinct centres. Fleshy polyp tissue forms a distinctive rim above the theca giving a subplocoid appearance. **Colour:** Steel grey. **Similar species:** Superficially resembles *Oulophyllia bennettae* underwater. See also *Favites paraflexuosa*. **Habitat:** Semi-exposed reef slopes. **Abundance:** Rare.

Taxonomic note: The generic designation is arbitrary as paliform lobes, usually characteristic of *Goniastrea*, are poorly developed. See 'New species described in *Corals of the World*' (Veron, in preparation) for further information.

x2

Group 3: Predominantly meandroid species

Goniastrea australensis
(Milne Edwards and Haime, 1857)

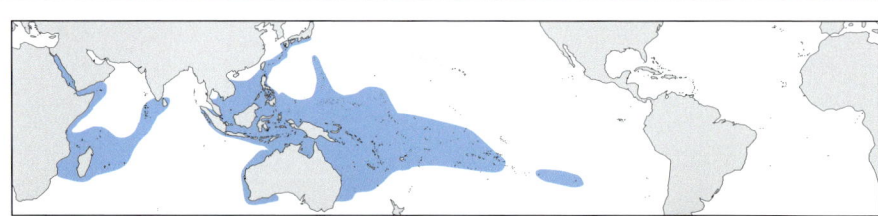

Characters: Colonies are submassive or encrusting, and meandroid, with sinuous valleys. Columella centres and paliform lobes are well developed. **Colour:** Very variable but usually a uniform dull green or brown or with walls and valley floors of contrasting dull or bright colours. **Similar species:** *Goniastrea australensis* is the only fully meandroid *Goniastrea*. Underwater, it may be difficult to distinguish from *Platygyra lamellina*, but skeletons are readily distinguished by their well developed columella centres and paliform lobes. Submeandroid colonies may resemble *Goniastrea pectinata*. See also *Oulophyllia crispa*. **Habitat:** Shallow or clear water. **Abundance:** Common, may be a dominant species of subtropical reefs.

Taxonomic references: Wijsman-Best (1972), Veron, Pichon and Wijsman-Best (1977). **Identification guides:** Veron (1986), Sheppard and Sheppard (1991), Nishihira and Veron (1995).

Family Faviidae | Genus *Goniastrea*

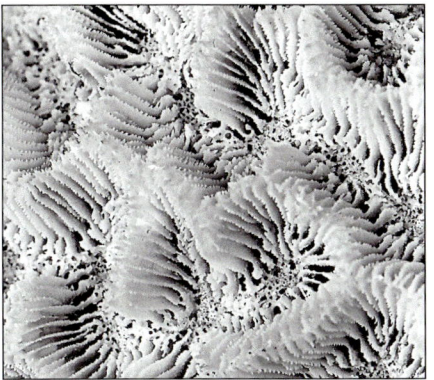

1 Encrusting a rock. GERALDTON, SOUTH-WEST AUSTRALIA *Photograph: Clay Bryce*

2 Two encrusting colonies overgrowing each other. SOLITARY ISLANDS, SOUTH-EAST AUSTRALIA *Photograph: author*

3, 4, 6, 7 Common variation in corallite shape and colour. **3, 4** GREAT BARRIER REEF, AUSTRALIA **6** HONSHU, JAPAN **7** NORFOLK ISLAND, WESTERN PACIFIC *Photographs: 3, 6, 7 author 4 Valerie Taylor*

5 Walls may be thick in shallow water colonies. GUAM *Photograph: Gustav Paulay*

Goniastrea favulus
(Dana, 1846)

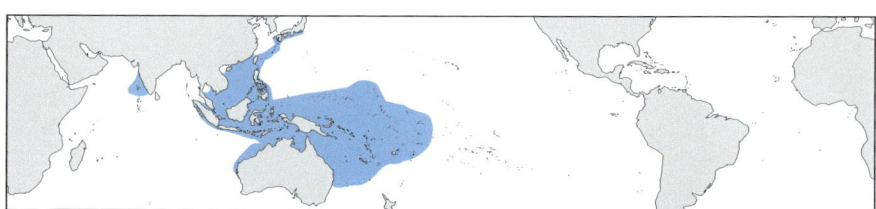

Characters: Colonies are mostly massive. Corallites are sometimes cerioid but are mostly submeandroid. Walls are thin, columellae are small and paliform lobes are well developed. **Colour:** Dull green or brown. **Similar species:** *Goniastrea retiformis*, which has similar corallite structures but is always cerioid. **Habitat:** Usually intertidal or just subtidal but also occurs in a wide range of other habitats. **Abundance:** Uncommon.

Taxonomic references: Wijsman-Best (1972), Veron, Pichon and Wijsman-Best (1977). **Identification guides:** Veron (1986), Nishihira and Veron (1995).

1 A colony on an intertidal reef flat. SCOTT REEF, WESTERN AUSTRALIA *Photograph: author*

2 *Goniastrea favulus* (bottom) being overgrown by and overgrowing *Goniastrea australensis* (top). These species are easily confused. NORFOLK ISLAND, WESTERN PACIFIC *Photograph: author*

3 Typical appearance of irregularly meandering valleys. NORFOLK ISLAND, WESTERN PACIFIC *Photograph: author*

4 An encrusting colony in a sheltered environment. Egg and sperm bundles are in the process of being released. LORD HOWE ISLAND, SOUTH-EAST AUSTRALIA *Photograph: Len Zell*

Goniastrea pectinata
(Ehrenberg, 1834)

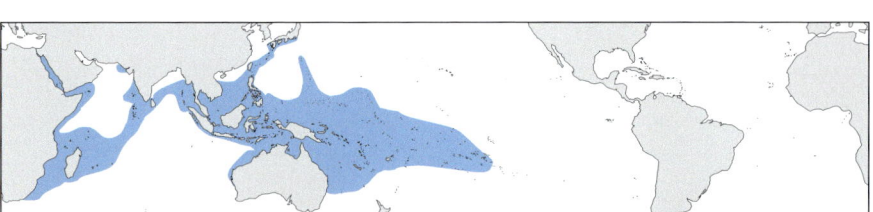

Characters: Colonies are submassive or encrusting. Corallites are cerioid to submeandroid. The latter usually have less than four centres. Walls are thick, paliform lobes are well developed. **Colour:** Usually pale brown or pink but may be dark brown in deep or turbid water. **Similar species:** *Goniastrea edwardsi*, which has markedly smaller corallites and *G. australensis* which has valleys of similar width but is usually fully meandroid. See also the merulinid *Merulina scheeri*. **Habitat:** Most shallow water environments. **Abundance:** Common.

Taxonomic references: Wijsman-Best (1972), Veron, Pichon and Wijsman-Best (1977). **Identification guides:** Randall and Myers (1983), Veron (1986), Sheppard and Sheppard (1991), Nishihira and Veron (1995).

Family Faviidae | Genus *Goniastrea*

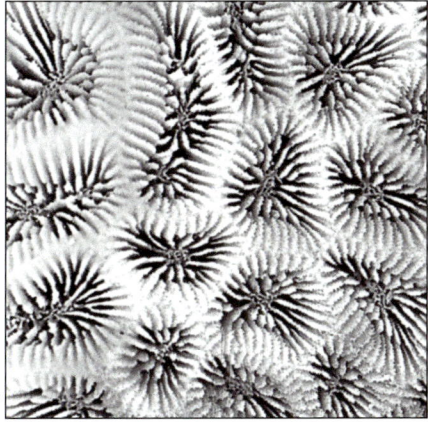

1 Corallite detail. SINAI PENINSULA, EGYPT *Photograph: author*

2 A submassive colony. GREAT BARRIER REEF, AUSTRALIA *Photograph: Ed Lovell*

3 Showing formation of nodular upgrowths in shallow water. GREAT BARRIER REEF, AUSTRALIA *Photograph: Len Zell*

GENUS *PLATYGYRA*
Ehrenberg, 1834

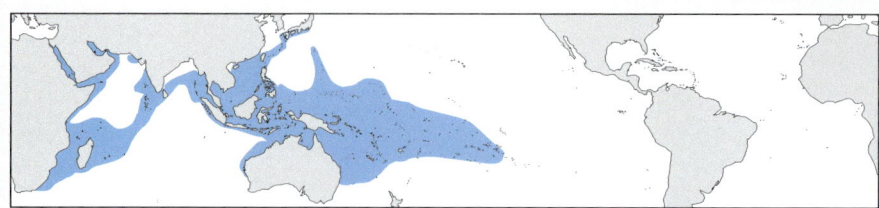

Characters: Colonies are massive and either flat or dome-shaped. Corallites are usually meandroid but may be cerioid. Paliform lobes are not developed; columellae seldom form centres, but rather form spongy wall-like structures. Tentacles are usually extended only at night. **Similar genera:** *Platygyra* is similar to *Goniastrea* and the Atlantic *Diploria*, and also to *Australogyra* and *Leptoria*. *Goniastrea* has well developed paliform lobes and columella centres. *Australogyra* has *Platygyra*-like skeletal structures but has a branching growth-form and no columellae. *Leptoria* is more meandroid than *Platygyra*, usually has distinctive and more solid wall-like columellae, and has uniformly spaced septa of equal size. **Earliest fossil record:** Eocene of the Pacific (doubtful record), Oligocene of the Caribbean and Tethys.

Taxonomic references: Chevalier (1975), Wijsman-Best (1976), Veron, Pichon and Wijsman-Best (1977). **Summary key:** See pp452, 454.

Identification of *Platygyra* species

Platygyra species all show similar skeletal modifications along environmental gradients and some, especially *P. daedalea* and *P. lamellina*, may sometimes be difficult to distinguish unless they occur together.

Platygyra can be divided into two groups. These groups aid identification, but do not indicate structural similarity other than in valley length.

Group 1: Species which are monocentric or form only short valleys, page 178.

Group 2: Species which are primarily meandroid, page 186.

Family Faviidae | Genus *Platygyra*

1 All *Platygyra* have similar growth-forms culminating in massive colonies such as this *P. daedalea*. GREAT BARRIER REEF, AUSTRALIA *Photograph: Neville Coleman*

2 *Platygyra* species are sometimes difficult to distinguish unless they are seen together underwater. Here *P. daedalea* (left) and *P. verweyi* (right) occur together. RYUKYU ISLANDS, JAPAN *Photograph: author*

3 The two most distinctive characters of *Platygyra* species are the absence of paliform lobes and poorly defined columella centres. These characters are often difficult to determine underwater, thus most meandroid Indo-Pacific faviids appear *Platygyra*-like. GREAT BARRIER REEF, AUSTRALIA *Photograph: Ed Lovell*

Family Faviidae | Genus *Platygyra*

> **Group 1: Species which are monocentric or form only short valleys**

Platygyra pini
Chevalier, 1975

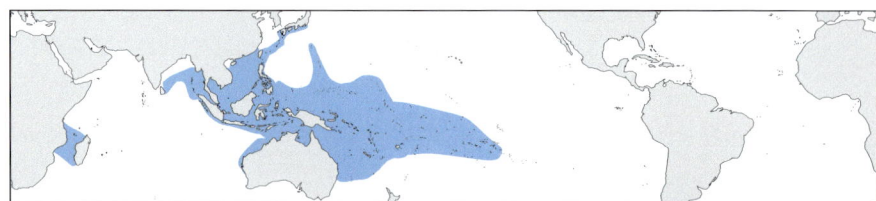

Characters: Colonies are massive to encrusting. Corallites are monocentric or form short valleys. Walls are thick, with rounded edges. Septa are thin and evenly spaced. There may be some development of columella centres and/or paliform lobes. **Colour:** Usually grey- or yellow-brown with green or cream valley floors. **Similar species:** *Platygyra ryukyuensis*, which has smaller valleys with thinner walls. See also *P. crosslandi*. **Habitat:** Shallow reef environments. **Abundance:** Usually uncommon.

Taxonomic references: Chevalier (1975), Veron, Pichon and Wijsman-Best (1977). **Identification guides:** Randall and Myers (1983), Veron (1986), Nishihira and Veron (1995).

x2

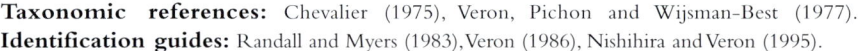

1

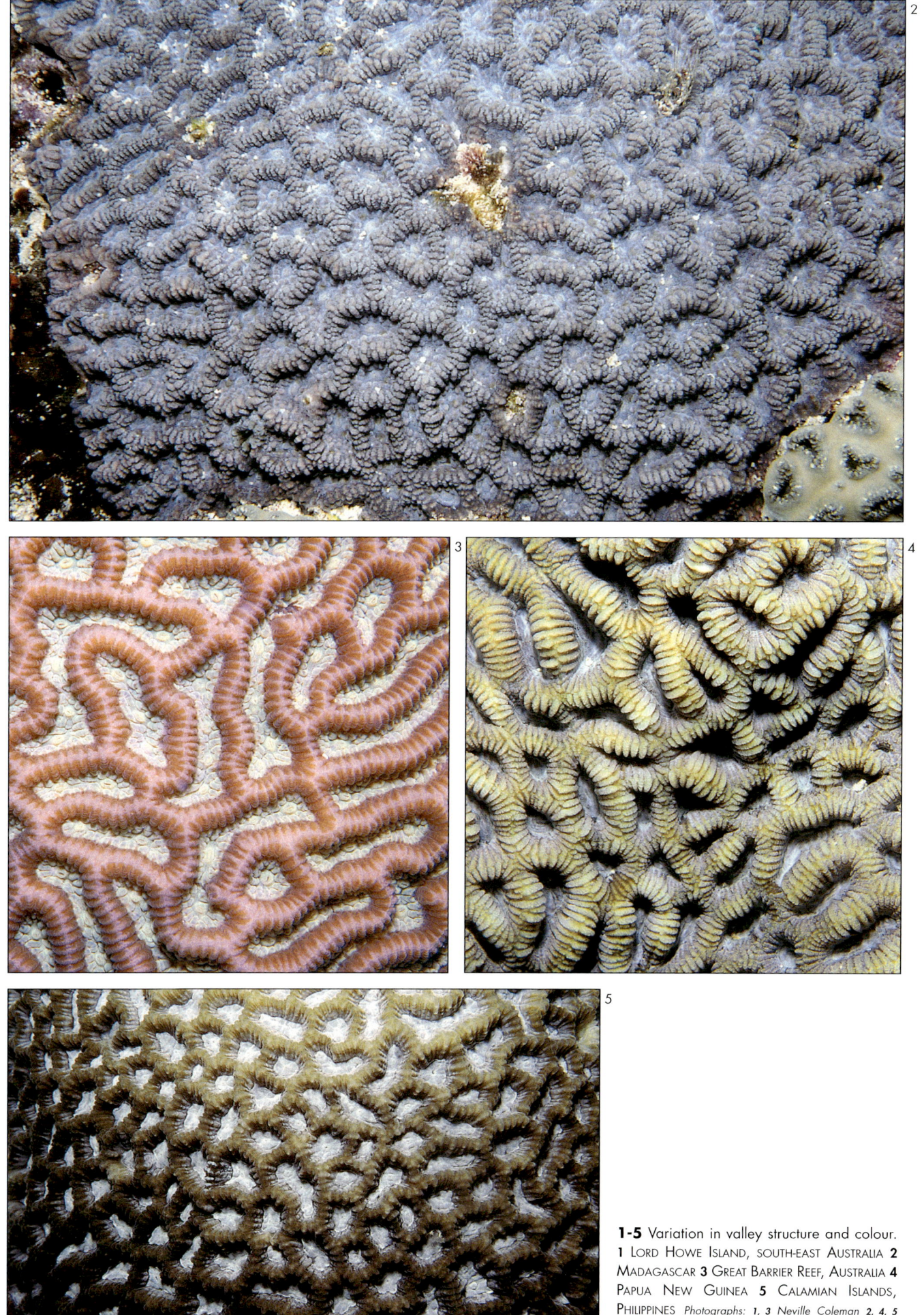

1-5 Variation in valley structure and colour. **1** Lord Howe Island, south-east Australia **2** Madagascar **3** Great Barrier Reef, Australia **4** Papua New Guinea **5** Calamian Islands, Philippines *Photographs: 1, 3 Neville Coleman 2, 4, 5 author*

Family Faviidae Genus *Platygyra*

Platygyra crosslandi
Matthai, 1928

Characters: Colonies are massive or form thick plates. Valleys are short. Walls are thick and rounded, with evenly exsert irregularly dentate septa. Columellae may be well developed. **Colour:** Brown, usually with greenish valleys. **Similar species:** *Platygyra pini*, which has shorter, more uniform valleys and less development of columellae. *Platygyra lamellina* has similar skeletal structures but is meandroid; *P. verweyi* has smaller corallites. **Habitat:** Most reef environments. **Abundance:** Common.

Taxonomic references: Matthai (1928), Scheer and Pillai (1983). **Identification guide:** Sheppard and Sheppard (1991).

x2

1, 3 *Platygyra crosslandi*. The usual appearance of a small encrusting colony. PEMBA ISLAND, TANZANIA *Photograph: author*

2 *Platygyra crosslandi*. Detail of valleys. PEMBA ISLAND, TANZANIA *Photographs: author*

4-6 *Platygyra verweyi*. Variation in colony surface structure and colour. **4** ASHMORE REEF, WESTERN AUSTRALIA **5** RYUKYU ISLANDS, JAPAN **6** CALAMIAN ISLANDS, PHILIPPINES *Photographs: author*

7 *Platygyra verweyi*. Detail of valleys. PAPUA NEW GUINEA *Photograph: author*

Platygyra verweyi
Wijsman–Best, 1976

Characters: Colonies are massive and cerioid to submeandroid with thin, acute walls. Septa are thin and uniformly spaced. Columellae are weakly developed or absent. **Colour:** Uniform or mottled grey or brown, usually with contrasting walls and valley floors. **Similar species:** *Platygyra sinensis*, which is more meandroid and has shallower valleys. See also *P. pini*, which has more widely spaced, more irregular septa and *P. carnosus*, which has well developed columellae. In general appearance *P. verweyi* has a neat *Goniastrea*-like skeletal structure. **Habitat:** Reef flats and upper slopes. **Abundance:** Usually uncommon.

Taxonomic reference: Wijsman-Best (1976). **Identification guide:** Nishihira and Veron (1995).

Platygyra ryukyuensis
Yabe and Sugiyama, 1936

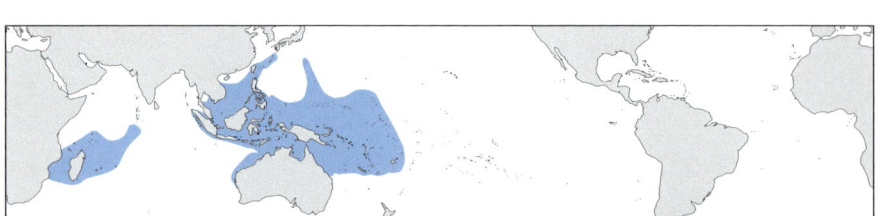

Characters: Colonies are massive with mostly short valleys. Some colonies are monocentric or nearly so. Valleys are narrow (3-4.5 mm wide) and walls are thin. Septo-costae are irregular and have irregular teeth. There are no paliform lobes. **Colour:** Dark brown, grey or green, usually with valley walls and floors of contrasting colours. **Similar species:** *Platygyra sinensis*, which is much more meandroid and has larger valleys. **Habitat:** Shallow reef environments. **Abundance:** Usually uncommon.

Taxonomic reference: Yabe and Sugiyama's original description/specimens. **Identification guide:** Nishihira and Veron (1995).

Family Faviidae — Genus *Platygyra*

1, 2 Colony surface. **1** SCOTT REEF, WESTERN AUSTRALIA **2** SEYCHELLES *Photographs: author*

3, 4 Variation in valley structure and colour. GREAT BARRIER REEF, AUSTRALIA *Photographs: 3 Ed Lovell 4 Valerie Taylor*

Platygyra carnosus
Veron, this publication

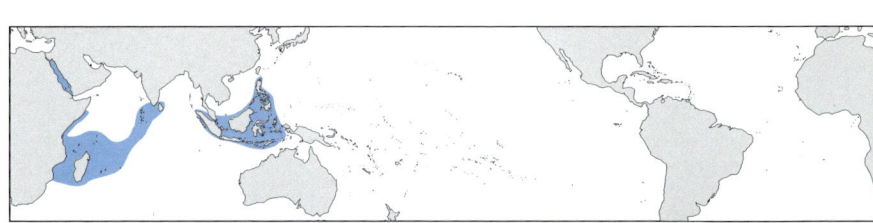

Characters: Colonies are massive and cerioid to submeandroid with thin, acute walls. Valleys are irregular in length in the same colony. Septa are thin and highly granulated. They converge and may fuse except where valleys are straight. Columellae are well developed. Polyps are fleshy. **Colour:** Uniform brown or red, with pale tops to walls. **Similar species:** No other *Platygyra* has such fleshy polyps. Skeletal structures are similar to those of *P. verweyi* and are somewhat *Goniastrea*-like. **Habitat:** Shallow reef environments. **Abundance:** Uncommon.

Taxonomic note: See 'New species described in *Corals of the World*' (in preparation) for further information.

Platygyra yaeyamaensis
Eguchi and Shirai, 1977

Characters: Colonies are encrusting or submassive. Most are monocentric or nearly so, especially towards the colony centre. Septa are exsert and have irregular teeth giving colonies a ragged appearance. Columellae form indistinct centres. Paliform lobes are weakly developed. **Colour:** Brown or cream, sometimes with green or cream valley floors. **Similar species:** *Favites stylifera*, which has smaller corallites. **Habitat:** Reef environments. **Abundance:** Rare.

Taxonomic reference: Shirai (1980). **Identification guides:** Veron (1986), Nishihira and Veron (1995).

Family Faviidae | Genus *Platygyra*

1 *Platygyra carnosus*. A submassive colony. SINAI PENINSULA, EGYPT *Photograph: author*

2 *Platygyra carnosus*. Surface of a small encrusting colony. MADAGASCAR *Photograph: author*

3 *Platygyra carnosus*. Corallite detail. SRI LANKA *Photograph: author*

4 *Platygyra yaeyamaensis*. Detail of valleys. PAPUA NEW GUINEA *Photograph: author*

5-7 *Platygyra yaeyamaensis*. Variation in valley structure and colour. **5** PAPUA NEW GUINEA **6** CALAMIAN ISLANDS, PHILIPPINES **7** RYUKYU ISLANDS, JAPAN *Photographs: author*

Family Faviidae — Genus *Platygyra*

> **Group 2: Species which are primarily meandroid**

Platygyra sinensis
(Milne Edwards and Haime, 1849)

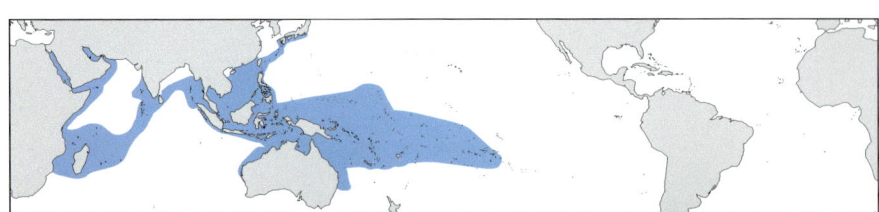

Characters: Colonies are massive or flat and usually fully meandroid, with thin walls. Septa are thin and slightly exsert. Columellae are weakly developed and there are no columella centres. **Colour:** Variable dull or bright colours. **Similar species:** *Platygyra ryukyuensis*. Resembles *Goniastrea favulus* underwater. **Habitat:** Most reef environments, especially back reef margins. **Abundance:** Usually uncommon.

Taxonomic references: Wijsman-Best (1972), Chevalier (1975), Veron, Pichon and Wijsman-Best (1977). **Identification guides:** Sheppard and Sheppard (1991), Nishihira and Veron (1995).

1 This species often forms large massive colonies. BORNEO, INDONESIA *Photograph: Gerry Allen*

2 Colony surface. GREAT BARRIER REEF, AUSTRALIA *Photograph: Mary Stafford-Smith*

3 On an intertidal reef flat. This combination of sinuous and straight valleys is not uncommon in large colonies. GREAT BARRIER REEF, AUSTRALIA *Photograph: Isobel Bennett*

4, 5 Detail of valleys. **4** MALDIVE ISLANDS **5** GREAT BARRIER REEF, AUSTRALIA *Photographs: 4 Neville Coleman 5 author*

Family Faviidae | Genus *Platygyra*

Platygyra contorta
Veron, 1990

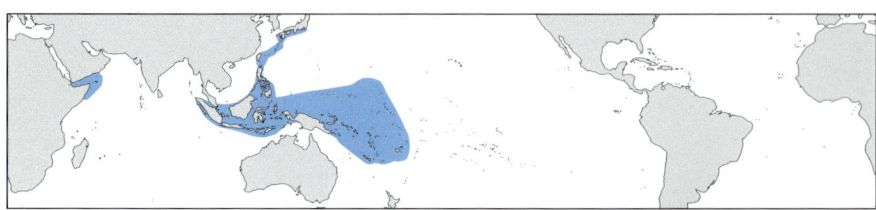

Characters: Colonies are massive, encrusting or columnar. Valleys are usually long and relatively straight at colony margins, becoming short, sinuous and contorted towards the colony centre. Walls are thin. Septa are highly irregular. **Colour:** Red, grey, pale yellow or green, with valley walls and floors of contrasting colours. **Similar species:** *Platygyra verweyi*, which has shorter valleys. See also *Goniastrea deformis*. **Habitat:** Rocky foreshores or shallow reef environments. **Abundance:** Common only around mainland Japan.

Taxonomic reference: Veron (1990a). **Identification guide:** Nishihira and Veron (1995).

Family Faviidae Genus *Platygyra*

1 A large lobed colony. Flores, Indonesia *Photograph: author*

2, 3 Variation in valley structure and colour. **2** Ryukyu Islands, Japan **3** Calamian Islands, Philippines *Photographs: author*

4 Plate-like lobes of a submassive colony. Calamian Islands, Philippines *Photograph: author*

5, 6 Valley detail. Valleys are usually fleshy in subtropical localities. **5** Great Barrier Reef, Australia **6** Honshu, Japan *Photographs: author*

Platygyra acuta
Veron, this publication

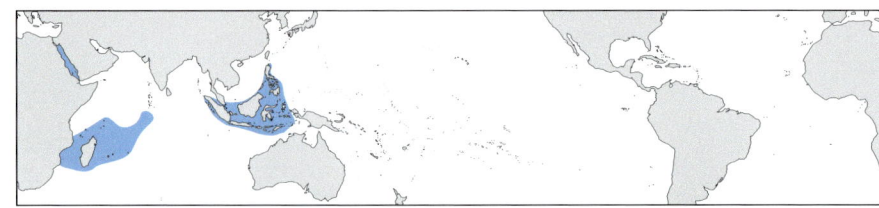

Characters: Colonies are massive and meandroid, with walls forming an acute or sharp edge. Septa are uniformly exsert and have ragged margins. Columellae are well developed but do not form centres. **Colour:** Walls are a uniform grey-brown with pale tops. Valley floors are greenish. **Similar species:** *Platygyra daedalea*, which has weakly developed columellae. Living colonies of *P. daedalea* do not have walls with such acute edges. See also *P. sinensis*, which has more meandroid valleys and walls of uniform thickness. **Habitat:** Most reef environments, especially shallow fringing reefs. **Abundance:** Sometimes common.

Taxonomic note: See 'New species described in *Corals of the World*' (Veron, in preparation) for further information.

1 *Platygyra acuta*. A small colony in a shallow fringing reef. SEYCHELLES *Photograph: author*

2 *Platygyra acuta*. Colony with short valleys. CALAMIAN ISLANDS, PHILIPPINES *Photograph: author*

3 *Platygyra acuta*. Valley detail. SEYCHELLES *Photograph: author*

4 *Platygyra acuta*. Colony with meandroid valleys and showing the acute edge to walls. SEYCHELLES *Photograph: author*

5 *Platygyra daedalea*. A small massive colony with typical characters of the species well developed. GREAT BARRIER REEF, AUSTRALIA *Photograph: Ed Lovell*

6, 8 *Platygyra daedalea*. Surface of large colonies. GREAT BARRIER REEF, AUSTRALIA *Photographs: author*

7 *Platygyra daedalea*. Characteristic appearance of valleys. CALAMIAN ISLANDS, PHILIPPINES *Photograph: author*

Platygyra daedalea
(Ellis and Solander, 1786)

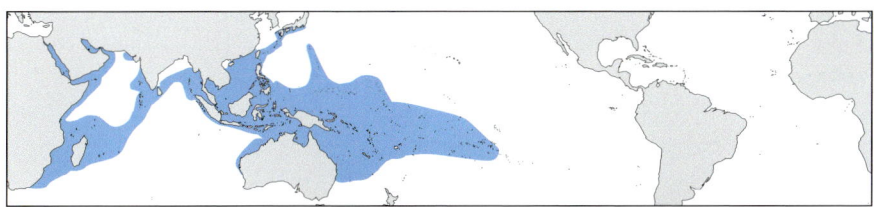

Characters: Colonies are massive to encrusting, and meandroid or submeandroid, with thick walls. Septa are exsert and have a characteristically ragged appearance. Columellae are weakly developed and centres are indistinct. **Colour:** Commonly brightly coloured, usually with brown walls and grey or green valleys. **Similar species:** *Platygyra lamellina*, which has thicker walls and neat rounded septa. **Habitat:** Most reef environments, especially back reef margins. **Abundance:** Common.

Taxonomic references: Wijsman-Best (1972), Chevalier (1975), Veron, Pichon and Wijsman-Best (1977). **Identification guides:** Randall and Myers (1983), Veron (1986), Sheppard and Sheppard (1991), Nishihira and Veron (1995), Coles (1996), Carpenter *et al.* (1997).

Platygyra lamellina
(Ehrenberg, 1834)

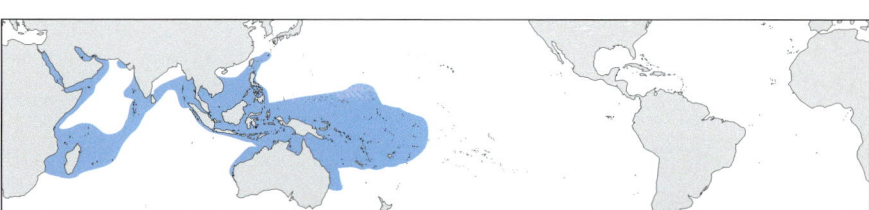

Characters: Colonies are massive and meandroid, with thick walls. Septa are uniformly exsert and are neat and rounded. Columellae may be well developed, but do not form distinct centres. **Colour:** Usually brown or with brown walls and grey or green valleys. **Similar species:** *Platygyra daedalea*. See also *P. crosslandi*. Resembles *Goniastrea australensis* underwater. **Habitat:** Most reef environments, especially back reef margins. **Abundance:** Usually uncommon.

Taxonomic references: Wijsman-Best (1972), Veron, Pichon and Wijsman-Best (1977). **Identification guides:** Veron (1986), Sheppard and Sheppard (1991), Nishihira and Veron (1995).

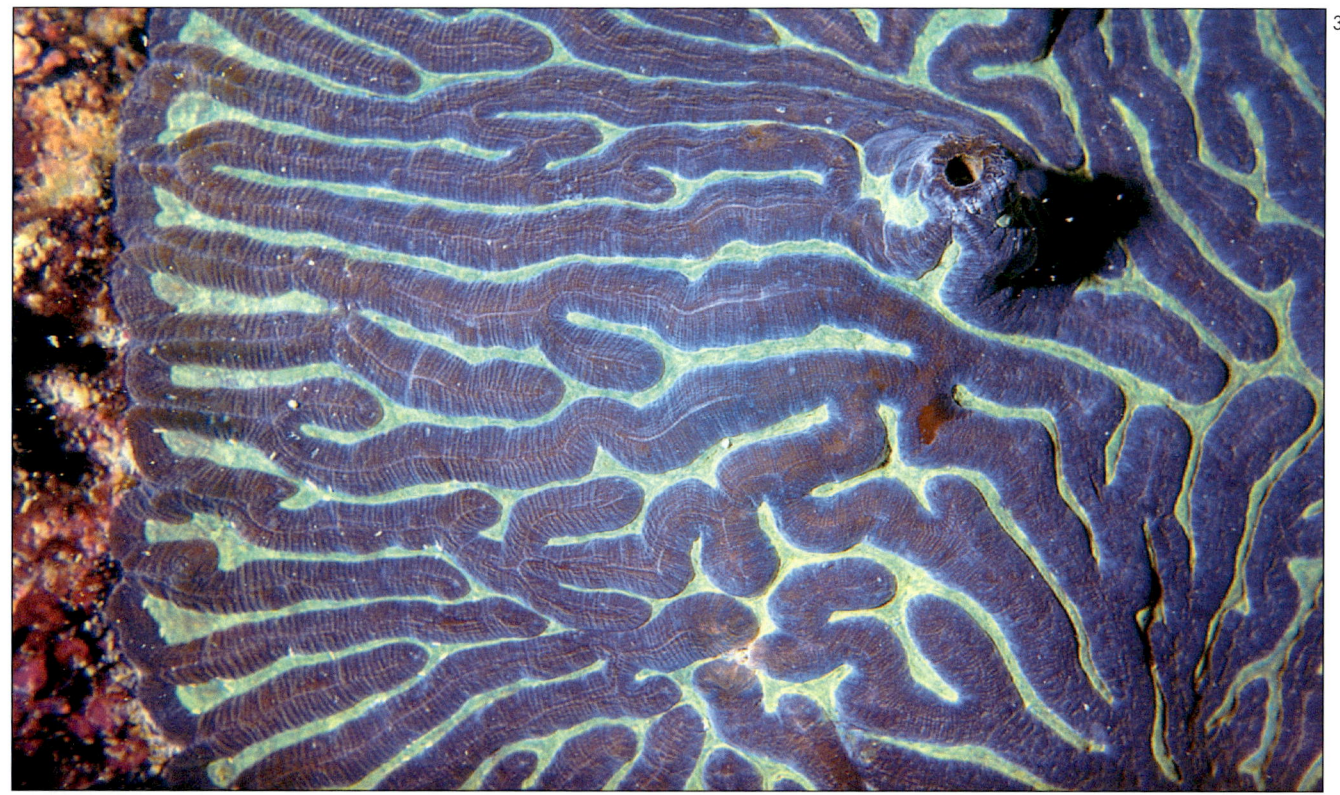

1 A massive colony showing a history of damage and recovery. CALAMIAN ISLANDS, PHILIPPINES *Photograph: author*

2, 3 This species usually forms flat plates with valleys approximately perpendicular to the margin. HOUTMAN ABROLHOS ISLANDS, SOUTH-WEST AUSTRALIA *Photographs: author*

4-6 Common variation in corallite shape and colour. GREAT BARRIER REEF, AUSTRALIA *Photographs: 4, 5 Mary Stafford-Smith 6 Neville Coleman*

Family Faviidae — Genus *Australogyra*

GENUS *AUSTRALOGYRA*
Veron and Pichon, 1982

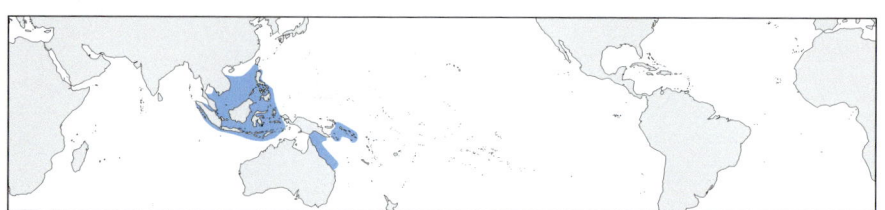

Characters, abundance and distribution: This genus has only one species, see *Australogyra zelli*. **Fossil record:** None.

Australogyra zelli
(Veron and Pichon, 1977)

Characters: Colonies have a branching growth-form. They may form compact hemispherical mounds or thickets over 2 metres across or occur as a few isolated branches. Corallites are monocentric or form short valleys. Walls are thick, rounded and smooth. Columellae are absent. Tentacles are extended only at night. **Colour:** Grey-green to brown. **Similar species:** *Australogyra* is similar to *Platygyra* in corallite structure but not growth-form. *Goniastrea ramosa* has a similar appearance but is much smaller and has paliform lobes. **Habitat:** Usually in turbid waters around high islands or in reef lagoons. **Abundance:** Uncommon.

Taxonomic reference: Veron, Pichon and Wijsman-Best (1977). **Identification guide:** Veron (1986).

1 *Australogyra zelli*. A compact hemispherical clump. Colonies as large as this are uncommon. PAPUA NEW GUINEA Photograph: author

2 *Australogyra zelli*. This species shows little variation: colonies consist of thick, interlocking, irregular branches. GREAT BARRIER REEF, AUSTRALIA Photograph: Ed Lovell

3 *Australogyra zelli*. Branch ends. GREAT BARRIER REEF, AUSTRALIA Photograph: Ed Lovell

GENUS *OULOPHYLLIA*
Milne Edwards and Haime, 1848

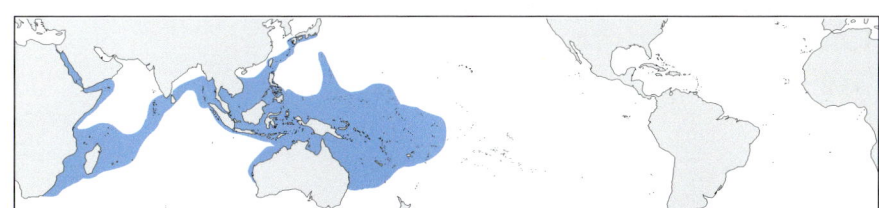

Characters: Colonies are massive and monocentric to meandroid, composed of large valleys with widely spaced, ragged septa and acute thin walls. Paliform lobes are usually present. Polyps are large and fleshy and tentacles are extended only at night. When tentacles are retracted, polyps have a coarse reptilian texture. Mouths are conspicuous. **Similar genera:** *Oulophyllia* is similar to the Atlantic *Colpophyllia*. Also similar to *Platygyra* and *Favites*, both of which have smaller skeletal structures. **Earliest fossil record:** Oligocene of the Tethys (doubtful record), Pleistocene of the Pacific.

Taxonomic references: Wijsman-Best (1976), Veron, Pichon and Wijsman-Best (1977).

4 *Oulophyllia* are often hemispherical as with this *Oulophyllia crispa*. GREAT BARRIER REEF Photograph: author

5 A rare undescribed *Oulophyllia*. PEMBA ISLAND, TANZANIA Photograph: author

6 Plate-like colonies like this *Oulophyllia crispa* may have *Symphyllia*-like radiating valleys. GREAT BARRIER REEF, AUSTRALIA Photograph: author

7 *Oulophyllia crispa* with polyps extended at night. Tentacles of all *Oulophyllia* species are long and have white knob-like tips. GREAT BARRIER REEF, AUSTRALIA Photograph: Neville Coleman

Oulophyllia crispa
(Lamarck, 1816)

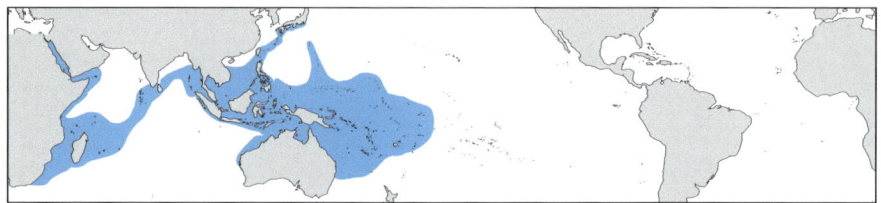

Characters: Colonies are thick plates or hemispherical and are frequently over one metre across. Valleys are short, broad (up to 20 mm), V-shaped and have sharp upper margins. Septa are usually thin and slope uniformly to the columellae, which usually form well defined centres. Paliform lobes may be present. Columellae are weakly developed. **Colour:** Uniform grey or with brown walls and pale cream or pink valley floors. **Similar species:** *Oulophyllia levis*. See also other *Oulophyllia* species. **Habitat:** Most reef environments, but especially in lagoons. **Abundance:** Uncommon.

Taxonomic references: Wijsman-Best (1972), Veron, Pichon and Wijsman-Best (1977). **Identification guides:** Veron (1986), Sheppard and Sheppard (1991), Nishihira and Veron (1995).

1 A small dome-shaped colony. RYUKYU ISLANDS, JAPAN *Photograph: author*

2 Valley detail at the centre of a large colony. WILLIS ISLAND, CORAL SEA *Photograph: Neville Coleman*

3 Sinuous valleys of a large colony. BOLINAO, PHILIPPINES *Photograph: author*

4-6 Variation in colour and surface detail of valleys. **4** PAPUA NEW GUINEA **5** PUERTO GALERA, PHILIPPINES **6** CALAMIAN ISLANDS, PHILIPPINES *Photographs: author*

Oulophyllia levis
(Nemenzo, 1959)

Characters: Colonies are thick plates or are hemispherical. Valleys are usually perpendicular to the margins of plates, and are sinuous towards colony centres. They are short, broad (up to 20 mm), V-shaped and have sharp upper margins. Columellae are weakly developed. **Colour:** Greenish- or yellowish-brown walls with yellow, green or pink valley floors. **Similar species:** *Oulophyllia crispa*, which is less colourful and has slightly larger valleys with less developed columellae. **Habitat:** Most reef environments. **Abundance:** Uncommon.

Taxonomic note: This species was previously considered a synonym of *Oulophyllia crispa*. **Taxonomic reference:** Nemenzo (1959). **Identification guide:** Nemenzo (1986).

1 A small massive colony. CALAMIAN ISLANDS, PHILIPPINES *Photograph: author*

2-4 Variation in colour and surface detail of valleys. **2** RYUKYU ISLANDS, JAPAN **3** VIETNAM **4** CALAMIAN ISLANDS, PHILIPPINES *Photographs: author*

Oulophyllia bennettae
(Veron and Pichon, 1977)

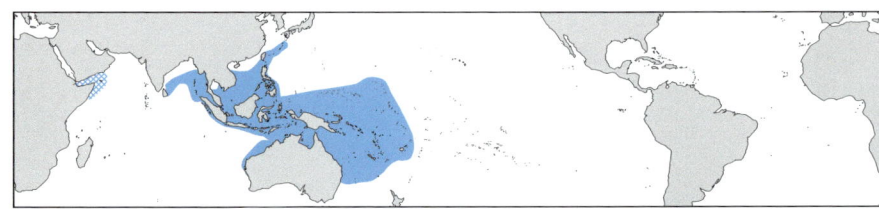

x2

Characters: Colonies are massive with large angular corallites, which may have up to three columellae. Septa are widely spaced, with large rounded teeth and some development of paliform lobes. Septa of adjacent corallites are aligned. **Colour:** Distinctive greenish-grey with green oral discs. **Similar species:** All other *Oulophyllia* species are more meandroid. See also *Goniastrea palauensis* and *Favites flexuosa*. **Habitat:** Upper reef slopes. **Abundance:** Uncommon but conspicuous.

Taxonomic reference: Veron, Pichon and Wijsman-Best (1977 as *Favites bennettae*). **Identification guides:** Veron (1986), Nishihira and Veron (1995).

1 This species shows little variation in the shape and colour of corallites. GREAT BARRIER REEF, AUSTRALIA *Photograph: author*

2 Surface of a small encrusting colony. VIETNAM *Photograph: author*

3 A large dome-shaped colony. GREAT BARRIER REEF, AUSTRALIA *Photograph: Ed Lovell*

4 Corallite detail. CALAMIAN ISLANDS, PHILIPPINES *Photograph: author*

Family Faviidae Genus *Leptoria*

GENUS *LEPTORIA*
Milne Edwards and Haime, 1848

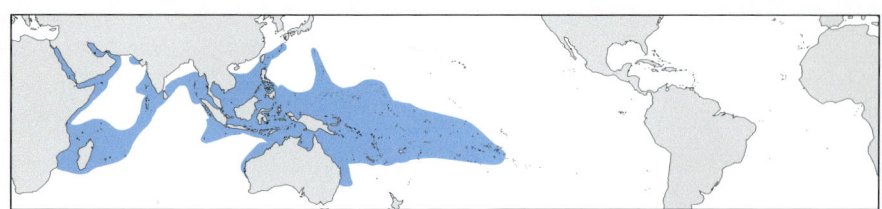

Characters: Colonies are massive or encrusting with sinuous valleys and neatly arranged equal septa and no paliform lobes. Tentacles are extended only at night. **Similar genera:** *Platygyra*, which is less sinuous; meandroid species also have coarser corallites. See also *Goniastrea*, which is less meandroid, has columellae forming distinct centres and well developed paliform lobes. **Earliest fossil record:** Eocene of the Caribbean and Pacific, Oligocene of the Tethys.

Taxonomic references: Chevalier (1975), Wijsman-Best (1976), Veron, Pichon and Wijsman-Best (1977).

Leptoria irregularis
Veron, 1990

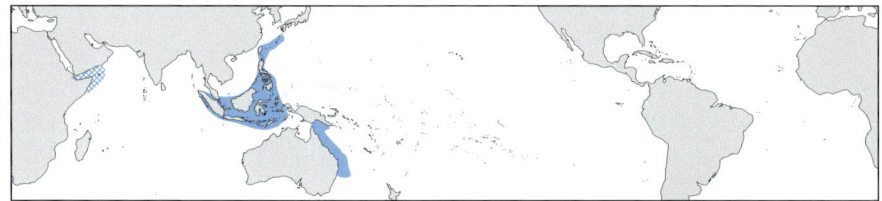

Characters: Colonies are submassive or laminar, up to 1.5 metres across. Valleys are 3-4 millimetres wide, usually perpendicular to colony margins, becoming sinuous towards the colony centre. Septa are irregular, with strong teeth. Columellae do not form centres and are not laminar. **Colour:** Pale blue-grey to translucent white. **Similar species:** *Leptoria phrygia*. See also the merulinids *Scapophyllia cylindrica* and *Merulina ampliata*. **Habitat:** Upper reef slopes. **Abundance:** Uncommon.

Taxonomic reference: Veron (1990a). **Identification guide:** Nishihira and Veron (1995).

x2

1

1 A large encrusting colony. RYUKYU ISLANDS, JAPAN *Photograph: author*

2 The common appearance of small colonies. RYUKYU ISLANDS, JAPAN *Photograph: author*

3 Valleys at the margin of an encrusting colony. CALAMIAN ISLANDS, PHILIPPINES *Photograph: author*

4 Detail of valleys. CALAMIAN ISLANDS, PHILIPPINES *Photograph: author*

Leptoria phrygia
(Ellis and Solander, 1786)

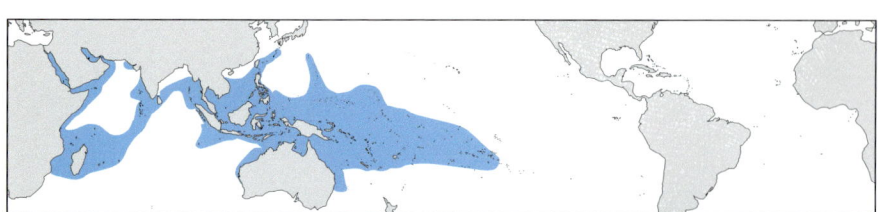

Characters: Colonies are massive, submassive or ridged, with an even surface and dense skeleton. Corallite valleys are sinuous and uniform. Septa are uniformly spaced and are of equal size. Columellae are plate-like with a lobed upper margin and do not form centres. **Colour:** Cream, brown or green, with walls and valleys of contrasting colours. **Similar species:** *Leptoria irregularis*, which has larger valleys which are straight at the colony margins, is distinctively coloured and has columellae which are not plate-like. **Habitat:** Occurs in most reef environments except where the water is turbid. **Abundance:** Common, especially on upper reef slopes.

Taxonomic references: Wijsman-Best (1972), Veron, Pichon and Wijsman-Best (1977). **Identification guides:** Randall and Myers (1983), Veron (1986), Sheppard and Sheppard (1991), Nishihira and Veron (1995), Coles (1996).

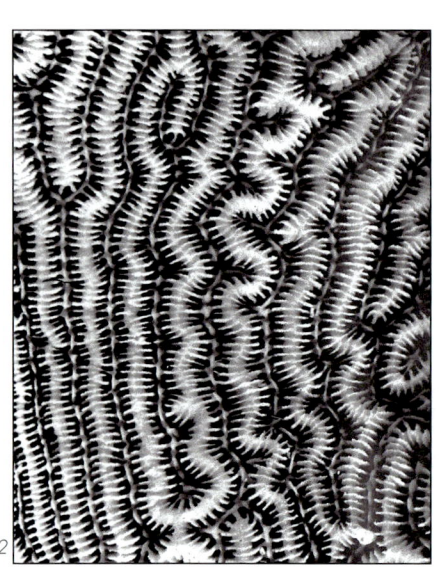

Family Faviidae | Genus *Leptoria*

1 A large colony of striking appearance. GREAT BARRIER REEF, AUSTRALIA *Photograph: Ed Lovell*

2 Colonies have even surfaces with uniform valleys forming characteristically sinuous patterns. GREAT BARRIER REEF, AUSTRALIA *Photograph: author*

3 Forming short thick columns. RYUKYU ISLANDS, JAPAN *Photograph: author*

4, 6 Variation in colour. **4** GREAT BARRIER REEF, AUSTRALIA **6** SOLOMON ISLANDS *Photographs: Roger Steene*

5 Small massive cushion-shaped colonies formed by regrowth after partial mortality are common. SINAI PENINSULA, EGYPT *Photograph: author*

Family Faviidae • Genus *Diploria*

GENUS *DIPLORIA*
Milne Edwards and Haime, 1848

Characters: Colonies are massive and meandroid. Columellae are interlinked. Paliform lobes are weakly developed or absent. Tentacles are extended only at night. **Similar genus:** *Platygyra*. **Earliest fossil record:** Eocene of the Tethys, Oligocene of the Caribbean.

Taxonomic reference: Walton Smith (1971).

Diploria labyrinthiformis
(Linnaeus, 1758)

Characters: Colonies are massive and usually hemispherical. Valleys are meandroid, parallel or sinuous, deep and 5-8 millimetres wide. Ambulacral grooves vary greatly in width within the same colony but may be wider than the valleys giving the superficial appearance of alternating valleys of two different sorts. Columellae are fine and do not form distinct centres. **Colour:** Tan to yellowish- or grey-brown. **Similar species:** Ambulacral grooves may reach the size of valleys of *Colpophyllia natans*. **Habitat:** Shallow reef environments. **Abundance:** Common.

Taxonomic references: Roos (1971), Cairns (1982), Zlatarski and Estalella (1982). **Identification guides:** Colin (1978), Humann (1993).

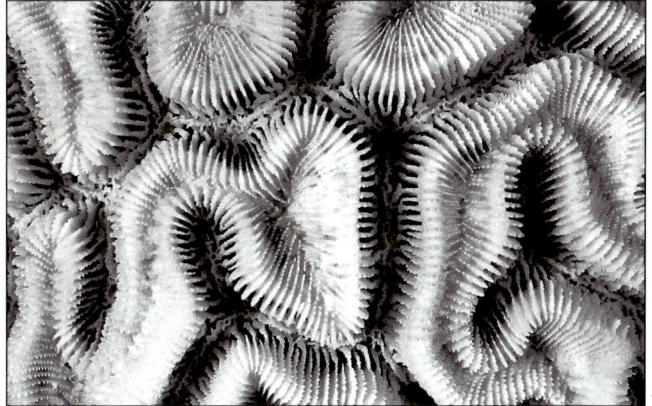

1 *Diploria labyrinthiformis* (above) and *Colpophyllia natans* (below) showing the clear size difference between these two genera. BELIZE *Photograph: Mary Stafford-Smith*

2 *Diploria labyrinthiformis*. This species forms conspicuous massive colonies. JAMAICA *Photograph: author*

3 *Diploria labyrinthiformis*. Forming an irregular colony, parts of which have wide ambulacral grooves. BELIZE *Photograph: author*

4 *Diploria labyrinthiformis*. Showing the alternation between valleys (which have retracted tentacles) and ambulacral grooves (wider than the valleys). BAHAMAS *Photograph: Neville Coleman*

5 *Diploria labyrinthiformis*. Surface detail of valleys and ambulacral grooves. BARBADOS *Photograph: author*

Family Faviidae | Genus *Diploria*

Diploria strigosa
(Dana, 1848)

Characters: Colonies are massive (especially in shallow water) or encrusting. The surface is even, with sinuous valleys 6-9 millimetres wide. Columellae are well developed although interlinked. Ambulacral grooves, if present, are fine. **Colour:** Purple-brown, grey or greenish with green or grey valley floors. **Similar species:** *Diploria clivosa*, which has a rough irregular colony surface, slightly smaller valleys and more numerous septa. Superficially resembles *Platygyra daedalea* of the Indo-Pacific. **Habitat:** Most reef environments, especially shallow slopes and lagoons. **Abundance:** Common.

Taxonomic references: Roos (1971), Cairns (1982), Zlatarski and Estalella (1982). **Identification guides:** Colin (1978), Wood (1983), Humann (1993).

x2

Family Faviidae | Genus *Diploria*

Diploria clivosa
(Ellis and Solander, 1786)

Characters: Colonies are massive (especially in shallow water) or encrusting. The surface is usually irregular, with sinuous valleys 4-8 millimetres wide. Columellae are well developed although interlinked. Ambulacral grooves, if present, are fine. **Colour:** Greys and browns, usually with walls and valleys of contrasting colours. **Similar species:** *Diploria strigosa*. **Habitat:** Most reef environments, especially shallow slopes and lagoons. **Abundance:** Common.

Taxonomic references: Roos (1971), Cairns (1982), Zlatarski and Estalella (1982). **Identification guides:** Colin (1978), Humann (1993).

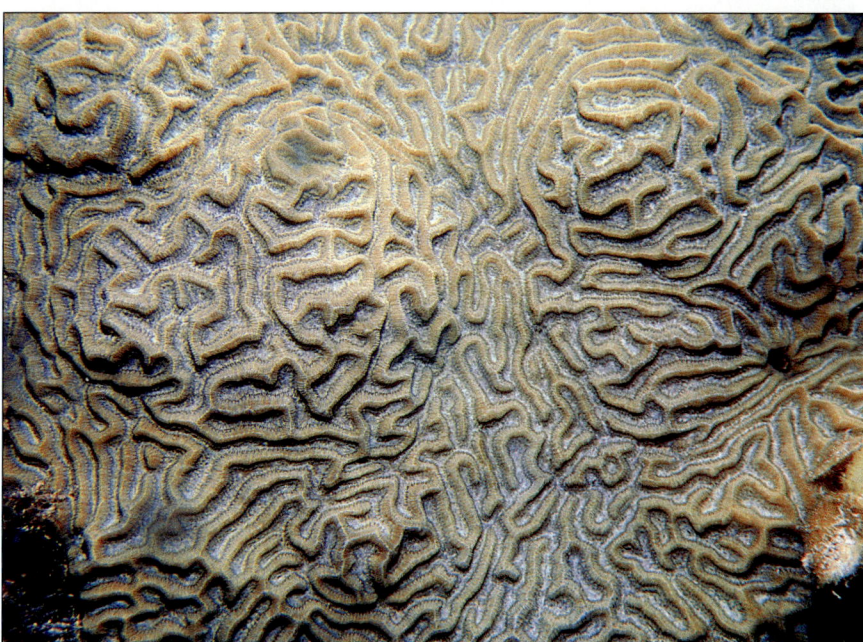

1 *Diploria strigosa*. Large boulder-like colonies with even surfaces in shallow water. BELIZE *Photograph: author*

2 *Diploria strigosa*. An encrusting plate. BELIZE *Photograph: author*

3 *Diploria strigosa*. Showing the straight valleys of a large colony. JAMAICA *Photograph: author*

4 *Diploria strigosa*. Showing lines of retracted tentacles halfway down valley walls. Colouration above and below this line is usually different. BELIZE *Photograph: author*

5 *Diploria clivosa*. A massive colony with a typically irregular surface. BELIZE *Photograph: author*

6 *Diploria clivosa*. Irregular valleys of a large colony. BELIZE *Photograph: author*

7 *Diploria clivosa*. Relatively straight valleys commonly found perpendicular to colony margins. BELIZE *Photograph: author*

8 *Diploria clivosa*. Surface of a large colony. BELIZE *Photograph: Mary Stafford-Smith*

GENUS *COLPOPHYLLIA*
Milne Edwards and Haime, 1848

Characters, abundance and distribution: This genus has only one species, see *Colpophyllia natans*.
Earliest fossil record: Eocene of the Caribbean and Tethys.

Colpophyllia natans
(Houttuyn, 1772)

Characters: Colonies are large and hemispherical or encrusting. Valleys are usually long and sinuous but may become short, with 1-3 centres and form *Hydnophora*-like (Merulinidae) monticules. Septa are exsert, equal, and finely toothed. A fine ambulacral groove occurs along the tops of walls. Tentacles are extended only at night.
Colour: Walls are brown or grey with valleys the same colour or green.
Similar species: *Manicina areolata*. Valleys are similar in size to *Oulophyllia* of the Indo-Pacific. *Diploria* species have smaller valleys and skeletal structures.
Habitat: Shallow reef environments.
Abundance: Sometimes common, conspicuous.

Taxonomic note: Colonies with 1-3 centres have been considered by some authors to be a distinct species, *Colpophyllia breviserialis*.
Taxonomic references: Roos (1971), Cairns (1982), Zlatarski and Estalella (1982), Fenner (1993). **Identification guides:** Colin (1978), Wood (1983), Humann (1993).

Family Faviidae Genus *Colpophyllia*

1 This species forms large distinctive colonies. BAHAMAS *Photograph: author*

2, 5, 6 Showing common variation in corallite shape and colour. **2** JAMAICA **5, 6** BELIZE *Photographs: author*

3 A colony with mostly monocentric valleys. This occurs primarily where colonies are semi-encrusting. JAMAICA *Photograph: author*

4 Colonies commonly have scalloped edges. BELIZE *Photograph: Mary Stafford-Smith*

211

GENUS *MONTASTREA*
Blainville, 1830

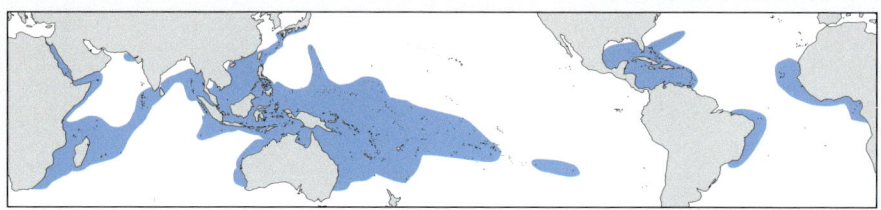

Characters: Colonies are massive, either flat or dome-shaped. Corallites are monocentric and plocoid. Daughter corallites are predominantly formed by extratentacular budding (from the walls of parent corallites). Some intratentacular budding may also occur. 'Groove and tubercle' formations (tiny tubes of tissue-thin epitheca formed by polychaete worms) are usually found in the second group of species below. **Similar genera:** *Montastrea* is a poorly defined genus, but the species within it are usually easily recognised. It is readily separated from the other massive faviid genera with extratentacular budding (*Plesiastrea*, *Diploastrea*, *Leptastrea* and *Cyphastrea*) because each of these has well defined characters. **Earliest fossil record:** Cretaceous of the Tethys (doubtful record), Eocene of the Caribbean.

Taxonomic references: Veron, Pichon and Wijsman-Best (1977), Wijsman-Best (1977). **Summary key:** See pp453-4.

Identification of *Montastrea* species

In the Indo-Pacific, *Montastrea* is a poorly defined genus although it contains mostly distinctive species within a given region. However, over wide geographic ranges, several species form well defined geographic subspecies of doubtful taxonomic affinities. The genus can be divided into three groups, based primarily on corallite size:

Group 1: Species with small corallites (less than 7 mm diameter) and no 'groove and tubercle' formation, opposite.

Group 2: Species with middle sized corallites, page 219.

Group 3: Species with large corallites (over 9 mm diameter), page 222.

Family Faviidae | Genus *Montastrea*

> Group 1: Species with small corallites (less than 7 mm diameter) and no 'groove and tubercle' formation

Montastrea serageldini
Veron, this publication

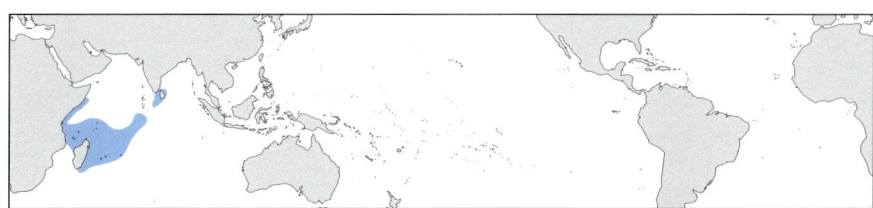

Characters: Colonies are massive, usually hemispherical, up to 0.8 metres across. Corallites are circular, and vary in size according to stage of growth. Mature corallites are 6-8 millimetres diameter, crowded and open, with a neat appearance. Septa are neatly arranged in two alternating orders. A paliform crown is formed from 6-8 septa. Columellae are small. **Colour:** Uniform grey, pinkish-brown or orange with pale oral discs. **Similar species:** Does not closely resemble any other *Montastrea*. *Plesiastrea versipora* has smaller, less compact corallites with thicker walls. See also *Plesiastrea devantieri*. **Habitat:** Shallow reef environments. **Abundance:** Rare.

Taxonomic note: See 'New species described in *Corals of the World*' (Veron, in preparation) for further information.

x2

1 All *Montastrea* species show predominantly extratentacular budding as in *M. curta* illustrated here. CALAMIAN ISLANDS, PHILIPPINES *Photograph: author*

2, 3 *Montastrea serageldini*. Variation in colour. **2** PEMBA ISLAND, TANZANIA **3** ZANZIBAR, TANZANIA *Photograph: author*

4 *Montastrea serageldini*. Surface of a large colony. PEMBA ISLAND, TANZANIA *Photograph: author*

Montastrea annularis
(Ellis and Solander, 1786)

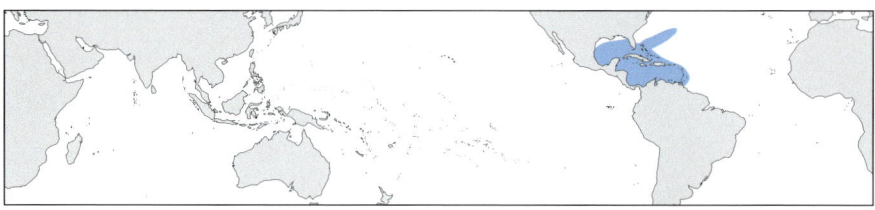

Characters: Colonies are massive, columnar or flat. Corallites are flush with the colony surface to conical, this variation mostly occurring between, rather than within, colonies. Septo-costae are neatly arranged. Long and short septa alternate. Columellae are small and compact. **Colour:** Brown or yellowish. **Similar species:** None in the Atlantic. *Montastrea salebrosa* and *M. curta* have corallites of similar size; the latter has similar morphological variations. See also *Favia stelligera*. **Habitat:** Most reef environments. **Abundance:** Common, frequently a dominant species of lagoons and upper reef slopes.

Taxonomic note: Some recent authors have divided *Montastrea annularis* into smaller, semi-distinct taxonomic units, *M. annularis*, *M. faveolata* and *M. franksi*. See 'What are species?' p425. **Taxonomic references:** Laborel (1969), Roos (1971), Zlatarski and Estalella (1982). **Identification guides:** Colin (1978), Humann (1993, illustrating the aforementioned smaller taxonomic units).

Family Faviidae											Genus *Montastrea*

1 Tiers of small plates and short columns which have a rough, irregular surface. BELIZE *Photograph: Mary Stafford-Smith*

2, 3, 8 *Montastrea annularis* showing variation in corallite detail. **2** BELIZE **3** CAYMAN ISLANDS **8** HONDURAS *Photographs: 2 author 3, 8 Paul Humann*

4 An irregular submassive colony with protuberant corallites. BAHAMAS *Photograph: author*

5 A colony of tiered plates with immersed corallites. BARBADOS *Photograph: author*

6 An irregularly divided colony with a smooth surface. BAHAMAS *Photograph: author*

7 Forming a massive boulder. JAMAICA *Photograph: author*

Family Faviidae | Genus *Montastrea*

Montastrea curta
(Dana, 1846)

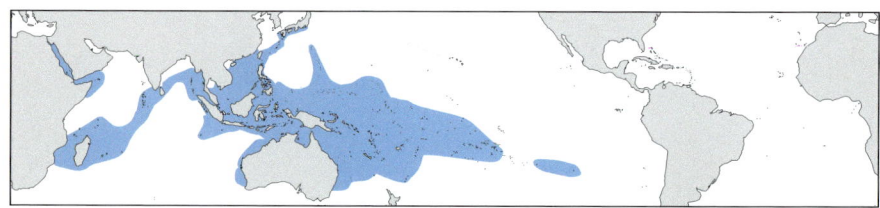

Characters: Colonies are spherical, columnar or flattened. Corallites are circular and widely spaced or closely compacted. The width of calices varies greatly, this variation mostly occurring between, rather than within, colonies. Long and short septa alternate. Small paliform lobes are usually developed. **Colour:** Cream or orange on reef flats, often with colours concentric to the oral discs. Usually dark brown when in shaded habitats. **Similar species:** *Montastrea salebrosa*, which has smaller more exsert corallites. See also *Favia stelligera*. **Habitat:** Shallow environments, especially reef flats. **Abundance:** Common.

Taxonomic note: This species is divisible into several smaller semi-distinct taxonomic units, see 'What are species?' p425. **Taxonomic references:** Chevalier (1971), Veron, Pichon and Wijsman-Best (1977). **Identification guides:** Veron (1986), Sheppard and Sheppard (1991), Nishihira and Veron (1995).

x2

1 Typical appearance of a small columnar colony. GUAM *Photograph: Gustav Paulay*

2 Tops of columns often have small compact corallites. CALAMIAN ISLANDS, PHILIPPINES *Photograph: author*

3 Sometimes septo-costae are distinctive, especially in colonies exposed to strong wave action. OSPREY REEF, CORAL SEA *Photograph: Neville Coleman*

4 A distinctive regional variant. NORFOLK ISLAND, WESTERN PACIFIC *Photograph: author*

5-7 Common variation in corallite shape and colour. **5** PAPUA NEW GUINEA **6** GREAT BARRIER REEF, AUSTRALIA **7** SCOTT REEF, WESTERN AUSTRALIA *Photographs: author*

8 Showing corallite detail and extratentacular budding. GREAT BARRIER REEF, AUSTRALIA *Photograph: Valerie Taylor*

‡ See also corallite detail, page 212.

Family Faviidae Genus *Montastrea*

217

Family Faviidae Genus *Montastrea*

Montastrea salebrosa
(Nemenzo, 1959)

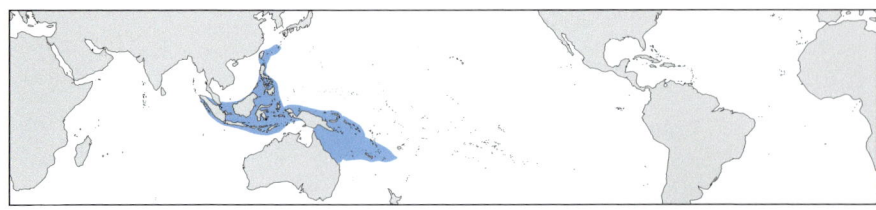

x2

Characters: Colonies are massive and mostly spherical. Corallites are circular and closely compacted. Corallites are small (3-4 mm diameter), exsert and have very thick walls. Corallites may face different directions according to place of budding. Septa are evenly spaced and strongly alternate. Costae are strongly beaded. Paliform crowns are well developed. Columellae are small. **Colour:** Tan or cream, sometimes with dark centres. **Similar species:** *Montastrea curta*. See also *Favia stelligera*. **Habitat:** Shallow environments, especially reef flats. **Abundance:** Rare.

Taxonomic note: Nemenzo (1959).

1 *Montastrea salebrosa*. A massive colony in shallow water. PAPUA NEW GUINEA *Photograph: author*

2 *Montastrea salebrosa*. Detail of a massive colony showing the extensive extratentacular budding. FLORES, INDONESIA *Photograph: author*

3-5 *Montastrea salebrosa*. Corallite detail. **3** CALAMIAN ISLANDS, PHILIPPINES **4** PAPUA NEW GUINEA **5** VANUATU *Photographs: author*

6 *Montastrea colemani* (left) and *M. valenciennesi* (right) have similar corallites. CEBU, PHILIPPINES *Photograph: author*

7 *Montastrea colemani*. Characteristic appearance of an encrusting colony. VIETNAM *Photograph: author*

8-10 *Montastrea colemani*. Variation in corallite shape and colour. CALAMIAN ISLANDS, PHILIPPINES *Photographs: author*

11 *Montastrea colemani*. Corallite detail. GUAM *Photograph: Gustav Paulay*

Family Faviidae | Genus *Montastrea*

> **Group 2: Species with middle sized corallites**
> 'Groove and tubercle' formations are a useful identification aid with this group, although they may not be present. The formations are tiny tubes or open grooves of very fine skeletal material (epitheca) made by minute polychaete worms. Underwater, they appear as irregular gaps or holes between some or all corallites. When fully developed, the corallites appear detached from each other.

Montastrea colemani
Veron, this publication

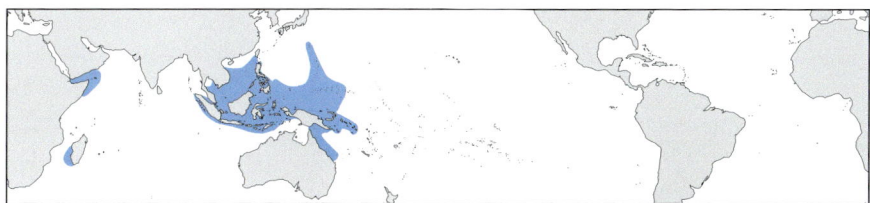

x2

Characters: Colonies are submassive to encrusting, with compact rounded corallites 5-8 millimetres diameter. 'Groove and tubercle' formations are well developed. Two cycles of septa clearly alternate; both are thickened over walls and are uniformly toothed. A paliform crown is well developed. **Colour:** Uniform brown or brown with green centres. **Similar species:** *Montastrea valenciennesi* has a similar appearance underwater and is distinguished by having larger, more irregular corallites. **Habitat:** Most reef environments. **Abundance:** Common.

Taxonomic note: See 'New species described in *Corals of the World*' (Veron, in preparation) for further information.

Montastrea annuligera
(Milne Edwards and Haime, 1849)

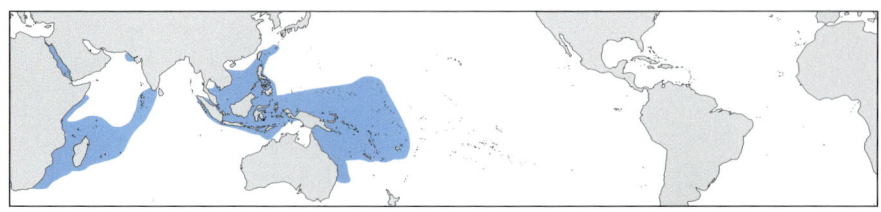

Characters: Colonies are irregular or encrusting. Corallites are circular. Septa taper from the wall to the columella and are in three cycles; those of the primary cycle are usually exsert, widely spaced, and have well developed paliform lobes. Most colonies have at least some development of 'groove and tubercle' structures. **Colour:** Mottled or uniform green and brown, with darker calices. **Similar species:** *Montastrea multipunctata*. *Montastrea curta* has no 'groove and tubercle' formation and septa are much less dentate. See also *M. valenciennesi*. **Habitat:** Protected reef backs and lagoons. **Abundance:** Uncommon.

Taxonomic reference: Veron, Pichon and Wijsman-Best (1977). **Identification guides:** Veron (1986), Nishihira and Veron (1995).

×2

Montastrea multipunctata
Hodgson, 1985

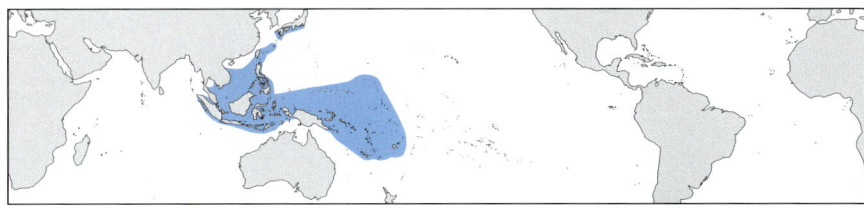

Characters: Colonies are encrusting and are less than 0.3 metres across. Corallites are 5-9 millimetres diameter and have thick fleshy mantles. Septa are in 3 or 4 distinct orders. Primary septa have 1 or 2 pointed teeth above the wall; this can sometimes be seen underwater. Paliform lobes are weakly developed. 'Groove and tubercle' formations are usually well developed. Sand, often containing turf algae, is usually compacted between the corallites and is probably held there by secretions of polychaete worms. **Colour:** Dark red (which may photograph dark brown as below), with green or white oral discs. **Similar species:** Resembles the mussid *Blastomussa wellsi* underwater. **Habitat:** Shallow water, especially adjacent to sandy substrates. **Abundance:** Rare.

Taxonomic reference: Hodgson (1985).

1 *Montastrea annuligera*. Colony with compact corallites and 'groove and tubercle' formations. PAPUA NEW GUINEA *Photograph: author*

2 *Montastrea annuligera*. Detail of corallites removing sediment. BALI, INDONESIA *Photograph: Roger Steene*

3 *Montastrea annuligera*. Detail of corallites. The species is best recognised by the irregular septa. PAPUA NEW GUINEA *Photograph: author*

4 *Montastrea multipunctata*. The usual appearance of a colony surface with sand between the corallites. CALAMIAN ISLANDS, PHILIPPINES *Photograph: author*

5 *Montastrea multipunctata*. Corallites with six primary septa clearly visible. CALAMIAN ISLANDS, PHILIPPINES *Photograph: author*

6 *Montastrea multipunctata*. Corallites with sand partly removed, revealing 'groove and tubercle' formations. CALAMIAN ISLANDS, PHILIPPINES *Photograph: author*

7 *Montastrea multipunctata*. Corallites with sand artificially removed. This seldom occurs naturally. CALAMIAN ISLANDS, PHILIPPINES *Photograph: author*

Family Faviidae | Genus *Montastrea*

> Group 3: Species with large corallites (over 9 mm diameter)

Montastrea cavernosa
(Linnaeus, 1766)

Characters: Colonies are massive, forming boulders or domes, or are flat plates. Corallites are variable, but are usually conical and exsert. Long and short septa strongly alternate, with alternate septa joined to the columella. **Colour**. Commonly green, brown, grey, or orange. **Similar species:** None. **Habitat:** All reef environments, especially lower slopes. **Abundance:** Common.

Taxonomic references: Laborel (1969), Zlatarski and Estalella (1982), Amaral (1994).
Identification guides: Colin (1978), Wood (1983), Humann (1993).

Family Faviidae / Genus *Montastrea*

1 Common appearance of a large massive colony. BAHAMAS *Photograph: author*

2 There is little significant geographic variation in this species. ABROLHOS ISLANDS, BRAZIL *Photograph: author*

3 Detail of corallites. BAHAMAS *Photograph: Pat Colin*

4 Corallites with tentacles partly extended. VIRGIN ISLANDS, CARIBBEAN *Photograph: Valerie Taylor*

5 Showing corallites of a small spherical colony. BELIZE *Photograph: author*

Family Faviidae   Genus *Montastrea*

Montastrea valenciennesi
(Milne Edwards and Haime, 1848)

Characters: Colonies are submassive to encrusting, with angular corallites 8-15 millimetres diameter. 'Groove and tubercle' formations are well developed. Long and short septa strongly alternate, are thickened over walls and are uniformly toothed. **Colour:** Usually green, brown or yellow with white septa and sometimes green oral discs. **Similar species:** *Montastrea colemani*. See also *M. magnistellata*. **Habitat:** Most reef environments. **Abundance:** Usually uncommon.

Taxonomic references: Veron, Pichon and Wijsman-Best (1977), Veron and Hodgson (1989). **Identification guides:** Veron (1986), Nishihira and Veron (1995).

1, 2, 4 *Montastrea valenciennesi*. Common variation in corallite shape and colour. **1, 2** GREAT BARRIER REEF, AUSTRALIA **4** GUAM *Photographs: 1, 2 author 4 Gustav Paulay*

3 *Montastrea valenciennesi*. A small submassive colony. GREAT BARRIER REEF, AUSTRALIA *Photograph: author*

5 *Montastrea magnistellata*. A massive colony. PAPUA NEW GUINEA *Photograph: author*

6 *Montastrea magnistellata*. An encrusting colony. PEMBA ISLAND, TANZANIA *Photograph: author*

7 *Montastrea magnistellata*. Undulating surface of a colony showing variation in corallite size. PEMBA ISLAND, TANZANIA *Photograph: author*

8 *Montastrea magnistellata*. Corallites are *Favia*-like, but have extratentacular budding. GREAT BARRIER REEF, AUSTRALIA *Photograph: Roger Steene*

Family Faviidae — Genus *Montastrea*

Montastrea magnistellata
Chevalier, 1971

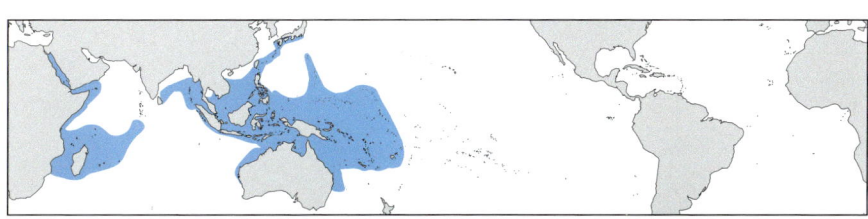

Characters: Colonies are massive, hemispherical or flattened and may be over 2 metres across. Corallites are round and of variable size, those on convex surfaces being larger than those on concave surfaces. Corallites are shallow, approximately 6-13 millimetres diameter, with tightly compacted septa. Columellae are large and small paliform lobes are usually developed. **Colour:** Commonly blue-grey. **Similar species:** *Montastrea valenciennesi* which has less compact, irregular septa. See also *Favia* species, especially *F. helianthoides*, which has corallites of similar size and shape. **Habitat:** Protected reef slopes. **Abundance:** Usually uncommon.

Taxonomic reference: Veron, Pichon and Wijsman-Best (1977). **Identification guides:** Veron (1986), Sheppard and Sheppard (1991), Nishihira and Veron (1995).

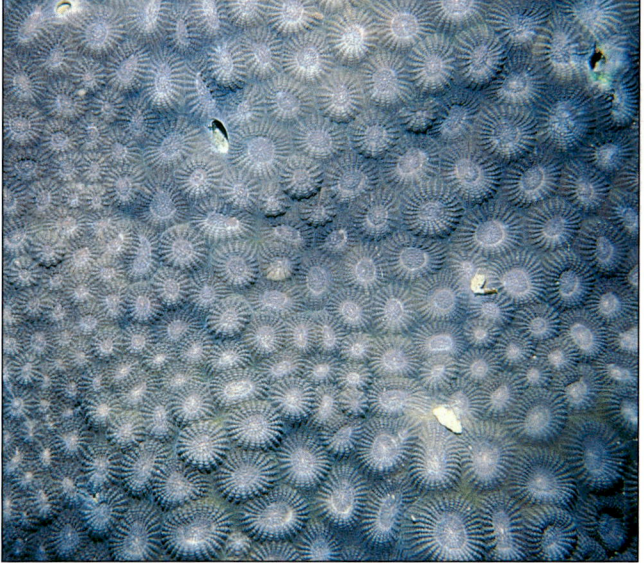

Family Faviidae | Genus *Plesiastrea*

GENUS *PLESIASTREA*
Milne Edwards and Haime, 1848

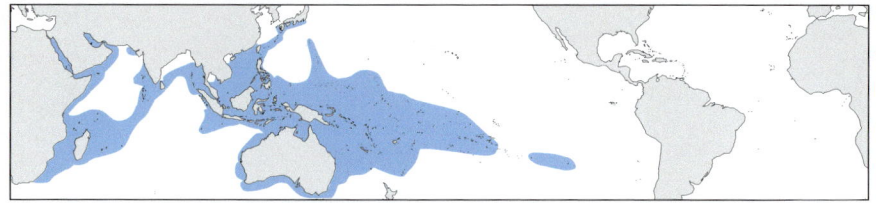

Characters: Colonies are massive, rounded or flattened. Corallites are small, rounded, plocoid and are formed by extratentacular budding. **Similar genera:** This is a poorly defined genus. *Plesiastrea versipora* is the only common, widespread species. **Earliest fossil record:** Miocene of the Tethys.

Taxonomic reference: Veron, Pichon and Wijsman-Best (1977).

Plesiastrea versipora
(Lamarck, 1816)

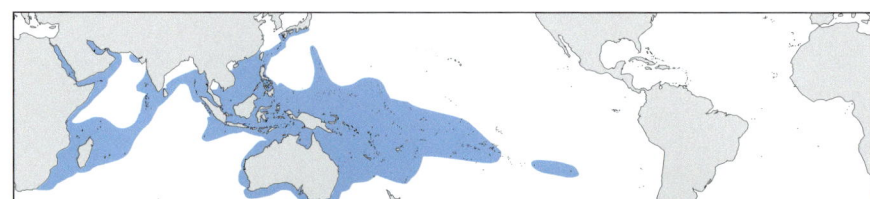

Characters: Colonies are flat and are frequently lobed, up to 3 metres across in high latitude localities, usually smaller in the tropics. Corallites are 2–4 millimetres diameter. Paliform lobes form a neat circle around small columellae. Tentacles are sometimes extended during the day; they are short and are of two alternating sizes. **Colour:** Yellow, cream, green or brown, usually pale colours in the tropics and bright colours (green or brown) in high latitude areas. **Similar species:** *Plesiastrea devantieri*. Sometimes confused with other faviids with corallites of similar size notably *Favia stelligera* and some *Montastrea* and *Cyphastrea* species. **Habitat:** Occurs in most reef environments but especially in shaded places such as under overhangs. Also occurs on rocky foreshores of temperate locations protected from strong wave action. **Abundance:** Seldom common.

Taxonomic references: Veron, Pichon and Wijsman-Best (1977), Wijsman-Best (1980). **Identification guides:** Randall and Myers (1983), Veron (1986), Sheppard and Sheppard (1991), Nishihira and Veron (1995), Coles (1996), Carpenter et al. (1997).

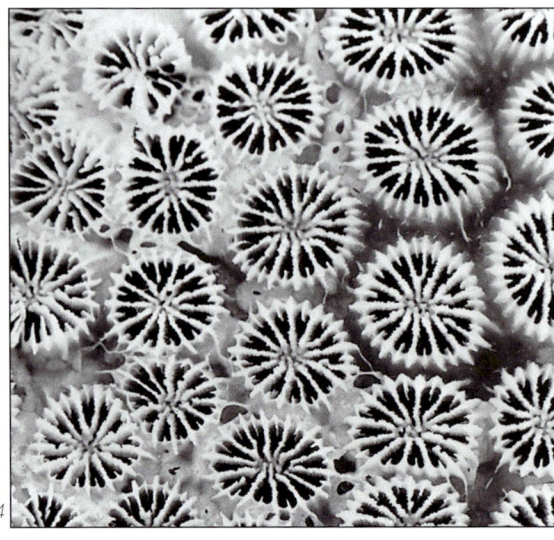

Family Faviidae — Genus *Plesiastrea*

1 A cushion-shaped colony, the most common growth-form. CARNAC ISLAND, WESTERN AUSTRALIA *Photograph: Clay Bryce*

2 Surface of a large massive colony. MADAGASCAR *Photograph: author*

3 A submassive colony. SOLITARY ISLANDS, SOUTH-EAST AUSTRALIA *Photograph: author*

4 The surface appearance of a massive colony. BASS STRAIT, SOUTHERN AUSTRALIA *Photograph: Neville Coleman*

5 Surface of an encrusting colony. PAPUA NEW GUINEA *Photograph: author*

6 With tentacles partly extended. CALAMIAN ISLANDS, PHILIPPINES *Photograph: author*

7 Corallite detail. VIETNAM *Photograph: author*

Family Faviidae / Genus *Plesiastrea*

Plesiastrea devantieri
Veron, this publication

Characters: Colonies are smooth massive boulders up to 1.5 metres across. Corallites are uniform, 3-5 millimetres diameter with well defined walls. Septa are in two distinct orders of 6-8 and are exsert. Paliform lobes are inconspicuous. Columellae are small. **Colour:** Red to orange, usually with green centres. **Similar species:** *Plesiastrea versipora*, which has more septa and conspicuous paliform lobes. **Habitat:** Shallow reef environments. **Abundance:** Seldom common.

Taxonomic note: See 'New species described in *Corals of the World*' (Veron, in preparation) for further information.

Family Faviidae

GENUS *OULASTREA*
Milne Edwards and Haime, 1848

Characters, abundance and distribution: This genus has only one species, see *Oulastrea crispata*. **Fossil record:** None.

Oulastrea crispata
(Lamarck, 1816)

Characters: Colonies are encrusting and grow to only a few centimetres across. Corallites are like a small *Montastrea*, are of uniform size and are closely compacted. Long and short septa alternate. Paliform lobes are well developed. Tentacles are sometimes extended during the day. **Colour:** Black with white upper margins to the septa. Dried skeletons are also black and white. **Similar species:** None. **Habitat:** Found only in subtidal turbid water, attached to wave washed rock. **Abundance:** Uncommon.

Taxonomic reference: Veron, Pichon and Wijsman-Best (1977). **Identification guide:** Nishihira and Veron (1995).

1 *Plesiastrea devantieri*. Colonies form large smooth boulders. SOCOTRA, YEMEN *Photograph: Lyndon DeVantier*

2 *Plesiastrea devantieri*. Colony surface. SOCOTRA, YEMEN *Photograph: Lyndon DeVantier*

3 *Oulastrea crispata*. The characteristic appearance of a small encrusting colony. FLORES, INDONESIA *Photograph: author*

4 *Oulastrea crispata*. With tentacles extended. HONSHU, JAPAN *Photograph: Takeshi Okamoto*

5 *Oulastrea crispata*. Corallite detail. GREAT BARRIER REEF, AUSTRALIA *Photograph: author*

GENUS *DIPLOASTREA*
Matthai, 1914

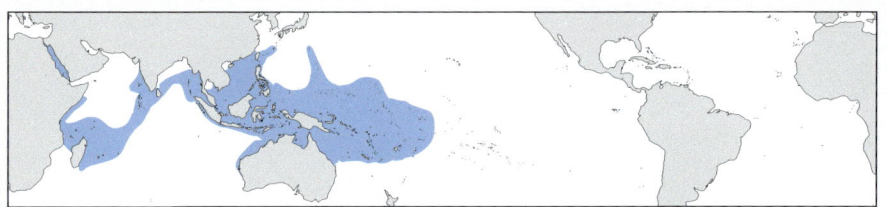

Characters, abundance and distribution: This genus has only one species, see *Diploastrea heliopora*.
Earliest fossil record: Eocene of the Indian Ocean (doubtful record) and Caribbean, Oligocene of the Tethys.

Diploastrea heliopora
(Lamarck, 1816)

Characters: Colonies are dome-shaped with an even surface and may be up to 2 metres high and 5 metres across. The skeleton is dense. Corallites form low cones with small openings and very thick walls. Columellae are large. Septa are equal and are thick at the wall and thin where joining the columellae. Tentacles are extended only at night.
Colour: Usually uniform cream or grey, sometimes greenish. **Similar species:** None. This is one of the most easily recognised and least variable of all massive corals. **Habitat:** Both exposed and protected reef environments.
Abundance: Sometimes common.

Taxonomic references: Chevalier (1975), Veron, Pichon and Wijsman-Best (1977). **Identification guides:** Randall and Myers (1983), Veron (1986), Sheppard and Sheppard (1991), Nishihira and Veron (1995).

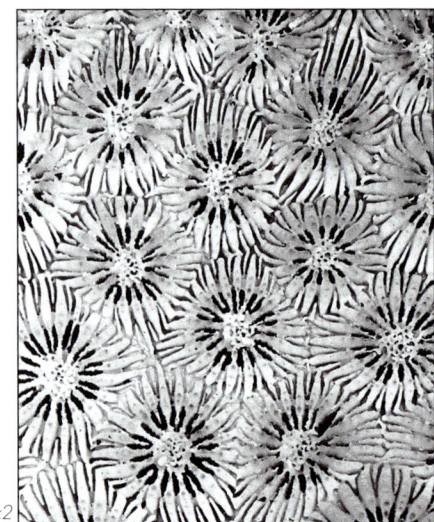

Family Faviidae | Genus *Diploastrea*

1 A typical hemispherical colony. BALI, INDONESIA *Photograph: Gerry Allen*

2, 4 Corallite detail. This species shows very little corallite variation. **2** GREAT BARRIER REEF, AUSTRALIA **4** SABAH, INDONESIA *Photographs: 2 Mary Stafford-Smith 4 Roger Steene*

3 Forming large spherical colonies. PEMBA ISLAND, TANZANIA *Photograph: author*

‡ See also corallite detail, page 84.

231

Family Faviidae

GENUS *LEPTASTREA*
Milne Edwards and Haime, 1848

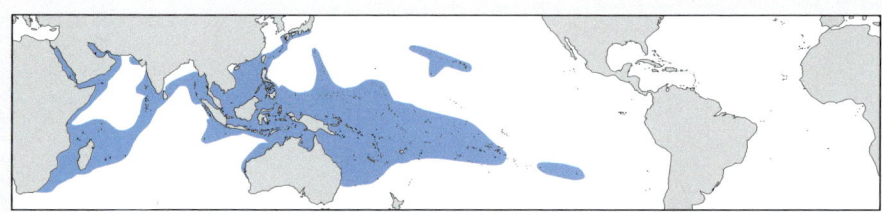

Characters: Colonies are massive, usually flat or dome-shaped. Corallites are cerioid to subplocoid. Costae are poorly developed or absent. Columellae consist of vertical pinnules. Septa have inward projecting teeth. The upper surfaces of colonies of most species growing in intertidal habitats are usually pale. Tentacles are sometimes extended during the day; in *Leptastrea pruinosa* they are always at least partly extended. **Similar genus:** *Leptastrea* is a well defined genus closest to *Cyphastrea* which is plocoid with widely separated corallites. **Earliest fossil record:** Oligocene of the Indo-Pacific, Miocene of the Tethys.

Taxonomic references: Chevalier (1975), Veron, Pichon and Wijsman-Best (1977), Wijsman-Best (1980). **Summary key:** See pp453-4.

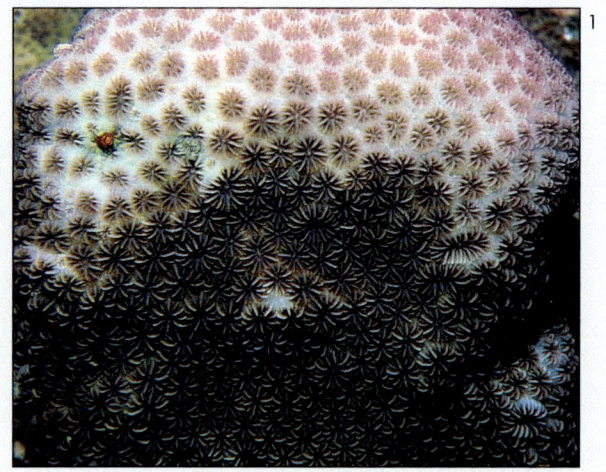

Leptastrea bewickensis
Veron and Pichon, 1977

Characters: Colonies are massive, flat or hillocky. Septa are in unequal cycles with approximately six widely spaced exsert primary septa plunging down to a small columella. Columellae are simple fused ridges. Tentacles are sometimes partly extended during the day. **Colour:** Dark grey, tan or brown. **Similar species:** *Parasimplastrea sheppardi*. **Habitat:** Sheltered reef environments. **Abundance:** Sometimes locally abundant.

Taxonomic references: Veron, Pichon and Wijsman-Best (1977), Wijsman-Best (1980).

x4

Leptastrea inaequalis
Klunzinger, 1879

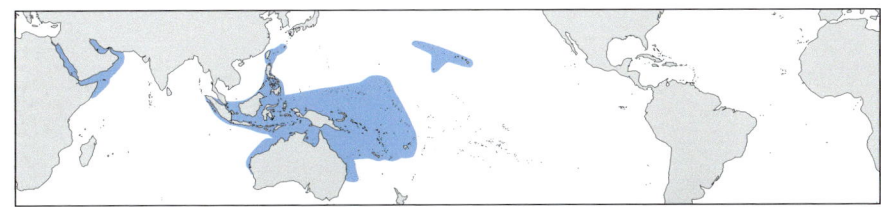

Characters: Colonies are massive and plocoid, with small barrel-shaped corallites (usually up to 3 mm diameter but occasionally 5 mm) of varying appearance, irregularly separated by deep grooves. Septa are in two (rarely three) unequal cycles, primary septa being wedge-shaped. 'Groove and tubercle' formations are usually well developed. **Colour:** Usually combinations of cream, green and yellow, with darker calices. **Similar species:** *Leptastrea bottae*, also *Cyphastrea* species with corallites of similar size. **Habitat:** A wide range of reef environments. **Abundance:** Uncommon.

Taxonomic reference: Veron, Pichon and Wijsman-Best (1977, as *Leptastrea* cf. *bottae*). **Identification guides:** Veron (1986, as *L.* cf. *bottae*), Nishihira and Veron (1995).

1 *Leptastrea* colonies in shallow water frequently have pale upper surfaces and dark sides, as seen in this colony of *L. transversa*. SINAI PENINSULA, EGYPT *Photograph: author*

2 *Leptastrea bewickensis*. Surface of a colony showing corallites with six primary septa. CALAMIAN ISLANDS, PHILIPPINES *Photograph: author*

3 *Leptastrea inaequalis*. Part of a large colony. PAPUA NEW GUINEA *Photograph: author*

4 *Leptastrea inaequalis*. Surface of a flat colony. GREAT BARRIER REEF, AUSTRALIA *Photograph: Neville Coleman*

5 *Leptastrea inaequalis*. Surface detail. PAPUA NEW GUINEA *Photograph: author*

6 *Leptastrea inaequalis*. Corallite detail. Openings to groove and tubercle formations are visible between corallites. GREAT BARRIER REEF, AUSTRALIA *Photograph: Valerie Taylor*

Leptastrea bottae
(Milne Edwards and Haime, 1849)

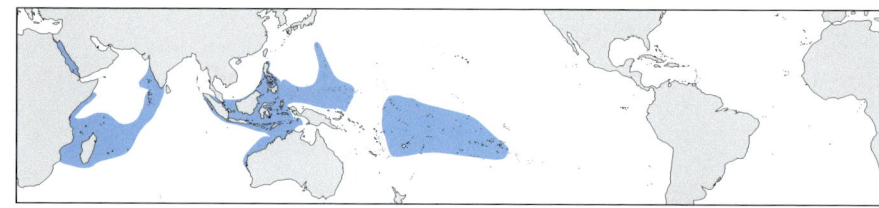

Characters: Colonies are massive to encrusting. Corallites are cylindrical, separated only by a fine groove. Septa are in three cycles, the longest being distinctive and exsert. 'Groove and tubercle' formations are sometimes well developed. **Colour:** White or cream with darker calices. **Similar species:** *Leptastrea inaequalis*, which has corallites of similar size, but these are barrel-shaped, more exsert, usually have extensive 'groove and tubercle' formations and have few, if any, third cycle septa. *Leptastrea transversa* has relatively compacted angular corallites and almost equal first and second cycle septa. **Habitat:** Shallow reef environments. **Abundance:** Uncommon.

Taxonomic reference: Milne Edwards and Haime's original description/specimens.

x4

Family Faviidae | Genus *Leptastrea*

Leptastrea aequalis
Veron, this publication

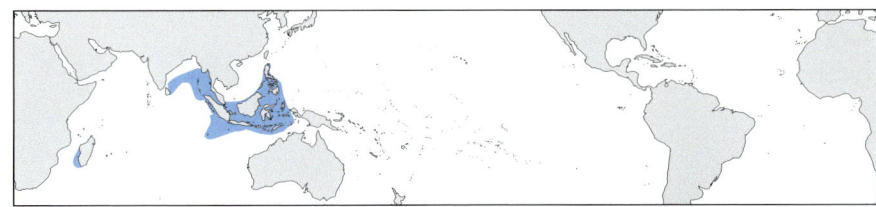

Characters: Colonies are massive to encrusting. Corallites are rounded and clearly plocoid, with thick walls. Corallites are often uniformly inclined on the colony surface. Septa are in two indistinct orders, primary septa being numerous and equal. 'Groove and tubercle' formations are rare. **Colour:** Cream with dark calices. **Similar species:** *Leptastrea inaequalis*, which has barrel-shaped corallites and six exsert primary septa and *L. bottae*, which has more compacted corallites and exsert primary septa. See also *L. transversa*. **Habitat:** Shallow reef environments. **Abundance:** Rare, except at Cocos (Keeling) Atoll.

Taxonomic note: See 'New species described in *Corals of the World*' (Veron, in preparation) for further information.

x4

1, 2 *Leptastrea bottae*. Surface of colonies showing variation in development of groove and tubercle formations. SINAI PENINSULA, EGYPT *Photographs: author*

3, 4 *Leptastrea bottae*. Variation in corallite detail. SINAI PENINSULA, EGYPT *Photographs: author*

5 *Leptastrea aequalis* (left) overgrowing *L. pruinosa* (right). COCOS (KEELING) ATOLL, WESTERN AUSTRALIA *Photograph: author*

6 *Leptastrea aequalis*. Corallite detail, showing the large open corallites. COCOS (KEELING) ATOLL, WESTERN AUSTRALIA *Photograph: author*

7 *Leptastrea aequalis*. Surface of a colony showing evenly spaced, compact corallites. MADAGASCAR *Photograph: author*

Leptastrea purpurea
(Dana, 1846)

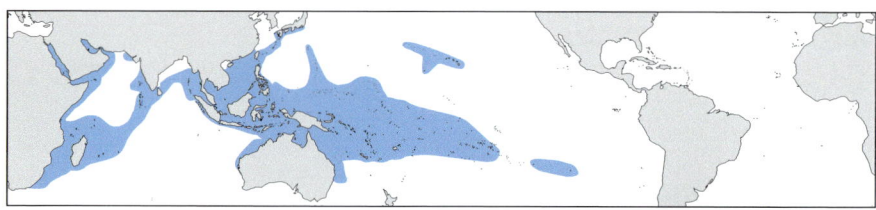

×4

Characters: Colonies are flat with angular, cerioid corallites which vary in size within the same colony. Colonies on reef flats may have several corallites in shallow valleys. Septa are tightly compact, approximately similar in size, and have margins that slope uniformly towards the corallite centre. Columellae are small and compact. **Colour:** Usually pale yellow, greenish or cream on the upper surface and dark sides. **Similar species:** *Leptastrea transversa*, which has more uniformly sized corallites and less compact septa with plunging inner margins. *Leptastrea pruinosa* has septa in more distinct orders and tentacles extended during the day. **Habitat:** A wide range of reef environments. **Abundance:** Common.

Taxonomic note: This species is divisible into several smaller semi-distinct taxonomic units, see 'What are species?' p425. **Taxonomic references:** Chevalier (1975), Veron, Pichon and Wijsman-Best (1977), Wijsman-Best (1980). **Identification guides:** Randall and Myers (1983), Veron (1986), Sheppard and Sheppard (1991), Nishihira and Veron (1995), Coles (1996).

1 *Leptastrea purpurea* (left) with *L. pruinosa* (right). ASHMORE REEF, WESTERN AUSTRALIA *Photograph: author*

2 *Leptastrea purpurea*. Surface of a large colony. PAPUA NEW GUINEA *Photograph: author*

3-5 *Leptastrea purpurea*. Common variation in corallite shape and colour. GREAT BARRIER REEF, AUSTRALIA *Photographs: author*

6 *Leptastrea pruinosa*. A submassive colony. CALAMIAN ISLANDS, PHILIPPINES *Photograph: author*

7, 8 *Leptastrea pruinosa*. Colony surface. **7** RYUKYU ISLANDS, JAPAN **8** SINAI PENINSULA, EGYPT *Photographs: author*

9-11 *Leptastrea pruinosa*. Common variation in corallite shape and colour. Note that tentacles are usually partly extended. **9** GREAT BARRIER REEF, AUSTRALIA **10** GUAM **11** VANUATU *Photographs:* **9** *Ed Lovell* **10** *Gustav Paulay* **11** *author*

Leptastrea pruinosa
Crossland, 1952

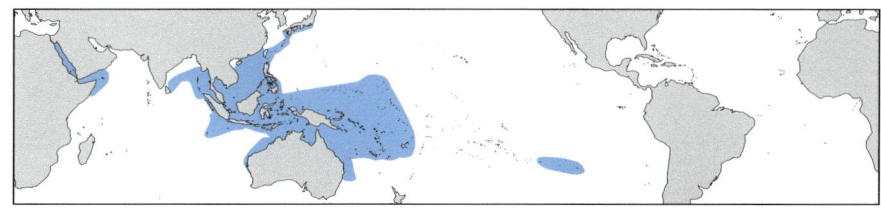

Characters: Colonies are flat with angular, cerioid corallites. Septa are in distinctive cycles and have granulated sides and margins. Tentacles are usually extended during the day. **Colour:** Commonly chocolate brown or pink with green or cream calices. **Similar species:** Readily identified underwater by having extended polyps during the day. Skeletons are differentiated from *Leptastrea purpurea* by having granulated septa giving a frosted appearance. **Habitat:** Shallow clear water. **Abundance:** Uncommon.

Taxonomic references: Veron, Pichon and Wijsman-Best (1977), Wijsman-Best (1980).
Identification guides: Veron (1986), Nishihira and Veron (1995).

x4

Family Faviidae   Genus *Leptastrea*

Leptastrea transversa
Klunzinger, 1879

Characters: Colonies are flat with angular, cerioid corallites. Septa are not tightly compact and plunge steeply near the columella. Columellae consist of a few pinnules aligned in a row and often fused. **Colour:** Usually grey, green or yellow with dark sides to the colony. **Similar species:** *Leptastrea purpurea*. **Habitat:** A wide range of reef environments. **Abundance:** Uncommon.

Taxonomic references: Chevalier (1975), Veron, Pichon and Wijsman-Best (1977), Wijsman-Best (1980). **Identification guides:** Randall and Myers (1983), Veron (1986), Sheppard and Sheppard (1991), Nishihira and Veron (1995), Carpenter *et al.* (1997).

GENUS
PARASIMPLASTREA
Sheppard, 1985

Characters, abundance and distribution: This genus has only one species, see *Parasimplastrea sheppardi*.
Earliest fossil record: Miocene of the Pacific (doubtful record).

Parasimplastrea sheppardi
Veron, this publication
(new name)

×2

Characters: Colonies are encrusting to submassive. Corallites are 4-6 millimetres diameter. A 'groove and tubercle' formation may be present, but corallites are cerioid to subplocoid, retaining individual walls. Septa have smooth or slightly serrated margins, are widely spaced, and plunge steeply within the corallite. Columellae are rudimentary, consisting of thickened inner septal margins. Polyps are fleshy. **Colour:** Brown with greenish calices. **Similar species:** *Leptastrea bewickensis*, which is fully cerioid, has more irregular septa and has small columellae. The mussid *Blastomussa merleti* has similar fleshy polyps. The smooth septa resemble those of the oculinids *Galaxea* and *Simplastrea*. **Habitat:** Reef environments. **Abundance:** Uncommon.

Taxonomic note 1: This genus is moved from Family Oculinidae because of general similarities with *Leptastrea bewickensis* and the occasional occurrence of septal serrations and rudimentary columellae. The species was formerly called *Parasimplastrea simplicitexta* (Umbgrove, 1939). This is an extinct fossil faviid with larger corallites and different septal structures. **Taxonomic note 2:** See 'New species described in *Corals of the World*' (Veron, in preparation) for further information. **Taxonomic reference:** Sheppard (1985, as *Parasimplastrea simplicitexta*). **Identification guide:** Sheppard and Sheppard (1991, as *Parasimplastrea simplicitexta*).

1 *Leptastrea transversa*. A massive colony. SINAI PENINSULA, EGYPT *Photograph: Mary Stafford-Smith*

2 *Leptastrea transversa*. An encrusting colony. CALAMIAN ISLANDS, PHILIPPINES *Photograph: author*

3-6 *Leptastrea transversa*. Surface of colonies. **3** GREAT BARRIER REEF, AUSTRALIA **4-6** PAPUA NEW GUINEA *Photographs: author*

7 *Parasimplastrea sheppardi*. A colony with tentacles withdrawn showing the exsert primary septa. OMAN *Photograph: Steve Coles*

8 *Parasimplastrea sheppardi*. A colony with tentacles extended. OMAN *Photograph: Steve Coles*

9 *Parasimplastrea sheppardi*. A submassive colony with tentacles partly extended. SOCOTRA, YEMEN *Photograph: Lyndon DeVantier*

10 *Parasimplastrea sheppardi*. An encrusting colony. OMAN *Photograph: Peter Harrison*

Family Faviidae   Genus *Cyphastrea*

GENUS *CYPHASTREA*
Milne Edwards and Haime, 1848

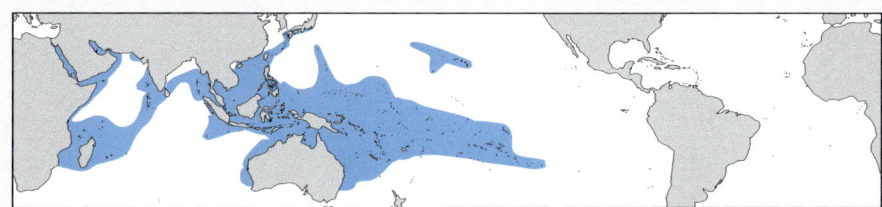

Characters: *Cyphastrea decadia* is arborescent with axial and radial corallites. All other species are massive or encrusting. Corallites are plocoid, with small calices. Costae are generally restricted to the corallite wall; the coenosteum is granulated. Tentacles are usually extended only at night. **Similar genera:** *Cyphastrea* is a well defined genus. It resembles some *Montastrea* with small corallites. See also *Plesiastrea versipora*, which is distinguished by having larger corallites with better developed paliform lobes and by having costae of adjacent corallites in contact, with no coenosteum granules. **Earliest fossil record:** Miocene of the Tethys.

Taxonomic references: Chevalier (1975), Veron, Pichon and Wijsman-Best (1977), Wijsman-Best (1980). **Summary key:** See pp452-3.

Cyphastrea japonica
Yabe and Sugiyama, 1932

Characters: Colonies are submassive or encrusting, with an irregular surface. Corallites are small and often crowded. Septa are in two unequal orders of 12 each, the first being exsert and irregularly toothed. Coenosteum spinules are prominent. 'Groove and tubercle' formations are often present. Colonies are often infested with parasitic barnacles. **Colour:** Mottled cream or yellowish-green, sometimes mottled grey. **Similar species:** *Cyphastrea chalcidicum*, which has strongly alternating costae. See also *C. ocellina*. **Habitat:** Shallow exposed reef environments. **Abundance:** Usually uncommon.

x4

Taxonomic references: Yabe and Sugiyama (1932), Veron (1992). **Identification guide:** Nishihira and Veron (1995).

Family Faviidae | Genus *Cyphastrea*

Cyphastrea chalcidicum
(Forskål, 1775)

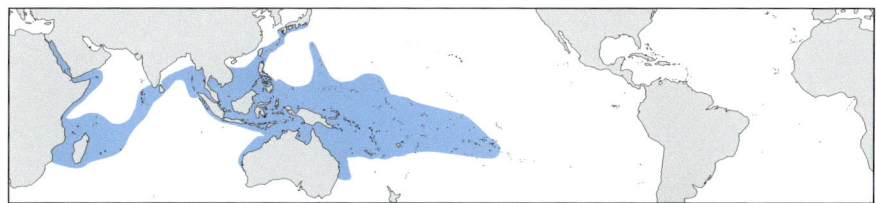

Characters: Colonies are encrusting to massive, with a tendency to form columns. Corallites are usually widely spaced and conical, with clearly alternating costae which are easily visible underwater. There are 12 primary septa. **Colour:** Usually uniform brown, green or cream with corallite walls and calices of contrasting colours. **Similar species:** *Cyphastrea serailia*. **Habitat:** A wide range of reefs, also rocky foreshores of subtropical locations. **Abundance:** Common, but less so than *C. serailia*.

Taxonomic references: Veron, Pichon and Wijsman-Best (1977), Wijsman-Best (1980).
Identification guides: Veron (1986), Nishihira and Veron (1995).

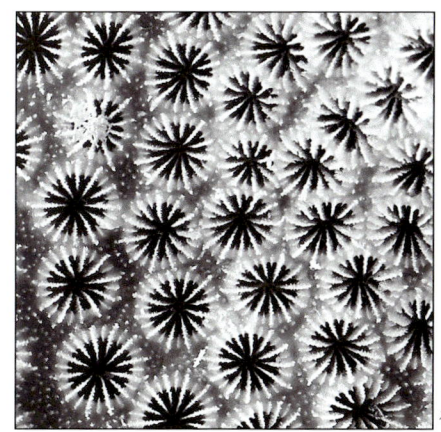
x4

1 *Cyphastrea japonica*. A large massive colony. VIETNAM *Photograph: author*

2, 3 *Cyphastrea japonica*. The surface appearance of massive colonies. **2** PUERTO GALERA, PHILIPPINES **3** RYUKYU ISLANDS, JAPAN *Photographs: author*

4, 5 *Cyphastrea japonica*. Variation in corallite shape and colour. **4** CALAMIAN ISLANDS, PHILIPPINES **5** RYUKYU ISLANDS, JAPAN *Photographs: author*

6 *Cyphastrea chalcidicum*. An encrusting colony. GREAT BARRIER REEF, AUSTRALIA *Photograph: author*

7, 9 *Cyphastrea chalcidicum*. Irregular surface of massive colonies. **7** PAPUA NEW GUINEA **9** SINAI PENINSULA, EGYPT *Photographs: author*

8 *Cyphastrea chalcidicum*. Showing extra-tentacular budding in exsert compact corallites. PAPUA NEW GUINEA *Photograph: author*

10 *Cyphastrea chalcidicum*. Corallite detail. SCOTT REEF, WESTERN AUSTRALIA *Photograph: author*

Cyphastrea serailia
(Forskål, 1775)

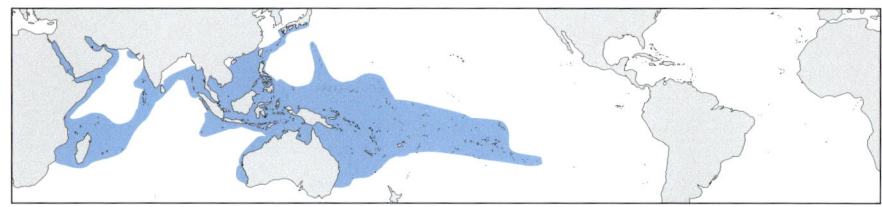

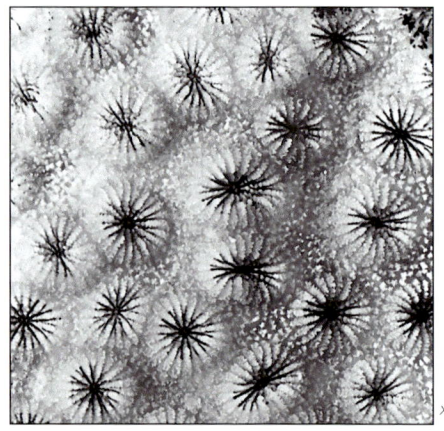

Characters: Colonies are massive to columnar with a smooth or hillocky surface. Corallites are rounded and equal in size. Costae do not alternate strongly. There are 12 primary septa. **Colour:** Usually uniform or mottled grey, brown or cream. **Similar species:** *Cyphastrea chalcidicum*, which has well developed alternating costae. *Cyphastrea microphthalma* has 10 primary septa. **Habitat:** All reef environments. **Abundance:** Common.

Taxonomic note: This is the most common and widespread *Cyphastrea*, yet is poorly defined. In isolated localities it may have little variation, while in central regions it is highly variable.
Taxonomic references: Chevalier (1975), Veron, Pichon and Wijsman-Best (1977), Wijsman-Best (1980). **Identification guides:** Veron (1986), Sheppard and Sheppard (1991), Nishihira and Veron (1995), Coles (1996), Carpenter *et al.* (1997).

x4

1

2

3

Family Faviidae | Genus *Cyphastrea*

1-3, 5, 6 Showing common variation in corallite structure and colour. **1** NORFOLK ISLAND, WESTERN PACIFIC **2** SCOTT REEF, WESTERN AUSTRALIA **3** VIETNAM **5, 6** GREAT BARRIER REEF, AUSTRALIA *Photographs: 1-3 author 5, 6 Ed Lovell*

4 Large colonies are usually massive in shallow water. PAPUA NEW GUINEA *Photograph: author*

Cyphastrea ocellina
(Dana, 1864)

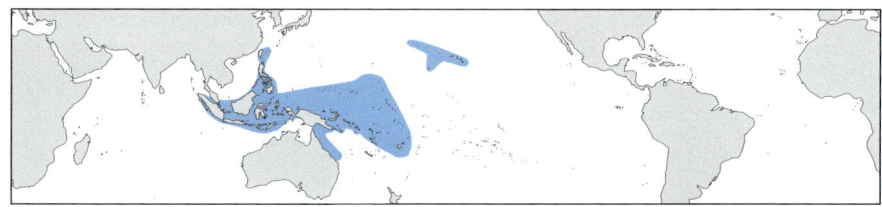

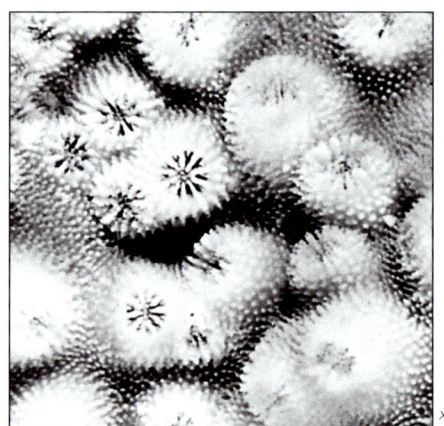

×4

Characters: Colonies are massive or encrusting, with an undulating surface. Corallites are less than 3 millimetres diameter and tightly compacted. Septa are in two unequal orders of 12 each and sometimes have a third order. Paliform lobes are small or absent. The coenosteum is covered with short spinules. **Colour:** Pale greenish-yellow or dark green. **Similar species:** *Cyphastrea japonica*, which forms less massive colonies, commonly has 'groove and tubercle' formations and has two very unequal orders of septo-costae. **Habitat:** Upper reef slopes. **Abundance:** Rare.

Taxonomic reference: Wijsman-Best (1980). **Identification guides:** Maragos (1977), Nishihira and Veron (1995).

Cyphastrea hexasepta
Veron, Turak and DeVantier, this publication

Characters: Colonies are massive, with a smooth or irregular surface. Corallites are widely spaced and immersed to tubular, the latter being strongly inclined on the colony surface. Calices are small. The coenosteum is covered with prominent spines which have elaborate surfaces. Septa are usually in two cycles. There are usually six primary septa although this varies among corallites. **Colour:** Mottled brown. **Similar species:** None. The small, widely spaced corallites are distinctive. *Cyphastrea microphthalma* is readily identified by its 10 primary septa. **Habitat:** Shallow reef environments. **Abundance:** Uncommon.

Taxonomic note: See 'New species described in *Corals of the World*' (Veron, in preparation) for further information.

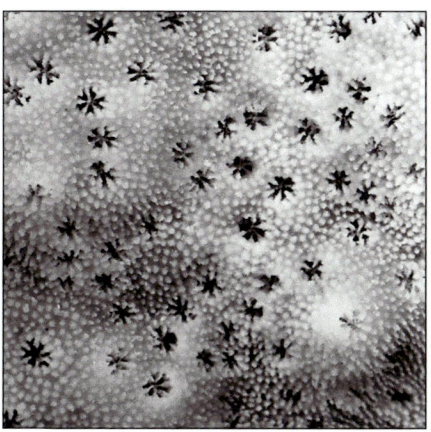

1 *Cyphastrea ocellina*. A massive colony. CALAMIAN ISLANDS, PHILIPPINES *Photograph: author*

2 *Cyphastrea ocellina*. Detail of a small encrusting colony. CALAMIAN ISLANDS, PHILIPPINES *Photograph: author*

3 *Cyphastrea ocellina*. With very exsert corallites. HAWAII *Photograph: Doug Fenner*

4 *Cyphastrea ocellina*. Corallite detail. CALAMIAN ISLANDS, PHILIPPINES *Photograph: author*

5 *Cyphastrea hexasepta*. A massive colony. SAUDI ARABIA, RED SEA *Photograph: Emre Turak*

Cyphastrea microphthalma
(Lamarck, 1816)

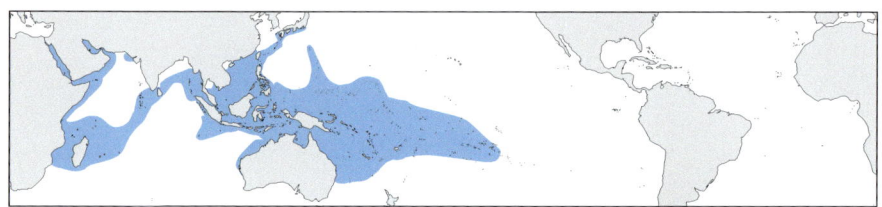

Characters: Colonies are massive, becoming thin encrusting plates where light levels are low. They commonly grow as mobile balls (coralliths). Corallites are tall and conical; compact in colonies exposed to strong light, widely spaced in encrusting colonies. They usually have 10 primary septa although this varies among corallites. **Colour:** Brown, cream or green, sometimes other colours. Septa are commonly white. **Similar species:** *Cyphastrea microphthalma* is readily identified by its 10 primary septa which are visible underwater. **Habitat:** Most reef environments. **Abundance:** Common, but less so than *C. serailia*.

Taxonomic references: Chevalier (1975), Veron, Pichon and Wijsman-Best (1977), Wijsman-Best (1980). **Identification guides:** Veron (1986), Sheppard and Sheppard (1991), Nishihira and Veron (1995), Coles (1996), Carpenter *et al.* (1997).

x4

Family Faviidae — Genus *Cyphastrea*

1 A small submassive colony. SINAI PENINSULA, EGYPT *Photograph: Mary Stafford-Smith*

2, 3, 6 Detail of encrusting colonies. **2** PEMBA ISLAND, TANZANIA **3** GUAM **6** ASHMORE REEF, WESTERN AUSTRALIA *Photographs: 2, 6 author 3 Gustav Paulay*

4, 5, 7 Common variation in corallite shape and colour. **4** VIETNAM **5** CALAMIAN ISLANDS, PHILIPPINES **7** SCOTT REEF, WESTERN AUSTRALIA *Photographs: author*

Cyphastrea agassizi
(Vaughan, 1907)

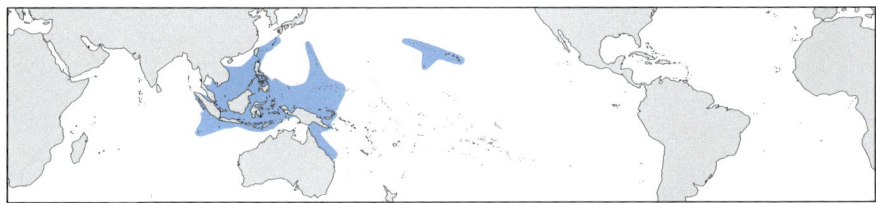

Characters: Colonies are massive. The surface is often deeply grooved. Corallites are widely spaced. Septa are in three unequal orders with the first order being exsert. The coenosteum between the corallites is usually smooth. Irregular 'groove and tubercle' formations may be present. **Colour:** The coenosteum is whitish, corallites are pale brown or green. Septa are sometimes orange. **Similar species:** *Cyphastrea ocellina*, which has smaller, more crowded corallites. See also *C. japonica* and *Leptastrea inaequalis*, which have more crowded corallites with thicker walls. **Habitat:** Shallow reef environments. **Abundance:** Uncommon.

Taxonomic reference: Vaughan's original description/specimens. **Identification guide:** Nishihira and Veron (1995).

1-4 *Cyphastrea agassizi*. Variation in corallite shape and colour. **1** CALAMIAN ISLANDS, PHILIPPINES **2** CEBU, PHILIPPINES **3** RYUKYU ISLANDS, JAPAN **4** GUAM Photographs: *1-3 author 4 Gustav Paulay*

Family Faviidae Genus *Cyphastrea*

Cyphastrea decadia
Moll and Borel–Best, 1984

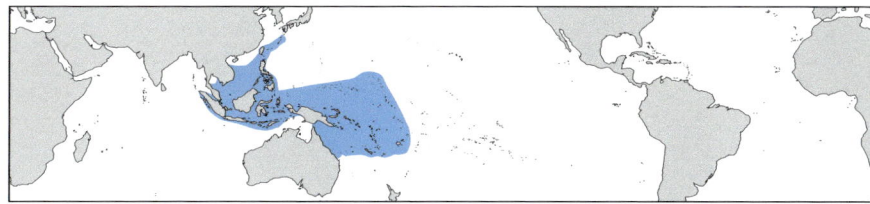

Characters: Colonies are branching, with *Acropora*-like axial corallites. On upper reef slopes, colonies have compact branches; on soft substrates, branches are more open and colonies are fragile. Primary septa vary from 10 to 12 in number. Costae are usually weakly developed. **Colour:** Brown, grey or cream. **Similar species:** None. **Habitat:** A wide range of reef environments, especially lagoons. **Abundance:** Uncommon.

Taxonomic references: Moll and Borel-Best (1984), Veron (1992). **Identification guide:** Nishihira and Veron (1995).

x4

5 *Cyphastrea decadia*. Corallite detail. CEBU, PHILIPPINES *Photograph:* Roger Steene

6 *Cyphastrea decadia*. A large stand in a sandy lagoon. RYUKYU ISLANDS, JAPAN *Photograph:* author

7 *Cyphastrea decadia*. A finely branched colony. SULAWESI, INDONESIA *Photograph:* Gerry Allen

8 *Cyphastrea decadia*. Detail of a colony exposed to moderate wave action. PAPUA NEW GUINEA *Photograph:* author

GENUS *SOLENASTREA*
Milne Edwards and Haime, 1848

Characters: Colonies are massive, hemispherical or with irregular upgrowths. Corallites are widely spaced, circular and equal in size. Long and short septo-costae slightly alternate. The coenosteum is smooth or blistery. Tentacles are sometimes extended during the day. **Similar genera:** None. **Earliest fossil record:** Oligocene of the Caribbean, Miocene of the Tethys.

Taxonomic reference: Zlatarski and Estalella (1982).

Solenastrea bournoni
Milne Edwards and Haime, 1849

Characters: Colonies are massive, forming hemispherical domes with smooth or slightly irregular surfaces. Corallite rims form conspicuous dark circles. Tentacles, which may be extended during the day, are long, giving colonies a ragged appearance. **Colour:** Cream or tan with colourless or green polyps. **Similar species:** *Solenastrea hyades*. **Habitat:** Shallow turbid environments including seagrass beds and reef lagoons. **Abundance:** Usually uncommon.

Taxonomic reference: Zlatarski and Estalella (1982). **Identification guide:** Humann (1993).

Family Faviidae   Genus *Solenastrea*

Solenastrea hyades
(Dana, 1846)

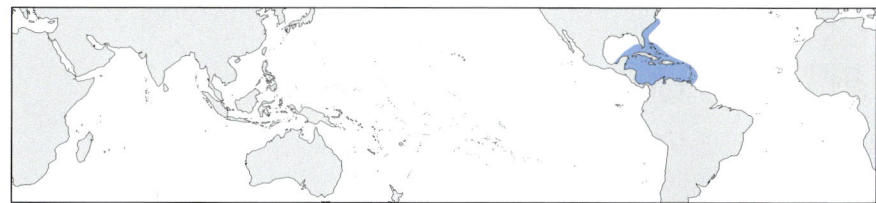

Characters: Colonies are submassive, with irregular upgrowths. Corallites may be irregularly spaced. Tentacles are commonly extended during the day and give colonies a furry appearance. **Colour:** Cream or tan with dark brown polyps. **Similar species:** *Solenastrea bournoni*, which forms even surfaced, hemispherical colonies with slightly smaller corallites. **Habitat:** A wide range of reef habitats and rocky foreshores, especially turbid environments. **Abundance:** Usually uncommon.

Taxonomic reference: Zlatarski and Estalella (1982). **Identification guide:** Humann (1993).

x5

1 *Solenastrea bournoni*. A massive colony in a turbid environment. FLORIDA *Photograph: Paul Humann*

2 *Solenastrea bournoni*. Detail of a massive colony. FLORIDA *Photograph: Paul Humann*

3 *Solenastrea hyades*. Colonies forming irregular columns as well as small columns and nodules. FLORIDA *Photograph: Paul Humann*

4 *Solenastrea hyades*. Detail of a small colony showing the usual appearance of dark corallites with tentacles extended during the day. GULF OF MEXICO *Photograph: Paul Humann*

GENUS *ECHINOPORA*
Lamarck, 1816

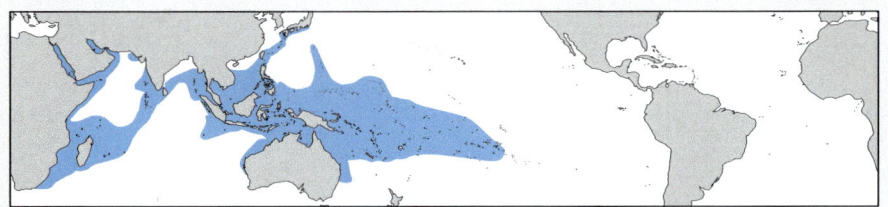

Characters: Colonies are massive, arborescent or laminar or mixtures of these forms. Corallites are plocoid (except *Echinopora fruticulosa* and *E. tiranensis* where corallites form branches), with calices up to 10 millimetres diameter. Septa are exsert and irregular. Columellae are usually prominent. Costae are mostly restricted to the corallite wall. The coenosteum is granulated (except *E. mammiformis*). Tentacles are extended only at night. **Similar genera:** *Cyphastrea*, which is distinguished by having a massive or encrusting growth-form (except *C. decadia*) and smaller corallites. *Echinopora* has a superficial resemblance to *Echinophyllia* species (Pectiniidae), especially *E. echinoporoides*. **Earliest fossil record:** Miocene of the Pacific.

Taxonomic references: Chevalier (1975), Veron, Pichon and Wijsman-Best (1977), Wijsman-Best (1980). **Summary key:** See p453.

Echinopora pacificus
Veron, 1990

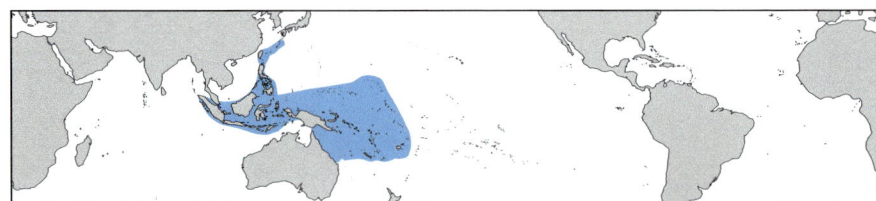

Characters: Colonies are usually unifacial plates with laminar margins and encrusting centres. Corallites are up to 10 millimetres diameter. Septo-costae are in two orders, the second seldom forming well developed septa. Septal teeth are exsert. Costae are beaded. Paliform lobes are not exsert. The coenosteum has tall spinules giving colonies a velvet-like appearance. **Colour:** Green, yellowish or grey-brown. **Similar species:** *Echinopora lamellosa*, which has similar but smaller corallites and forms larger colonies with more development of plates. **Habitat:** Most shallow reef environments. **Abundance:** Usually uncommon.

Taxonomic reference: Veron (1990a). **Identification guide:** Nishihira and Veron (1995).

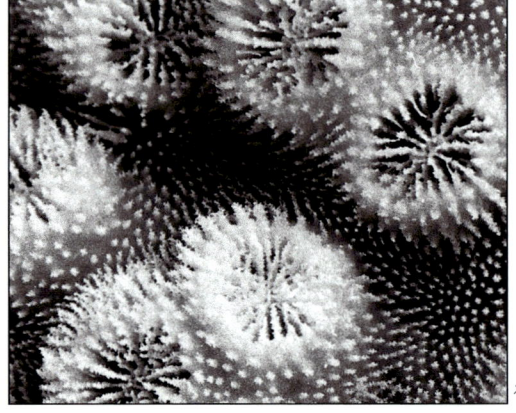

x4

1

1 A large colony of unifacial plates. Papua New Guinea *Photograph: author*

2 The plate-like edge of an encrusting colony. This is the most common colour of the species. Bolinao, Philippines *Photograph: author*

3, 5 Common variation in corallite shape and colour. **3** Bolinao, Philippines **5** Ryukyu Islands, Japan *Photographs: author*

4 Corallite detail. Bolinao, Philippines *Photograph: author*

Echinopora lamellosa
(Esper, 1795)

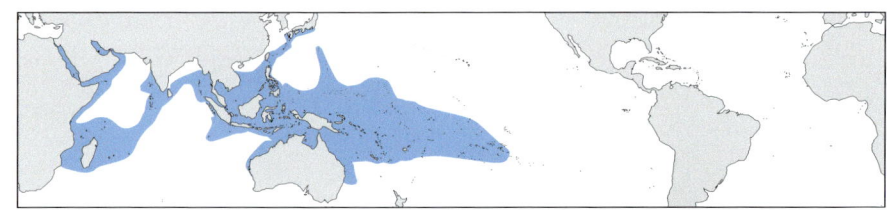

Characters: Colonies are thin laminae arranged in whorls or tiers or, rarely, forming tubes. Stands over 5 metres across are not unusual. Corallites are relatively thin walled and small (2.5-4 mm diameter). Columellae are small and compact, and paliform lobes are well developed. **Colour:** Amber, pale to dark brown or greenish, often with darker brown or green calices. **Similar species:** *Echinopora ashmorensis*, *E. pacificus* and *E. gemmacea*. **Habitat:** May be a dominant species in shallow water habitats with flat substrates. **Abundance:** Common.

Taxonomic references: Chevalier (1975), Veron, Pichon and Wijsman-Best (1977). **Identification guides:** Veron (1986), Sheppard and Sheppard (1991), Nishihira and Veron (1995), Coles (1996).

x4

Family Faviidae Genus *Echinopora*

1 A complex of plates of different colonies. DAMPIER ARCHIPELAGO, WESTERN AUSTRALIA *Photograph: author*

2 A large colony of tiered plates. PEMBA ISLAND, TANZANIA *Photograph: author*

3 Forming flat plates. ZANZIBAR, TANZANIA *Photograph: author*

4 This species sometimes forms upright tubes, superficially resembling those of *E. ashmorensis*. GREAT BARRIER REEF, AUSTRALIA *Photograph: author*

5-7 Common variation in corallite shape and colour. GREAT BARRIER REEF, AUSTRALIA *Photographs: author*

Echinopora ashmorensis
Veron, 1990

Characters: Colonies may be over 2 metres across and are composed of irregularly contorted tubes with hollow centres or, rarely, solid branches. Corallites are 3-5 millimetres diameter and are usually conical. They have three cycles of septa, the first two cycles reaching the columellae, which are well developed. Paliform lobes are prominent. **Colour:** Uniform brown or dark blue-grey. **Similar species:** *Echinopora lamellosa*, which may also form tubular branches. See also *E. horrida*. **Habitat:** Shallow protected reef environments. **Abundance:** Rare.

Taxonomic reference: Veron (1990a).

x4

1 *Echinopora ashmorensis*. Part of a large colony composed entirely of tubes. Ashmore Reef, Western Australia *Photograph: author*

2 *Echinopora ashmorensis*. Corallite detail. Ningaloo Reefs, Western Australia *Photograph: Clay Bryce*

3 *Echinopora ashmorensis*. Surface detail of tubes. Ningaloo Reefs, Western Australia *Photograph: Clay Bryce*

4 *Echinopora fruticulosa*. The typical growth-form of a large colony. Sinai Peninsula, Egypt *Photograph: author*

5 *Echinopora fruticulosa*. Interlocking branches of which colonies are made. Sinai Peninsula, Egypt *Photograph: author*

6 *Echinopora fruticulosa*. Corallite detail. Sinai Peninsula, Egypt *Photograph: Mary Stafford-Smith*

Echinopora fruticulosa
(Ehrenberg, 1834)

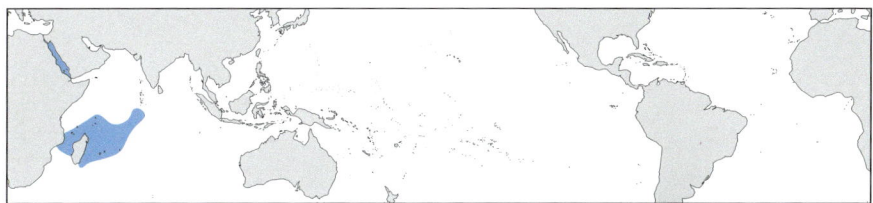

Characters: Colonies are dome-shaped clumps of interlocking branches up to 2 metres across. Branches are formed of single tubular corallites (or axial corallites) with lateral buds. Corallites are 5-8 millimetres diameter. Costal spines are widely spaced and not exsert. There is little tendency to form laminae or solid bases. **Colour:** Pinkish-brown with pale corallite ends. **Similar species:** Corallites are similar to those of *Echinopora gemmacea*. The growth-form is more like that of *E. horrida*, which has corallites with exsert costal and coenosteum spines. **Habitat:** Most shallow reef environments. **Abundance:** Common.

Taxonomic reference and identification guide: Sheppard and Sheppard (1991).

Family Faviidae Genus *Echinopora*

Echinopora gemmacea
Lamarck, 1816

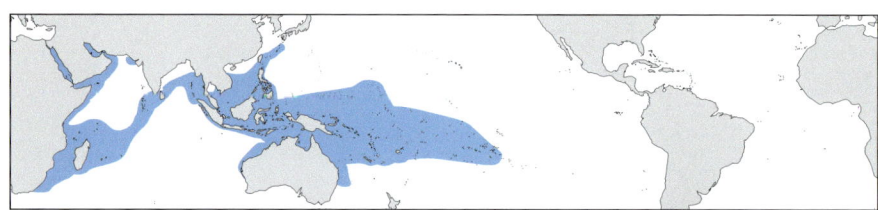

Characters: Colonies are laminar and bifacial, sometimes forming contorted branches. Corallites are 3.5-4.5 millimetres diameter. Columellae are large, and paliform lobes are not well developed. Primary septa may be thick and are always exsert. **Colour:** Usually grey, sometimes pale cream to dark brown or green. **Similar species:** *Echinopora horrida*, which has similar corallite characters but is usually branching whereas *E. gemmacea* is usually laminar. See also *E. hirsutissima*. **Habitat:** Shallow protected reef environments. **Abundance:** Usually uncommon.

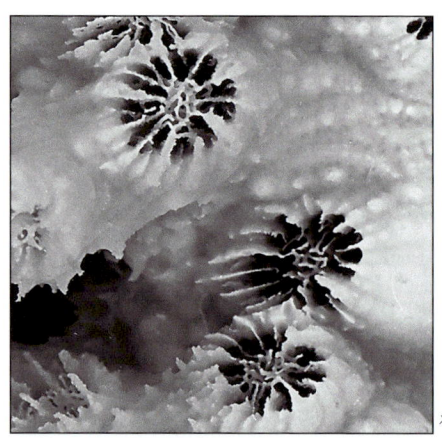

Taxonomic references: Chevalier (1975), Veron, Pichon and Wijsman-Best (1977). **Identification guides:** Veron (1986), Sheppard and Sheppard (1991), Nishihira and Veron (1995).

1 Encrusting a pile of rubble. RYUKYU ISLANDS, JAPAN *Photograph: author*

2 *Echinopora gemmacea* (right) being overgrown by *E. lamellosa* (left). ASHMORE REEF, WESTERN AUSTRALIA *Photograph: author*

3 Corallites on a submassive colony. RYUKYU ISLANDS, JAPAN *Photograph: author*

4 Corallites at the edge of a plate. RYUKYU ISLANDS, JAPAN *Photograph: author*

5 Corallite detail. GREAT BARRIER REEF, AUSTRALIA *Photograph: Mary Stafford-Smith*

Echinopora hirsutissima
Milne Edwards and Haime, 1849

Characters: Colonies are submassive or encrusting. Corallites are 4-7 millimetres diameter and are thick walled with prominent skeletal structures. Costae are strongly beaded. The coenosteum is densely covered with thick, finely elaborated spinules.
Colour: Mustard, green, brown or purple, often with pale septo-costae.
Similar species: *Echinopora irregularis*. See also *E. horrida* and *E. gemmacea*, which have less coarse corallite structures and smaller corallites.
Habitat: Shallow reef environments.
Abundance: Sometimes common in the western Indian Ocean, rare elsewhere.

Taxonomic references: Chevalier (1975), Veron, Pichon and Wijsman-Best (1977).
Identification guide: Veron (1986).

1 The common appearance of an encrusting colony. PEMBA ISLAND, TANZANIA *Photograph: author*

2 Colonies often have irregular shapes. SEYCHELLES *Photograph: author*

3-7 Variation in corallite colour and shape. **3** MADAGASCAR **4, 5** PEMBA ISLAND, TANZANIA **6** ZANZIBAR, TANZANIA **7** SINAI PENINSULA, EGYPT *Photographs: author*

x4

Family Faviidae — Genus *Echinopora*

Echinopora irregularis
Veron, Turak and DeVantier, this publication

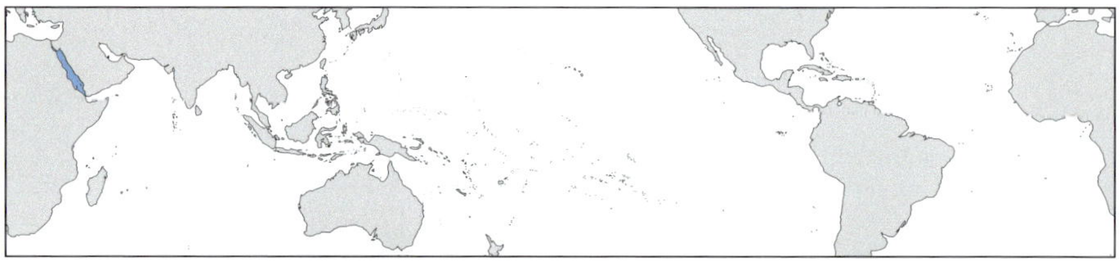

Characters: Colonies are composed of irregularly encrusting bases with short branches of irregularly fused corallites. Corallites are 4–7 millimetres diameter and are thick walled forming an interlocking clump. Septa are in three cycles, the first being exsert with prominent elaborated spines. Costae are irregularly contorted, also with elaborated spines which extend across the coenosteum. **Colour:** Cream. **Similar species:** *Echinopora hirsutissima*, which has much less irregular corallites. **Habitat:** Shallow partly protected reef environments. **Abundance:** Rare.

Taxonomic note: See 'New species described in *Corals of the World*' (Veron, in preparation) for further information.

x4

1

Echinopora robusta
Veron, this publication

Characters: Colonies are massive or are thick plates. Corallites are up to 8 millimetres diameter, closely compacted, and thick walled. Septo-costae are in two alternating cycles, those of the first cycle being very exsert and coarsely beaded. Columellae are large and consist primarily of fused paliform lobes. **Colour:** Grey- or greenish-brown. **Similar species:** *Echinopora forskaliana*. See also *E. hirsutissima*, which has relatively widely spaced corallites, less conspicuous first cycle septo-costae and smaller columellae. Colonies with closely compacted corallites are *Favia*-like. **Habitat:** Shallow reef environments. **Abundance:** Rare.

Taxonomic note: See 'New species described in *Corals of the World*' (Veron, in preparation) for further information.

1 *Echinopora irregularis*. Colonies are composed of a complex of irregularly fused corallites. SAUDI ARABIA, RED SEA *Photograph: Lyndon DeVantier*

2 *Echinopora robusta*. Colony surface showing the compact corallites. SRI LANKA *Photograph: author*

3 *Echinopora robusta*. Colony surface. The six septa of the first cycle are very exsert. SRI LANKA *Photograph: author*

4 *Echinopora robusta*. Corallite detail. SRI LANKA *Photograph: author*

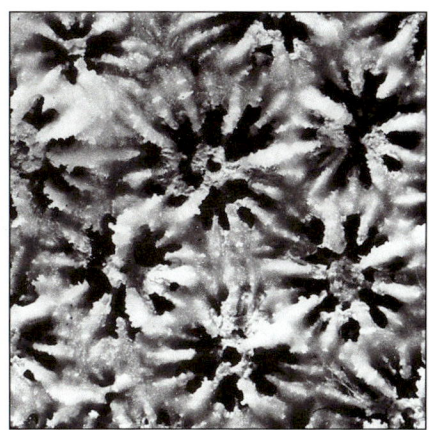
x4

Echinopora forskaliana
(Milne Edwards and Haime, 1850)

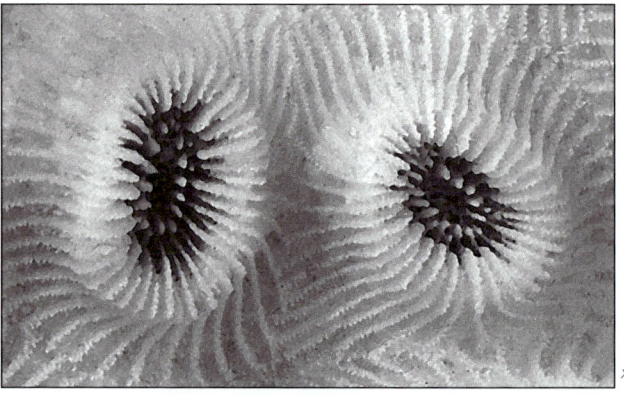

Characters: Colonies are massive, with little tendency to form plates. Corallites are tall and conical, up to 10 millimetres diameter. Long and short septo-costae slightly alternate and are neatly and uniformly beaded. Paliform lobes form a neat crown. The columella has spines twisted into a spiral. **Colour:** Pink, sometimes brown. **Similar species:** *Echinopora robusta*. See also *E. gemmacea* which has less exsert corallites and tends to form plates rather than massive colonies. **Habitat:** Shallow reef environments. **Abundance:** Common.

Taxonomic reference: Wijsman-Best (1980).

Echinopora tiranensis
Veron, Turak and DeVantier, this publication

Characters: Colonies are composed of basal plates with short branches of irregularly fused corallites. Corallites are 3-4 millimetres diameter and are thick walled. Those on basal plates are up to 8 millimetres exsert and are strongly inclined. Septa are in four orders and may fuse according to Pourtalès plan. Primary septa have paliform lobes. Columellae are compact. Costae are in two orders and are mostly smooth. **Colour:** Pale brown and cream. **Similar species:** *Echinopora gemmacea*, which does not have a smooth coenosteum. *Echinopora mammiformis* has larger, less exsert corallites with the columella twisted into a spiral and no fusion of septa. **Habitat:** Shallow reef environments. **Abundance:** Rare.

Taxonomic note: The type series of *Echinopora ehrenbergi* Milne Edwards and Haime (1849) is a mixture of this species and *E. gemmacea*. See 'New species described in *Corals of the World*' (Veron, in preparation) for further information.

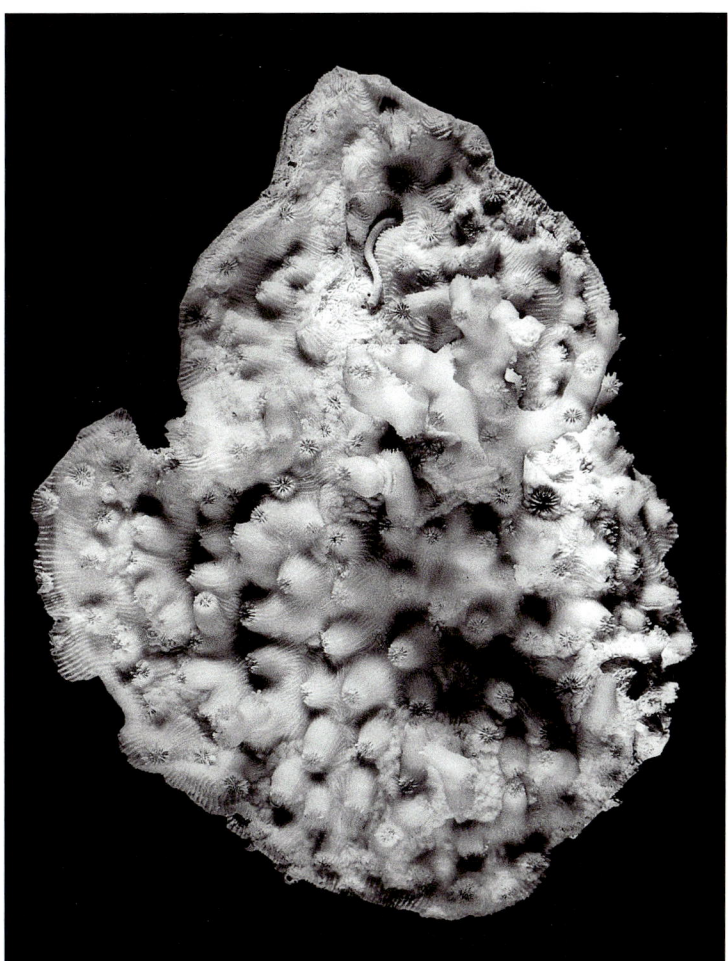

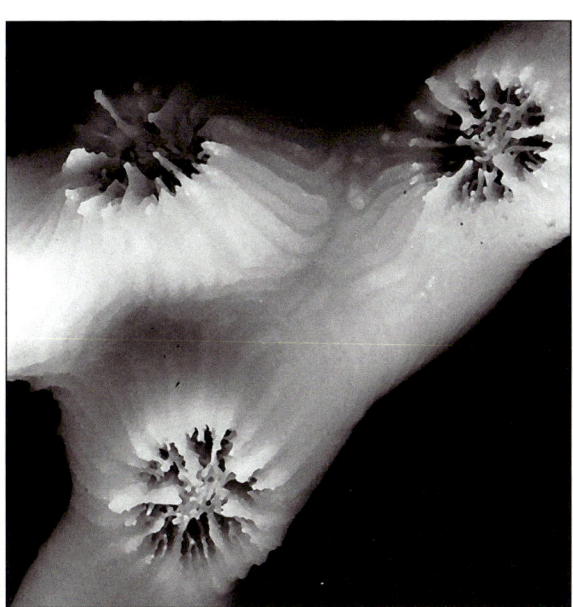

1 *Echinopora forskaliana*. A massive colony. SINAI PENINSULA, EGYPT *Photograph: Mary Stafford-Smith*

2 *Echinopora forskaliana*. Colonies of flat plates. SINAI PENINSULA, EGYPT *Photograph: author*

3 *Echinopora forskaliana*. Surface of a massive colony. SINAI PENINSULA, EGYPT *Photograph: author*

4, 5 *Echinopora forskaliana*. Variation in the structure of corallites. SINAI PENINSULA, EGYPT *Photographs: author*

Echinopora mammiformis
(Nemenzo, 1959)

Characters: Colonies are commonly over 5 metres across and are composed of contorted branches and flat plates. Rarely submassive. Corallites are conical, 7-10 millimetres diameter. The coenosteum is smooth or nearly so. Costae are smooth or slightly beaded. Columellae are twisted into a spiral. **Colour:** Distinctive cream with blue (which may photograph purple) corallites. **Similar species:** *Echinopora horrida* and *E. gemmacea* may both have similar growth-forms but both have spines down their costae and a spiny coenosteum. **Habitat:** Shallow water, especially lagoons and back reef margins. **Abundance:** Common.

Taxonomic reference: Veron, Pichon and Wijsman-Best (1977). **Identification guides:** Veron (1986), Nishihira and Veron (1995).

Family Faviidae Genus *Echinopora*

1 A large colony composed of basal plates and upright branches. APO ISLAND, PHILIPPINES *Photograph: author*

2 Plates and branches commonly occur in adjacent colonies as seen here, or within the same colony. CALAMIAN ISLANDS, PHILIPPINES *Photograph: author*

3 Colony composed of short branches in a contorted basal plate. GREAT BARRIER REEF, AUSTRALIA *Photograph: Len Zell*

4 Corallite detail of a plate. PAPUA NEW GUINEA *Photograph: Valerie Taylor*

5 Corallite detail of a branch. GREAT BARRIER REEF, AUSTRALIA *Photograph: Roger Steene*

Family Faviidae | Genus *Echinopora*

Echinopora horrida
Dana, 1846

Characters: Colonies are composed of contorted branches, sometimes with flat laminar bases. Stands over 5 metres across and one metre high are not unusual. Corallites are 4-6 millimetres diameter, thick walled and have six thick primary septa. The coenosteum is covered with tall spinules. **Colour:** Dark brown, cream or green. **Similar species:** *Echinopora gemmacea*, which does not form extensive branches. *Echinopora mammiformis* has a smooth coenosteum with distinctively coloured corallites. **Habitat:** Forms large stands on protected horizontal substrates including shallow reefs and lagoons. **Abundance:** Uncommon.

Taxonomic references: Chevalier (1975), Veron, Pichon and Wijsman-Best (1977). **Identification guide:** Veron (1986).

1 *Echinopora horrida*. Typical appearance of a thicket. PAPUA NEW GUINEA *Photograph: author*

2 *Echinopora horrida*. Contorted branches. Polyps are partly extended. GREAT BARRIER REEF, AUSTRALIA *Photograph: author*

3 *Echinopora horrida*. Branch tips of a compact colony. PAPUA NEW GUINEA *Photograph: Neville Coleman*

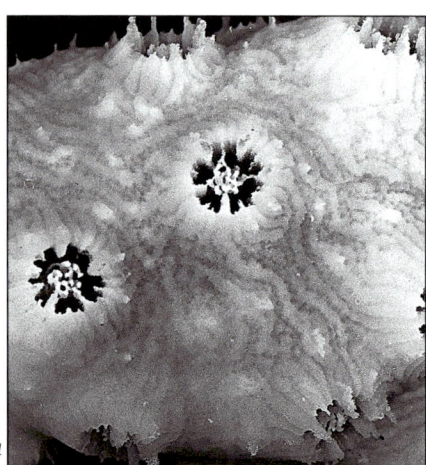

x4

Family Faviidae — Genus *Moseleya*

GENUS *MOSELEYA*
Quelch, 1884

Characters, abundance and distribution: This genus has only one species, see *Moseleya latistellata*. **Fossil record:** None.

Moseleya latistellata
Quelch, 1884

Characters: Colonies are flat, submassive, usually disc-like, and sometimes free-living. Corallites are cerioid with a large central corallite (up to 35 mm diameter) surrounded concentrically with angular daughter corallites. Septa have fine teeth and usually exsert paliform lobes. Tentacles are extended only on dark nights. **Colour:** Pale to deep green or brown. **Similar species:** May resemble the mussid *Acanthastrea* (especially *A. bowerbanki*), which can have the same colony and corallite shapes. *Acanthastrea* has more fleshy polyps, much larger septal teeth and never has large paliform lobes. See also *Goniastrea palauensis*. **Habitat:** Restricted to turbid water with muddy substrates. Also occurs in muddy areas exposed at low tide. **Abundance:** Uncommon.

Taxonomic reference: Veron, Pichon and Wijsman-Best (1977). **Identification guides:** Veron (1986), Nishihira and Veron (1995).

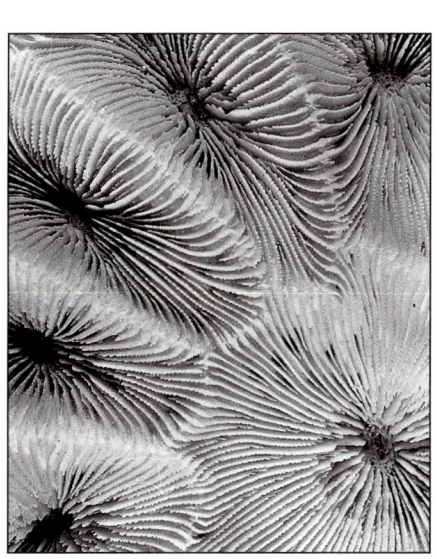
×1

4 *Moseleya latistellata*. The usual appearance of a small colony. KOMODO, INDONESIA *Photograph: Valerie Taylor*

5 *Moseleya latistellata*. Corallite detail. GREAT BARRIER REEF, AUSTRALIA *Photograph: Ed Lovell*

6 *Moseleya latistellata*. The central corallite. GREAT BARRIER REEF, AUSTRALIA *Photograph: author*

Family Trachyphylliidae

FAMILY TRACHYPHYLLIIDAE
Verrill, 1901

Characters: Solitary to colonial and zooxanthellate. The family is separated from the Faviidae by growth-form, the presence of large paliform lobes and fine teeth on the septa. **Similar family:** Trachyphylliidae is close to the Faviidae, especially to genus *Moseleya*, so much so that its status is somewhat arbitrary. **Genera:** One genus, *Trachyphyllia*.

Taxonomic references: Chevalier (1975), Veron, Pichon and Wijsman-Best (1977)
Summary key: See p459.

Opposite: The Trachyphylliidae appears to be an ancient family but the fossil record is unclear due to confusion between *Trachyphyllia* (illustrated here) and the extinct mussid genus *Antillia*, which has an identical growth-form. BALI INDONESIA *Photograph: Gerry Allen*

Family Trachyphylliidae   Genus *Trachyphyllia*

GENUS *TRACHYPHYLLIA*
Milne Edwards and Haime, 1848

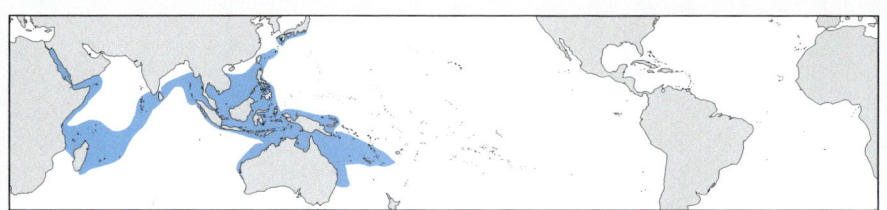

Characters, abundance and distribution: This genus has only one species, see *Trachyphyllia geoffroyi*. **Earliest fossil record:** Eocene of the Tethys and Indian Ocean, Oligocene of the Caribbean.

Trachyphyllia geoffroyi
(Audouin, 1826)

Characters: Colonies are flabello-meandroid and free-living. They are usually hourglass shaped, up to 80 millimetres in length with one to three separate mouths. Large, fully flabello-meandroid colonies are uncommon. Valleys have large regular septa and paliform lobes and a large columella of tangled spines. Polyps are fleshy. When tentacles are retracted during the day a large mantle extends well beyond the perimeter of the skeleton. This retracts if disturbed. At night tentacles in several rows are extended from the expanded oral disc inside the mantle. The mouth is approximately 10 millimetres across. **Colour:** Polyps, especially the mantles, are often brightly coloured, usually yellow, brown, blue or green. **Similar species:** None. **Habitat:** Inter-reef environments and on soft substrates around continental islands. Frequently found with other free-living corals: *Heteropsammia* (Dendrophylliidae), *Heterocyathus* (Caryophylliidae) and the fungiids, *Cycloseris* and *Diaseris*. Large colonies are found only in certain protected, shallow island embayments. **Abundance:** Rare on reefs, common around continental islands and some inter-reef areas.

Taxonomic references: Chevalier (1975), Veron, Pichon and Wijsman-Best (1977), Veron and Hodgson (1989). **Identification guides:** Veron (1986), Sheppard and Sheppard (1991), Nishihira and Veron (1995).

1 A fully developed flabello-meandroid colony. CALAMIAN ISLANDS, PHILIPPINES *Photograph: author*

2 Typical shape of a flabello-meandroid colony. BALI, INDONESIA *Photograph: Gerry Allen*

3 An hourglass shaped colony. GREAT BARRIER REEF, AUSTRALIA *Photograph: Roger Steene*

4 The mantle of colonies may be very large and fleshy. AMBON, INDONESIA *Photograph: Valerie Taylor*

5 A voracious feeder at night. GREAT BARRIER REEF, AUSTRALIA *Photograph: Ed Lovell*

‡ See also polyp detail, page 270.

Family Poritidae

FAMILY
PORITIDAE
Gray, 1842

Characters: Colonial, zooxanthellate and mostly extant. Colonies are usually massive, laminar or branching. Corallites have a wide size range but are usually compacted, with little or no coenosteum. Walls and septa are porous. **Similar families:** None. **Genera:** Five genera, each of which is distinct. *Stylaraea* and *Poritipora* are rare, monospecific, and clearly related to *Porites*. *Porites* and *Goniopora* are very different but are related by their patterns of septal fusion. *Alveopora* has tenuous affinities with *Goniopora*. Family Poritidae, therefore, is essentially a heterogeneous assembly of distantly related genera.

Taxonomic reference: Veron and Pichon (1982).
Summary key: See p457.

Opposite: The Poritidae, represented here by a majestic colony of *Porites rus* with *P. cylindrica* (foreground), is a well defined family. The Poritidae are the most common fossil corals of the Cenozoic and were the dominant family of the Tethys Sea for most of its history. GREAT BARRIER REEF, AUSTRALIA *Photograph: Valerie Taylor*

GENUS *PORITES*
Link, 1807

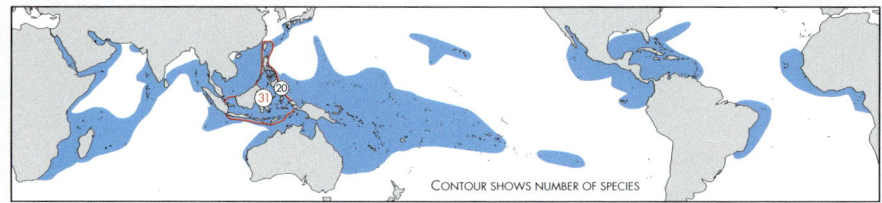

Characters: Colonies are flat (laminar or encrusting), massive or branching. Massive colonies are spherical or hemispherical when small and helmet or dome-shaped when large, and may be over 5 metres across. Corallites are small, immersed, and filled with septa. Tentacles of most, but not all, species are extended only at night. **Similar genera:** *Porites* superficially resembles *Montipora* (Acroporidae) and also *Stylaraea*. *Porites* differs from *Montipora* by many differences in growth-form and corallites are usually larger and more compacted and lack the elaborate papillae and tuberculae which characterise *Montipora*. Underwater, *Porites* can readily be seen to have corallites filled with septa, whereas corallites of *Montipora* contain only inward projecting septal teeth. **Earliest fossil record:** Eocene of the Caribbean and Tethys.

Taxonomic reference: Veron and Pichon (1982). **Summary key:** pp457-8.

Identification of *Porites* species

Porites species are the most difficult of all the major genera to identify, partly because their corallites are both small and variable and partly because species with massive growth-forms are difficult to recognise underwater and thus common species may mask the presence of uncommon species. Taxonomic difficulties are also due to geographic variation. Despite the wide geographic range of the genus as a whole, latitudinal attenuation, especially between reef and non-reef environments, occurs more abruptly in *Porites* than in all other major genera except *Fungia*.

Family Poritidae — Genus *Porites*

Key diagrams summarising the most useful diagnostic characters are given with each species. As there may be wide variation in corallite characters within a single colony (illustrated, p278), as well as environmental and geographic variation, these diagrams are only representative of an average corallite on a colony not exposed to extreme environments.

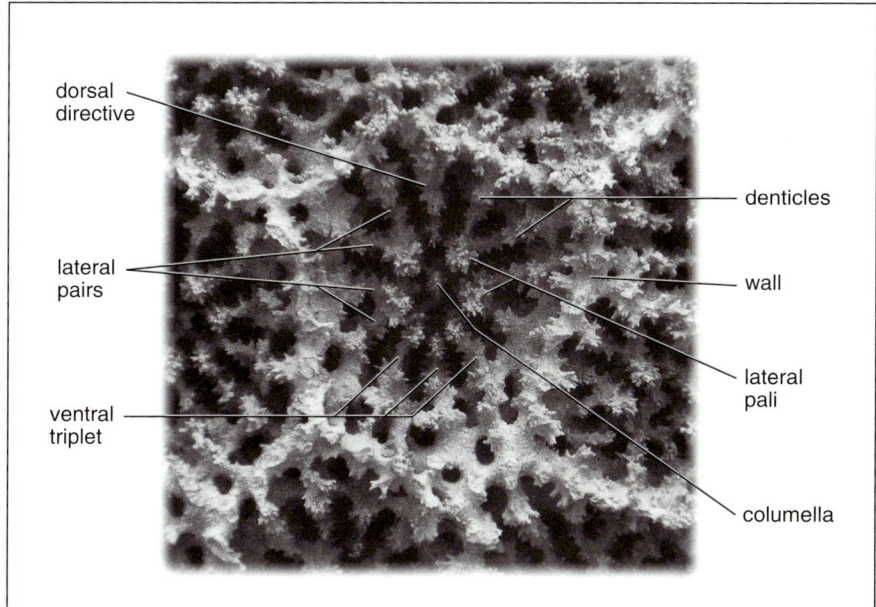

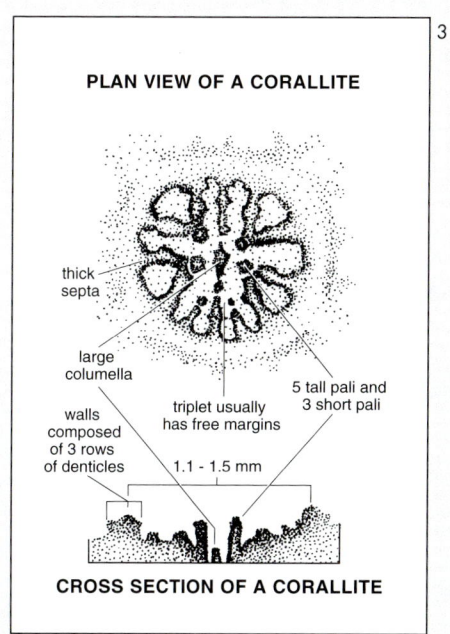

1 Comparison of corallites of *Porites* (left) and *Montipora* (right). GREAT BARRIER REEF, AUSTRALIA *Photograph: Ed Lovell*

2 Branching *Porites* colonies (foreground) are usually easier to identify than massive colonies (background). ZANZIBAR, TANZANIA *Photograph: author*

3 A scanning electron microscope image of an 'average' corallite of *Porites australiensis* (left) with the key diagram of corallite characters of the same species (right). The orientation is the same. This orientation, as well as degree of development of individual skeletal structures, varies among adjacent corallites.

The corallite structures commonly used in species identification are:

septa: These are named according to a convention and have variable development in different species. As illustrated (right), each corallite has one dorsal directive and one ventral directive septum, four lateral pairs of septa arranged symmetrically and two more septa, one either side of the ventral directive which, together with the ventral directive, form the triplet. The inner margins of the triplet may be free or fused. In the latter case they may be fused along their inner margins or each outer septum may be fused to the sides of the ventral directive by a cross bar forming a trident.

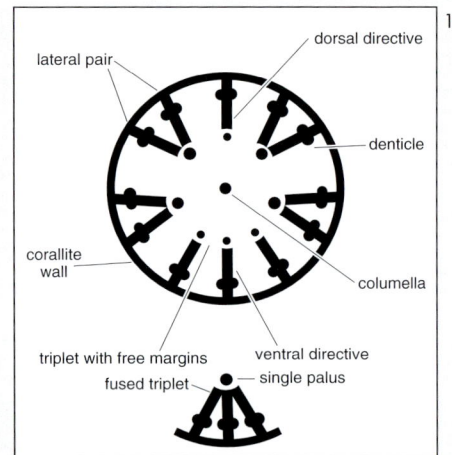

pali: The pali are vertical pillars positioned as illustrated right. These have variable development in different species. The four pali associated with the lateral pairs of septa are usually the largest. A fifth palus is usually associated with the dorsal directive septum. One, two or three pali may be associated with the triplet; one palus if the triplet is fused, two pali if they are fused by a cross bar and three pali if the triplet is not fused.

denticles: These are vertical pillars resembling pali and are arranged along the top of the septa at fixed intervals. The pali and denticles may form concentric circles.

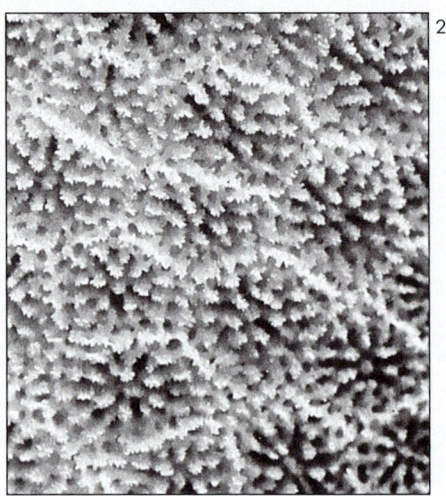

walls: These may be thick but commonly consist of three rows of denticles as in *P. australiensis*, illustrated page 277.

columellae: Some species do not have columellae. (Some specimens may have the columella missing from some or all corallites: this appears to be due to the activities of parasitic worms occupying the corallites and removing all skeletal elements from their centres.)

Other corallite structures (not illustrated above) sometimes used in species identification are:

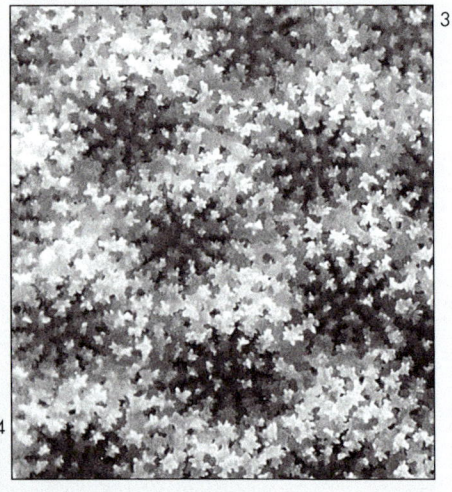

radii: These usually occur deep within the corallite and connect the pali to the columella. The columella, pali and denticles are all covered with granules and may be similar in appearance (see, for example, page 287).

synapticular rings: There may be two synapticular rings deep within the corallite. The inner ring links the pali and is joined to the columella by the radii. The outer ring is usually less visible and occurs near the corallite wall.

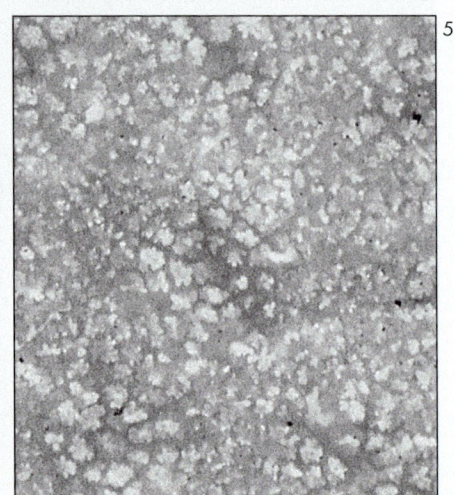

Family Poritidae — Genus *Porites*

Porites can be divided into six groups, based on colony size and growth-form.

Group 1: Species forming large massive colonies, page 280.

Group 2: Species forming small massive colonies, page 292.

Group 3: Species forming thick columns or thick plates, page 296.

Group 4: Species forming composites of columns, laminae and branches, page 303.

Group 5: Species forming composites of laminae and branches, page 312.

Group 6: Species forming primarily branching colonies, page 326.

1 Diagrammatic representation of a *Porites* corallite, showing septal structures commonly used in species identification.

2, 3, 5 Corallite variation within a colony of *Porites lutea* (×10). The corallites occur within 300 mm of each other, around the lip of the base of a helmet-shaped colony.

4 Most massive *Porites* species are very tolerant of sedimentary environments partly because they protect themselves with a thick film of mucous. This film must be removed before colonies can be identified. GREAT BARRIER REEF, AUSTRALIA *Photograph: Neville Coleman*

6 Corallite size is sometimes important to identification. Here *P. rus*, which has the smallest corallites of any *Porites* species, is overgrowing *P. lobata*, which has middle sized corallites. *Porites rus* and its allies are sometimes placed in a separate subgenus, *Synaraea*. SEYCHELLES *Photograph: author*

7 Massive *Porites* are very common in shallow water. Both growth-form and corallite structures are modified by aerial exposure and wave action. GREAT BARRIER REEF, AUSTRALIA *Photograph: Neville Coleman*

8 *Porites* colonies commonly house a wide variety of other fauna. SOLOMON ISLANDS *Photograph: Jim Maragos*

Group 1: Species forming large massive colonies

It requires experience to identify massive *Porites* species because corallite characters are variable, even within a single colony. They can be identified underwater with the aid of a hand lens as most corallite characters can be seen if tentacles are fully retracted.

Porites astreoides
Lamarck, 1816

Characters: Colonies are generally massive but are often encrusting. They are usually small but may be several metres across. The surface is usually lumpy but is sometimes smooth or nodular. Tentacles are commonly extended during the day. **Colour:** Usually bright yellow to dull grey-brown, sometimes green. **Similar species:** *Porites branneri*. **Habitat:** All reef environments. **Abundance:** Common.

Taxonomic references: Laborel (1969), Roos (1971), Zlatarski and Estalella (1982). **Identification guides:** Colin (1978), Humann (1993).

Family Poritidae | Genus *Porites*

1 A small encrusting colony. CAYMAN ISLANDS
Photograph: Nancy Sefton

2 Forming a boulder in shallow water. BELIZE
Photograph: author

3 Surface of a colony with tentacles extended. BELIZE *Photograph: Don Potts*

4 A small encrusting plate. VIRGIN ISLANDS
Photograph: Don Potts

5 Detail of extended tentacles. ABROLHOS ISLANDS, BRAZIL *Photograph: author*

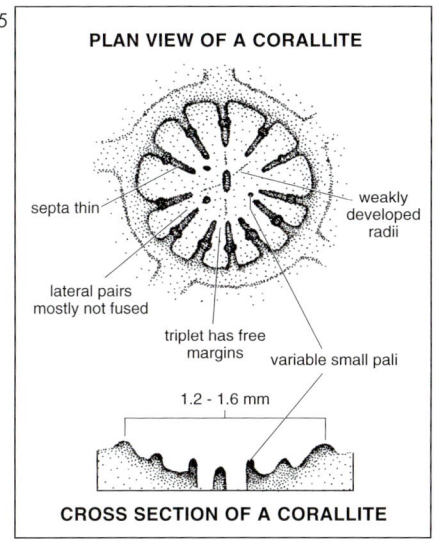

281

Porites solida
(Forskål, 1775)

Characters: Colonies are massive, usually hemispherical, and may be several metres across. The surface is smooth to undulating. Corallites are conspicuously large. **Colour:** Brown or greenish yellow. **Similar species:** *Porites lobata*, which has weakly developed pali. Relatively easily recognised underwater. **Habitat:** Shallow reef environments. **Abundance:** Common.

Taxonomic reference: Veron and Pichon (1982). **Identification guides:** Veron (1986), Sheppard and Sheppard (1991), Nishihira and Veron (1995).

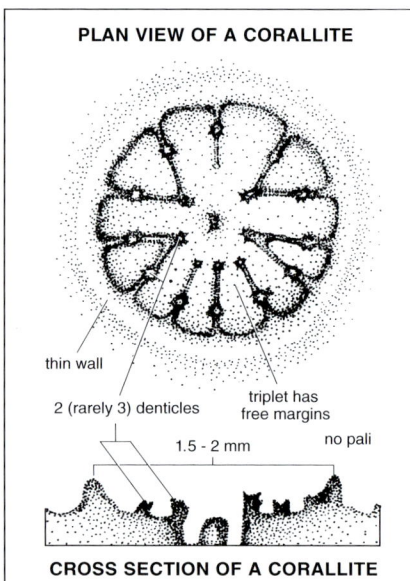

Porites panamensis
Verrill, 1866

Characters: Colonies are lobed or form short branches. The surface is smooth to undulating. **Colour:** Greenish-grey.
Similar species: *Porites lobata*, which has a more massive growth-form. *Porites mayeri* has similar corallite characters.
Habitat: Rocky foreshores. **Abundance:** Locally common.

Taxonomic note: *Porites sverdrupi* Durham, 1947 may be a separate species, primarily characterised by some development of branches and a rough coenosteum. **Taxonomic reference:** Verrill's original description/specimens.

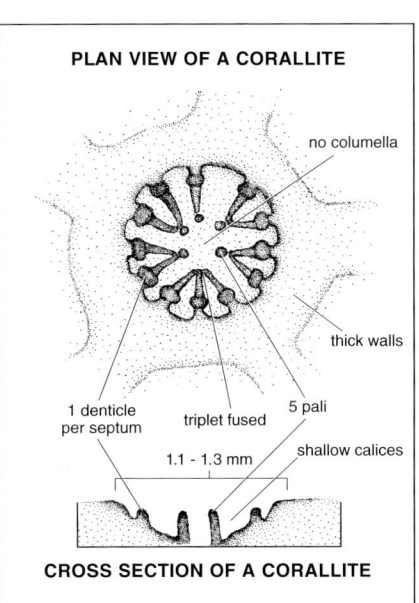

1 *Porites solida*. Common appearance of colonies in shallow water. FLORES, INDONESIA *Photograph: author*

2 *Porites solida*. A small colony with clearly recognisable corallite characters. SINAI PENINSULA, EGYPT *Photograph: author*

3 *Porites solida*. Corallite detail. The absence of pali is readily seen underwater. GREAT BARRIER REEF, AUSTRALIA *Photograph: author*

4 *Porites panamensis*. Columnar colony. GULF OF CALIFORNIA *Photograph: Dan Gotshall*

5 *Porites panamensis*. Edge of a hillocky plate. GULF OF CALIFORNIA *Photograph: Dan Gotshall*

Porites lobata
Dana, 1846

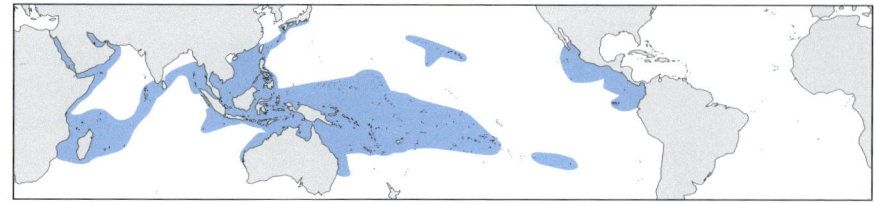

Characters: Colonies are usually hemispherical or helmet-shaped and may be over 4 metres across. They commonly form 'micro-atolls' in intertidal habitats. The surface is usually smooth. **Colour:** Usually cream or pale brown but may be bright blue, purple or green in shallow water. **Similar species:** *Porites solida*. See also *P. australiensis*, which is distinguished by having taller pali, especially on the lateral pairs of septa. **Habitat:** Frequently a dominant species of back reef margins, lagoons and some fringing reefs. **Abundance:** Probably the most common *Porites*.

Taxonomic note: This species is divisible into several smaller semi-distinct taxonomic units, see 'What are species?' p425. **Taxonomic references:** Wells (1954), Veron and Pichon (1982). **Identification guides:** Veron (1986), Sheppard and Sheppard (1991), Nishihira and Veron (1995).

1 This species commonly forms helmet-shaped colonies with lobed upper surfaces. GREAT BARRIER REEF, AUSTRALIA *Photograph: Peter Isdale*

2 An intertidal 'micro-atoll'. GREAT BARRIER REEF, AUSTRALIA *Photograph: Ed Lovell*

3 An unusual growth-form in a shallow turbulent habitat. CLIPPERTON ATOLL, FAR EASTERN PACIFIC *Photograph: author*

4 A massive colony. SINAI PENINSULA, EGYPT *Photograph: Mary Stafford-Smith*

5 Surface detail. Corallites are open and have relatively few skeletal elements. GREAT BARRIER REEF, AUSTRALIA *Photograph: Ed Lovell*

Family Poritidae Genus *Porites*

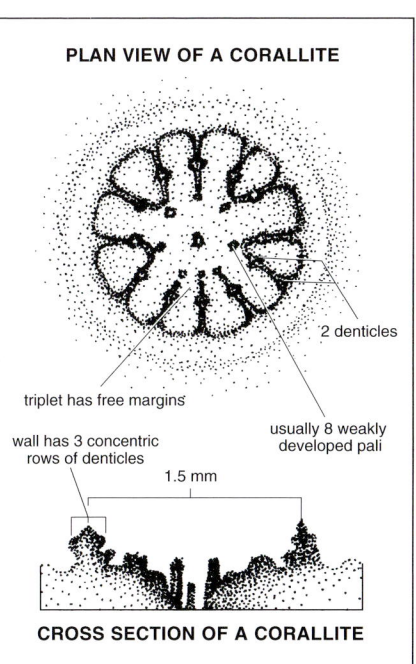

Family Poritidae   Genus *Porites*

Porites australiensis
Vaughan, 1918

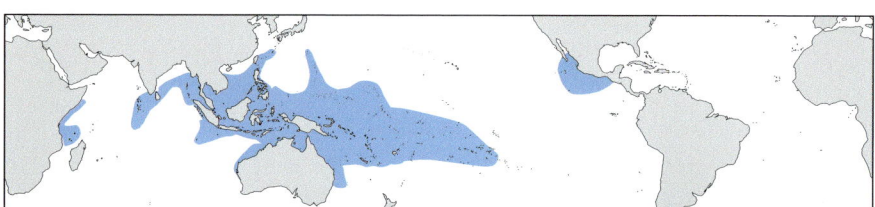

Characters: Colonies are hemispherical to helmet-shaped and may be large. The surface is smooth or has irregular humps and nodules. **Colour:** Usually cream or yellow but may be bright colours in shallow water. **Similar species:** *Porites lobata*. **Habitat:** Occurs with *P. lutea* and *P. lobata* on back reef margins, lagoons and fringing reefs. **Abundance:** Common.

Taxonomic references: Wells (1954), Veron and Pichon (1982). **Identification guides:** Veron (1986), Nishihira and Veron (1995).

1 *Porites australiensis*. This species forms large, usually helmet-shaped colonies. GREAT BARRIER REEF, AUSTRALIA Photograph: Ed Lovell

2, 3 *Porites australiensis*. Variation in surface detail of colonies in shallow water. **2** ZANZIBAR, TANZANIA **3** PAPUA NEW GUINEA Photographs: author

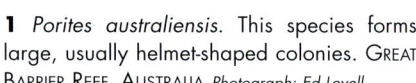

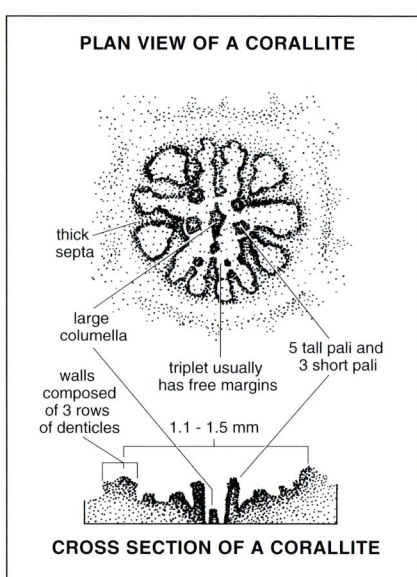

Porites lutea
Milne Edwards and Haime, 1851

Characters: Colonies are hemispherical or helmet-shaped and may be over 4 metres across. They usually form 'micro-atolls' in intertidal habitats. The surface is usually smooth. **Colour:** Usually cream or yellow but may be bright colours in shallow water. **Similar species:** *Porites australiensis*, which has thicker walls and five tall and three short pali. Underwater, the corallites of *P. lutea* are filled with skeletal elements, whereas the corallites of *P. lobata* appear to have fewer elements and thus look more open. *Porites somaliensis* has similar corallite characters. **Habitat:** Occurs with *P. lobata* and *P. australiensis* on back reef margins, lagoons and fringing reefs. **Abundance:** Common.

Taxonomic references: Wells (1954), Veron and Pichon (1982), Veron (1992). **Identification guides:** Veron (1986), Sheppard and Sheppard (1991), Nishihira and Veron (1995), Carpenter *et al.* (1997).

4 *Porites lutea*. A large helmet-shaped colony. CARTIER REEF, WESTERN AUSTRALIA *Photograph: author*

5 *Porites lutea*. An intertidal 'micro-atoll'. GREAT BARRIER REEF, AUSTRALIA *Photograph: author*

6 *Porites lutea*. Surface detail. In contrast with *P. lobata*, corallites appear filled with skeletal elements. GREAT BARRIER REEF, AUSTRALIA *Photograph: Ed Lovell*

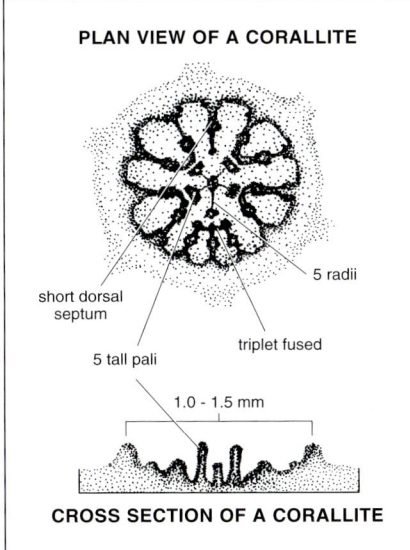

Family Poritidae — Genus *Porites*

Porites myrmidonensis
Veron, 1985

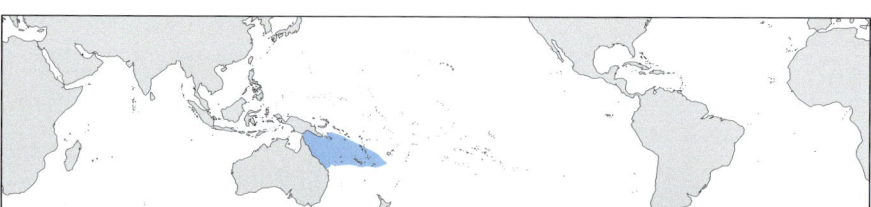

Characters: Colonies are massive and may be over 4 metres across. The surface is usually nodular. Corallites are in deeply excavated pits and are distinctive underwater. **Colour:** Uniform or mottled green or brown. **Similar species:** *Porites solida*, which has superficially similar corallites. **Habitat:** Upper reef slopes exposed to strong wave action. **Abundance:** Usually uncommon.

Taxonomic reference: Veron (1985). **Identification guide:** Veron (1986).

1 *Porites myrmidonensis*. Large colonies. GREAT BARRIER REEF, AUSTRALIA *Photograph: Terry Done*

2 *Porites myrmidonensis*. Surface detail. GREAT BARRIER REEF, AUSTRALIA *Photograph: author*

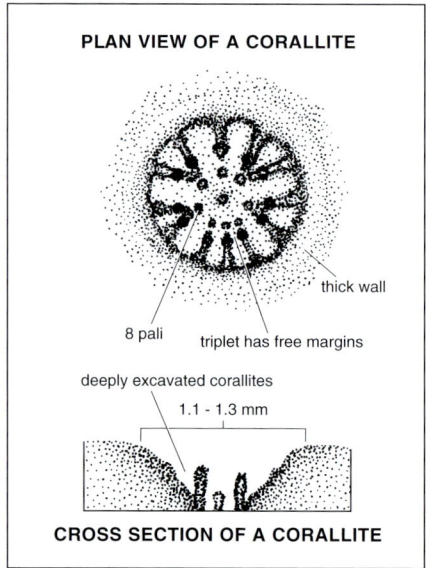

Porites mayeri
Vaughan, 1918

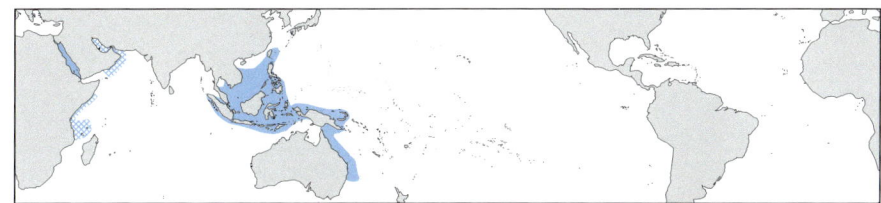

Characters: Colonies are hemispherical and may be over 4 metres across. The surface is smooth or lobed. **Colour:** Cream or brown, sometimes purple or blue (which may photograph pink). **Similar species:** *Porites stephensoni* and *P. murrayensis*, which have similar corallites. **Habitat:** Back reef margins, lagoons and fringing reefs. **Abundance:** Usually uncommon.

Distribution note: Records from the Indian Ocean and the Persian/Arabian Gulf are doubtful. **Taxonomic reference:** Veron and Pichon (1982). **Identification guide:** Veron (1986).

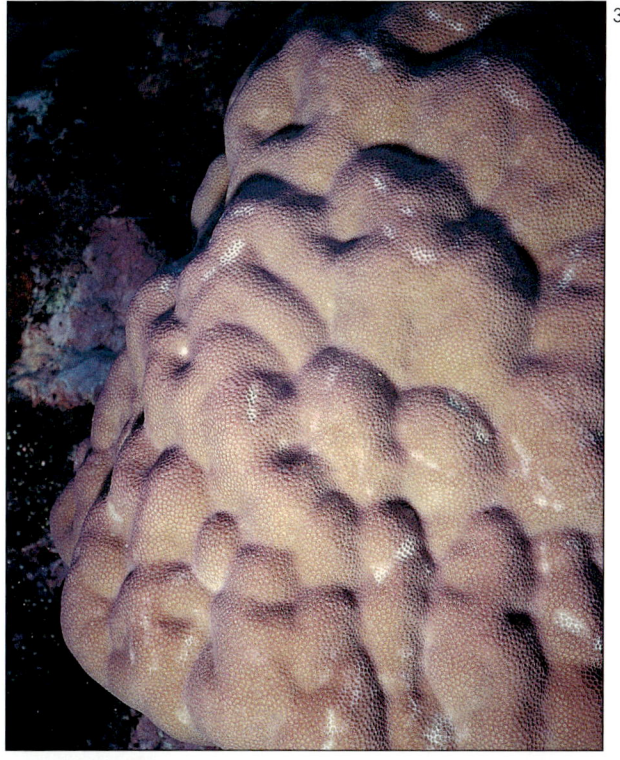

3 *Porites mayeri*. This species forms large colonies with a smooth surface and small corallites. GREAT BARRIER REEF, AUSTRALIA Photograph: author

4 *Porites mayeri*. Detail of a lobed colony. PAPUA NEW GUINEA Photograph: Neville Coleman

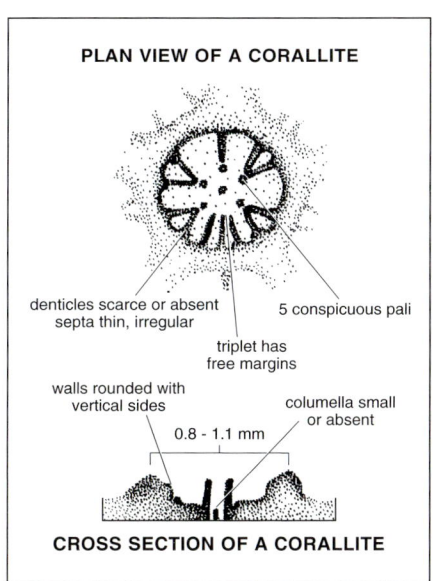

Family Poritidae | Genus *Porites*

Porites echinulata
Klunzinger, 1879

Characters: Colonies are massive, flat or consisting of flat-topped columns. The coenosteum between corallites is granulated giving a rough appearance which can be seen underwater. **Colour:** Grey or tan. **Similar species:** *Porites lichen*, which has thick walls and a tendency for corallites to form valleys. Corallites are similar to those of *P. australiensis*. **Habitat:** Shallow reef environments. **Abundance:** Common.

Distribution note: Records from the Indian Ocean are doubtful. **Taxonomic reference:** Scheer and Pillai (1983). **Identification guide:** Sheppard and Sheppard (1991).

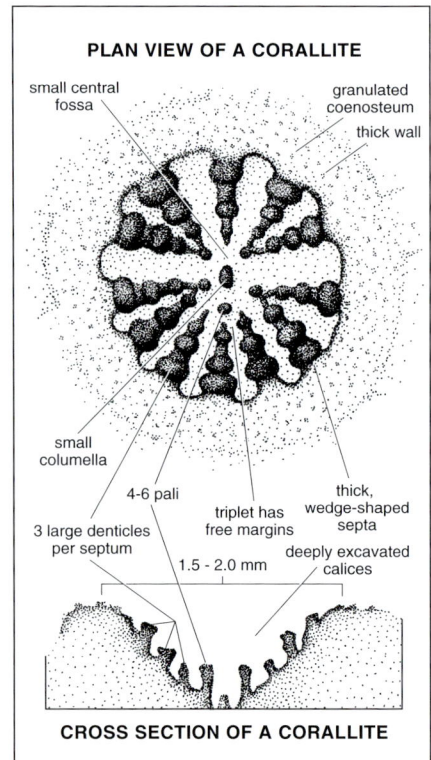

Family Poritidae | Genus *Porites*

Porites somaliensis
Gravier, 1911

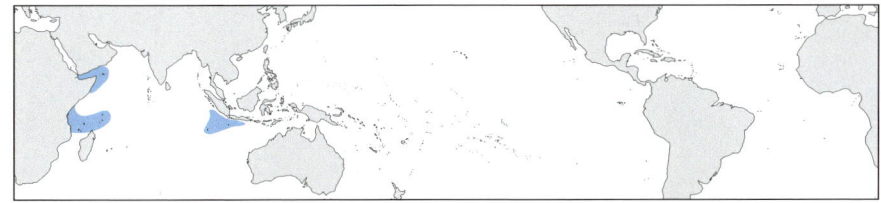

Characters: Colonies are massive, hemispherical, with a knobbly surface and may be several metres across. **Colour:** Pale brown. **Similar species:** *Porites stephensoni*, which has similar but smaller corallites. *Porites lutea* has similar corallite characters. **Habitat:** Shallow reef environments. **Abundance:** Rare, except at Cocos (Keeling) Atoll.

Taxonomic reference: Vaughan (1918).

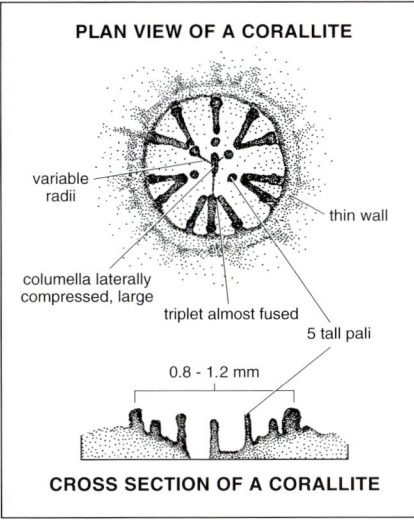

1 *Porites echinulata*. A large columnar colony. SINAI PENINSULA, EGYPT *Photograph: author*

2 *Porites echinulata*. Colony surface. SINAI PENINSULA, EGYPT *Photograph: author*

3 *Porites echinulata*. Surface detail. SINAI PENINSULA, EGYPT *Photograph: author*

4 *Porites somaliensis*. Characteristically knobbly surface of a massive colony. SEYCHELLES *Photograph: author*

Family Poritidae — Genus *Porites*

Group 2: Species forming small massive colonies

Colony size is an unreliable character in identifying massive *Porites*. Nevertheless, the following species seldom, if ever, form large colonies. Species of this group mostly brood larvae whereas those of the previous group are mostly separately sexed spawners.

Porites murrayensis
Vaughan, 1918

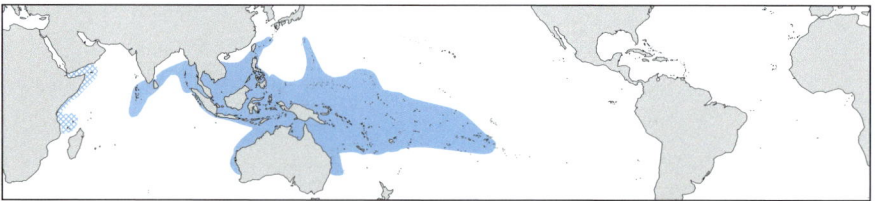

Characters: Colonies are hemispherical or spherical, up to 0.2 metres across. The colony surface is smooth and corallites are evenly spaced. **Colour:** Usually cream or brown, but may be bright colours in shallow water. **Similar species:** *Porites lobata*, which is distinguished by its larger corallites and longer septa. *Porites stephensoni* has similar corallite characters. **Habitat:** Shallow clear water, especially reef flats. **Abundance:** Sometimes common.

Distribution note: Records from the western Indian Ocean are doubtful. **Taxonomic references:** Wells (1954), Veron and Pichon (1982). **Identification guides:** Veron (1986), Nishihira and Veron (1995).

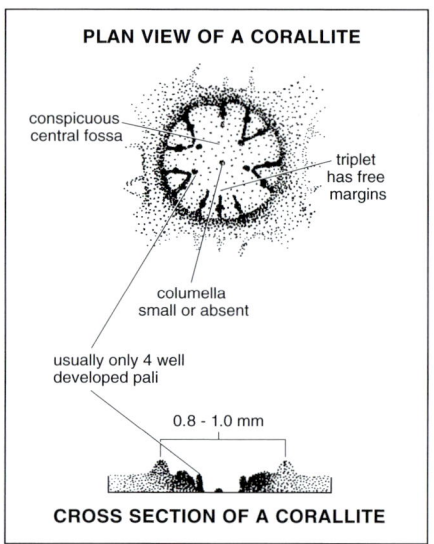

Porites stephensoni
Crossland, 1952

Characters: Colonies are encrusting, hemispherical or columnar and are usually less than 0.1 metre across. **Colour:** Cream or brown. **Similar species:** *Porites murrayensis*. **Habitat:** Reef flats. **Abundance:** Uncommon.

Taxonomic reference: Veron and Pichon (1982). **Identification guide:** Veron (1986).

1 *Porites murrayensis*. A small colony with a typically smooth surface. GREAT BARRIER REEF, AUSTRALIA Photograph: author

2 *Porites stephensoni*. Surface of a massive colony. PAPUA NEW GUINEA Photograph: Neville Coleman

3 *Porites stephensoni*. General appearance of corallites. GREAT BARRIER REEF, AUSTRALIA Photograph: author

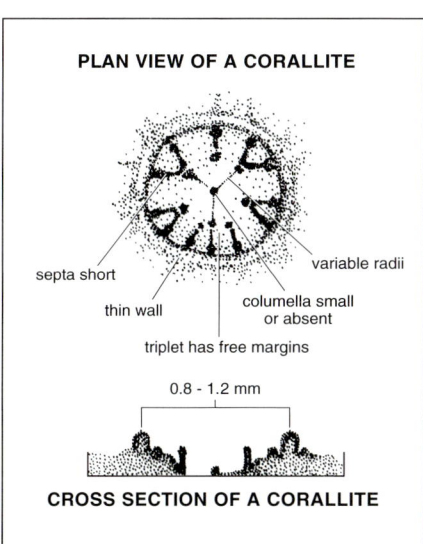

Porites densa
Vaughan, 1918

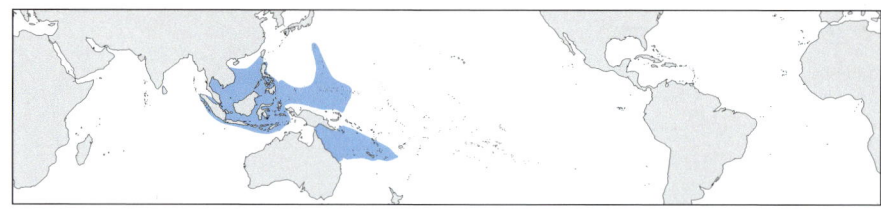

Characters: Colonies are hemispherical, less than 0.15 metres across, with an even surface. **Colour:** Cream, grey or brown. **Similar species:** *Porites echinulata*, which has more widely spaced corallites and a columnar growth-form. The deeply excavated calices of *P. densa* are readily recognisable underwater. See also *P. echinulata* and *Poritipora paliformis* which have corallites of similar size. **Habitat:** Back reef margins. **Abundance:** Sometimes common.

Taxonomic reference: Veron and Pichon (1982). **Identification guide:** Veron (1986).

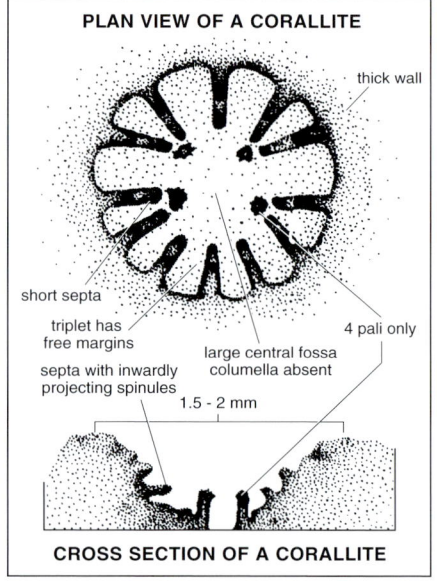

Porites brighami
Vaughan, 1907

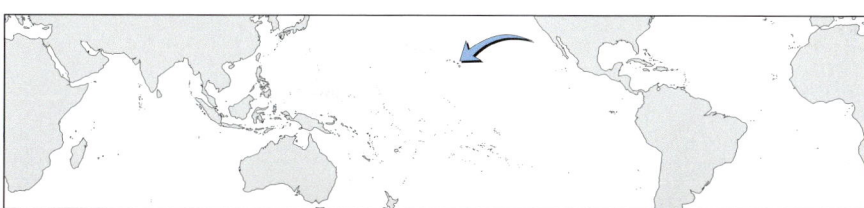

Characters: Colonies are small nodules, usually less than 0.1 metre across. Corallites are deep, funnel-shaped, and polygonal. Septa are thickened near the wall. **Colour:** Grey, brown or green. **Similar species:** *Porites astreoides*, which forms massive colonies. **Habitat:** Shallow reef environments. **Abundance:** Common.

Taxonomic reference: Vaughan's original description/specimens. **Identification guide:** Maragos (1977).

1 *Porites densa*. Corallites are large and conspicuous. Tentacles are partially extended. CALAMIAN ISLANDS, PHILIPPINES *Photograph: author*

2 *Porites brighami*. An encrusting colony. HAWAII *Photograph: Jim Maragos*

3 *Porites brighami*. Colonies usually consist of small lumpy nodules. HAWAII *Photograph: Jim Maragos*

4 *Porites brighami*. Corallite detail. HAWAII *Photograph: Keoki Stender*

Family Poritidae | Genus *Porites*

Group 3: Species forming thick columns or thick plates

Porites nodifera
Klunzinger, 1879

Characters: Colonies are compact clusters of columns, which have a smooth surface. Individual columns tend to be isolated subcolonies. Tops of columns tend to be squared off. Small colonies usually consist of nodular subcolonies. **Colour:** Pale brown. **Similar species:** Well developed colonies have a distinctive growth-form. Small colonies are similar to those of *Porites columnaris*. Corallites are similar to those of *P. myrmidonensis* and *P. pukoensis*. **Habitat:** Shallow water. Tolerant of high salinities. **Abundance:** Common and may form patch reefs.

Taxonomic reference: Scheer and Pillai (1983). **Identification guide:** Sheppard and Sheppard (1991).

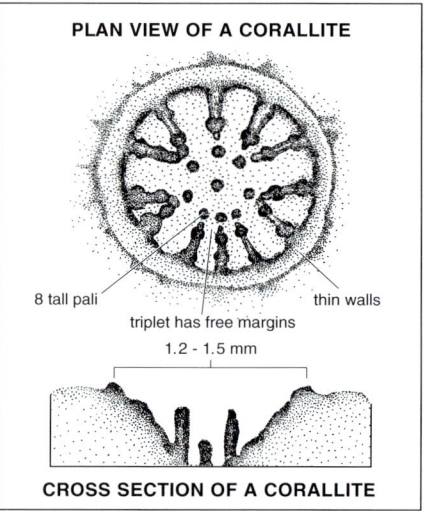

1, 2 Large colonies composed of columns. SINAI PENINSULA, EGYPT *Photographs: author*

3 Small colony, showing how individual columns are separated into subcolonies. SINAI PENINSULA, EGYPT *Photograph: Mary Stafford-Smith*

4 *Porites nodifera* (left) forming nodules similar in shape to *P. echinulata* (right). SINAI PENINSULA, EGYPT *Photograph: author*

5 Colony surface. SINAI PENINSULA, EGYPT *Photograph: author*

6 Corallite detail. SINAI PENINSULA, EGYPT *Photograph: author*

Family Poritidae　　　　　　　　　　　　　　　　　　　　　　　　　　　　　　　Genus *Porites*

Porites evermanni
Vaughan, 1907

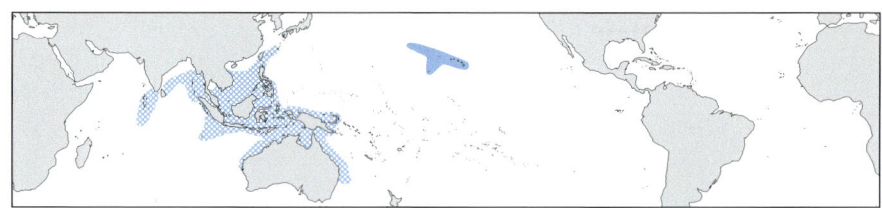

Characters: Colonies are massive with a tendency to form columns. Tentacles are usually extended during the day. **Colour:** Usually mustard, sometimes brown. **Similar species:** Readily distinguished by extended polyps giving colonies a distinctive colour and a furry appearance. Corallites are similar to those of *Porites australiensis* and *P. columnaris*. **Habitat:** Shallow protected reef environments, especially lagoons. **Abundance:** Usually uncommon.

Taxonomic note: There are unresolved taxonomic issues between Hawaiian and central Indo-Pacific occurrences of this species. **Taxonomic reference:** Vaughan's original description/specimens. **Identification guides:** Maragos (1977), Nishihira and Veron (1995).

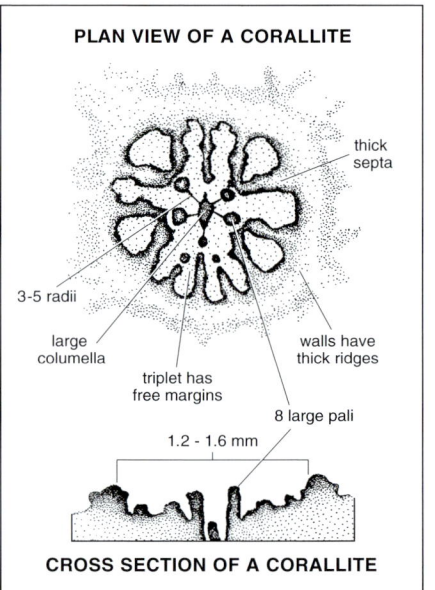

1 *Porites evermanni*. The columnar growth-form is typical of the species. SCOTT REEF, WESTERN AUSTRALIA *Photograph: author*

2 *Porites evermanni*. Tops of columns. MALDIVE ISLANDS *Photograph: Neville Coleman*

3 *Porites evermanni*. Detail of a colony with polyps extended during the day. This species is relatively easily recognised underwater. PAPUA NEW GUINEA *Photograph: author*

4 *Porites pukoensis*. Colonies have a submassive to columnar growth-form. HAWAII *Photograph: Don Potts*

5 *Porites pukoensis*. Colony surface with polyps partly extended during the day. HAWAII *Photograph: Don Potts*

Family Poritidae | Genus *Porites*

Porites pukoensis
Vaughan, 1907

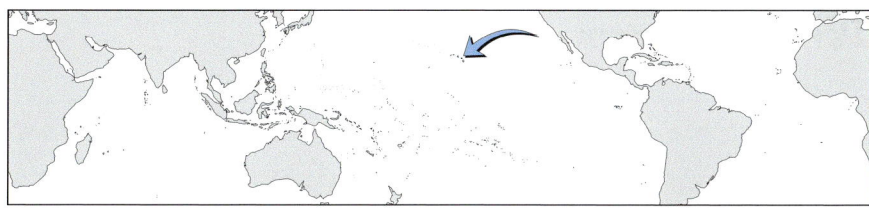

Characters: Colonies are massive with a tendency to form columns. Tentacles are usually extended during the day.
Colour: Brown or tan. **Similar species:** *Porites evermanni*, which lacks denticles at the bases of septa. See also *P. nodifera*.
Habitat: Shallow protected reef environments, especially lagoons. **Abundance:** Usually uncommon.

Taxonomic reference: Vaughan's original description/specimens. **Identification guide:** Maragos (1977).

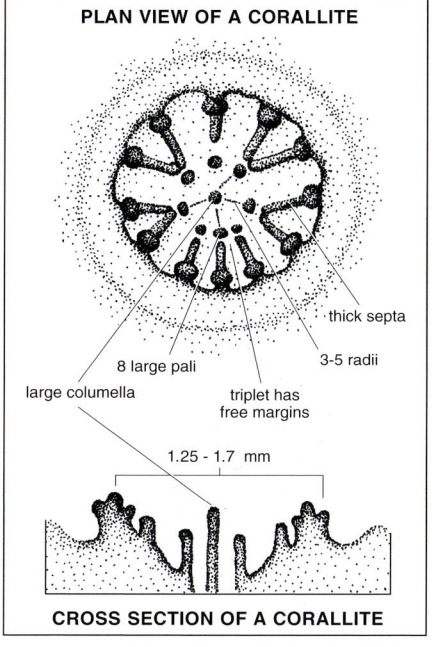

Porites columnaris
Klunzinger, 1879

Characters: Colonies are submassive or form short columns which may branch. These have a smooth surface. **Colour:** Pale brown. **Similar species:** *Porites nodifera*, which has a similar growth-form, but corallite characters are distinct. Corallites of *P. columnaris* are similar to those of *P. lutea* and *P. evermanni*. **Habitat:** Shallow reef environments. **Abundance:** Uncommon.

Taxonomic reference: Scheer and Pillai (1983).

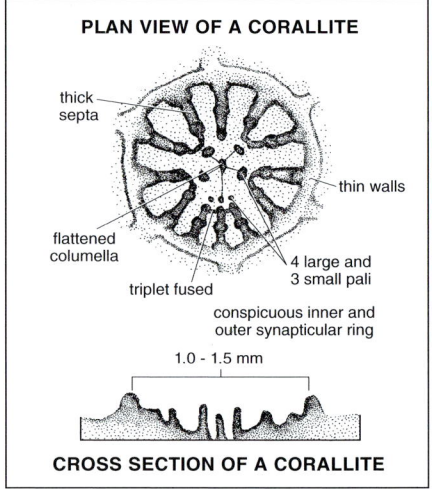

Porites arnaudi
Reyes-Bonilla and Carricart-Ganivet, 2000

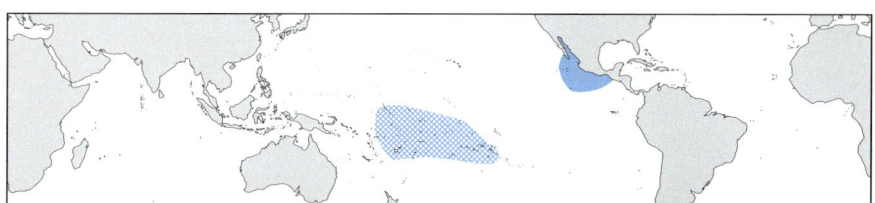

Characters: Colonies are flat plates which may be arranged as tiers or whorls. Corallites are compact and deeply excavated. **Colour:** Pale brown or greenish-grey, usually with pale margins to plates. **Similar species:** None. Corallites are similar to those of *Porites lobata*. **Habitat:** Upper reef slopes or rocky foreshores exposed to strong wave action. **Abundance:** Common.

Distribution note: Colonies in the central Pacific are doubtfully this species. **Taxonomic reference:** Reyes-Bonilla and Carricart-Ganivet (2000).

1 *Porites columnaris.* A prostrate columnar colony. SINAI PENINSULA, EGYPT *Photograph: author*

2 *Porites arnaudi.* Forming tiers of plates that are characteristic of the species. CLIPPERTON ATOLL, FAR EASTERN PACIFIC *Photograph: author*

3 *Porites arnaudi.* Side view of compact tiers. KIRIBATI, WESTERN PACIFIC *Photograph: Len Zell*

4 *Porites arnaudi.* Detail of a plate. CLIPPERTON ATOLL, FAR EASTERN PACIFIC *Photograph: author*

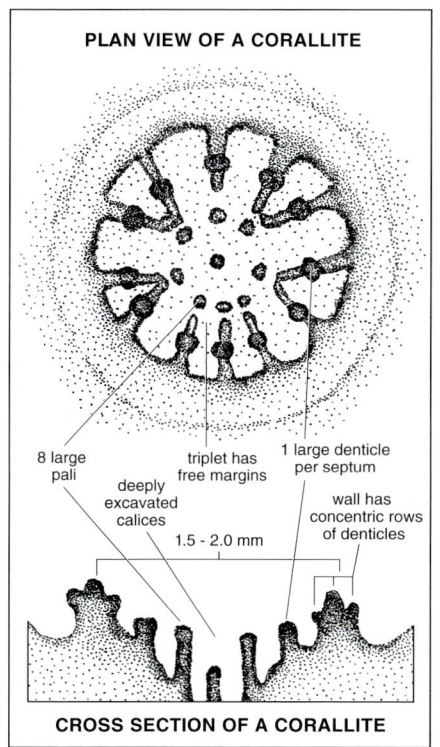

Porites colonensis
Zlatarski, 1990

Characters: Colonies are thin plates, sometimes in tiers. Surfaces of plates are smooth or undulating. **Colour:** Brown with white polyps. **Similar species:** *Porites astreoides* where forming thin plates, which is distinguished by having weakly developed paliform lobes and by colour. **Habitat:** Reef slopes. **Abundance:** Locally common.

Taxonomic reference: Zlatarski (1990). **Identification guide:** Humann (1993).

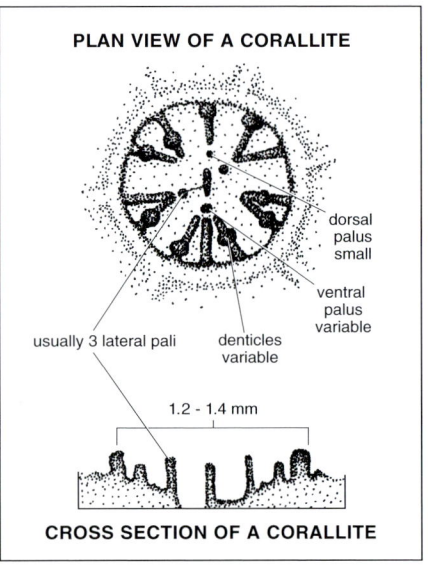

1 *Porites colonensis*. Common appearance of a flat plate projecting beneath *Agaricia*. HONDURAS *Photograph: Paul Humann*

2 *Porites colonensis*. Surface of a flat plate. COZUMEL, MEXICO *Photograph: Doug Fenner*

3 *Porites aranetai*. Surface of a submassive colony. VIETNAM *Photograph: author*

| Family Poritidae | Genus *Porites* |

> **Group 4: Species forming composites of columns, laminae and branches**

Porites aranetai
Nemenzo, 1955

Characters: Colonies are encrusting with a smooth or nodular surface forming short branches. The septal pattern is highly variable. **Colour:** Mottled green or pale cream. **Similar species:** *Porites lichen*, which commonly has corallites aligned in rows. Corallites are similar to those of *P. lutea* and *P. cocosensis*. **Habitat:** Shallow reef environments. **Abundance:** Locally common.

Taxonomic reference: Nemenzo (1955). **Identification guide:** Nishihira and Veron (1995).

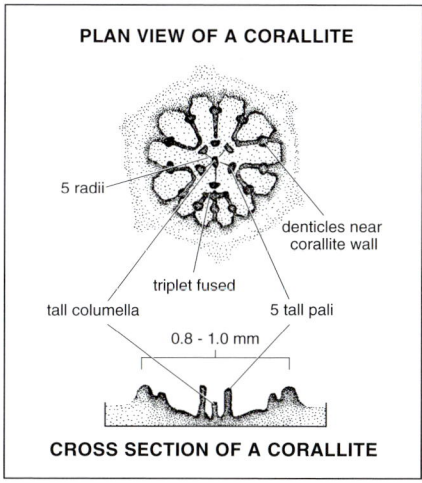

Family Poritidae | Genus *Porites*

Porites lichen
Dana, 1846

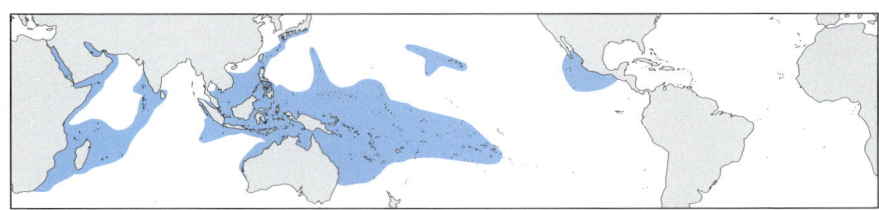

Characters: Colonies are flat laminae or plates, or fused nodules and columns. Corallites are commonly aligned in irregular rows separated by slight ridges. Septal structures are variable and irregular. **Colour:** Usually bright yellowish-green, sometimes brown. **Similar species:** *Porites annae* and *P. heronensis*. **Habitat:** Frequently a dominant species of lagoons and reef slopes. **Abundance:** Common and usually conspicuous.

Taxonomic references: Wells (1954), Veron and Pichon (1982). **Identification guides:** Veron (1986), Nishihira and Veron (1995).

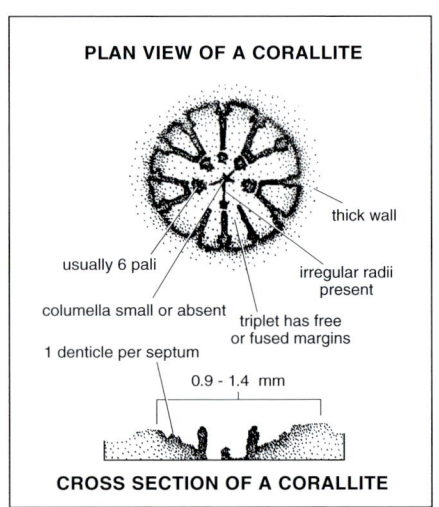

1 A typical growth-form and common colour. GREAT BARRIER REEF, AUSTRALIA *Photograph: author*

2 Corallite detail. GREAT BARRIER REEF, AUSTRALIA *Photograph: Ed Lovell*

3 Colony of small tiered plates. PAPUA NEW GUINEA *Photograph: Neville Coleman*

4 Encrusting colonies. GREAT BARRIER REEF, AUSTRALIA *Photograph: author*

Family Poritidae Genus *Porites*

305

Porites heronensis
Veron, 1985

Characters: Colonies are massive, encrusting or columnar. Corallites are shallow giving colonies a smooth surface. **Colour:** Cream, green, brown or mottled. **Similar species:** *Porites lichen*, which has a similar growth-form but has larger corallites and is usually bright yellowish-green in colour. See also *P. desilveri*. **Habitat:** Semi-protected shallow reef or terrigenous rock substrates. **Abundance:** Common in subtropical localities, rare in the tropics.

Taxonomic reference: Veron (1985). **Identification guides:** Veron (1986), Nishihira and Veron (1995).

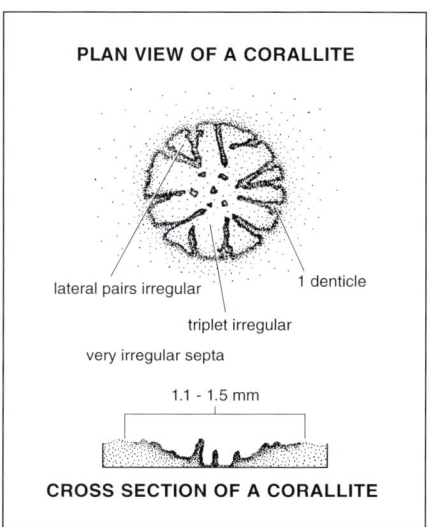

1, 2 *Porites heronensis*. Colonies with characteristically developed columns. **1** LORD HOWE ISLAND, SOUTH-EAST AUSTRALIA **2** NORFOLK ISLAND, WESTERN PACIFIC *Photographs: 1 James Brown 2 author*

3 *Porites heronensis*. Surface appearance of columns. Polyps remain extended during the day. NORFOLK ISLAND, WESTERN PACIFIC *Photograph: author*

4 *Porites okinawensis*. Surface of a small colony. RYUKYU ISLANDS, JAPAN *Photograph: Masanori Nonaka*

Porites okinawensis
Veron, 1990

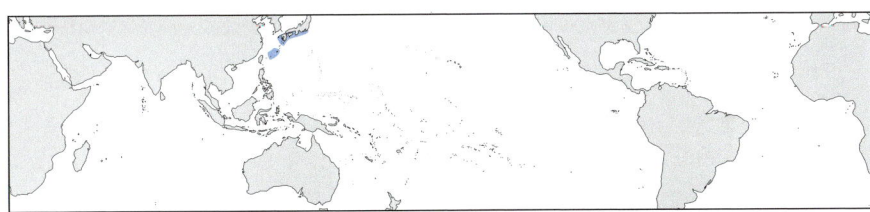

Characters: Colonies are encrusting to massive, with an irregular surface. Corallites are superficial and angular. The coenosteum between corallites is granulated giving a rough appearance which can be seen underwater. **Colour:** Pale brown or cream, sometimes green in high latitudes. **Similar species:** A distinctive species. Corallites are sometimes similar to those of *P. lutea*. **Habitat:** Semi-protected shallow reef or terrigenous rock substrates. **Abundance:** Uncommon.

Taxonomic reference: Veron (1990a).

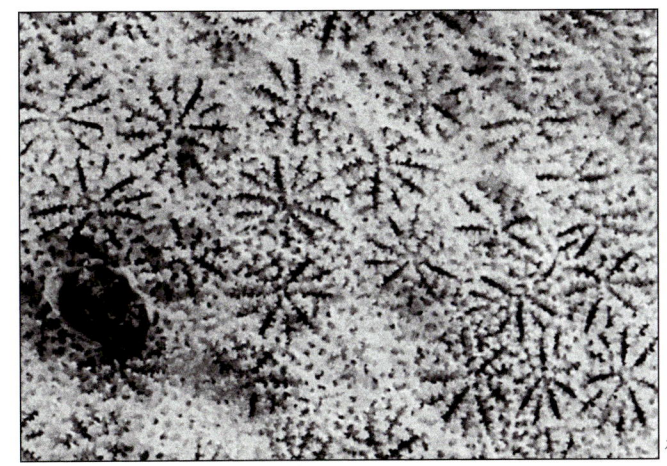

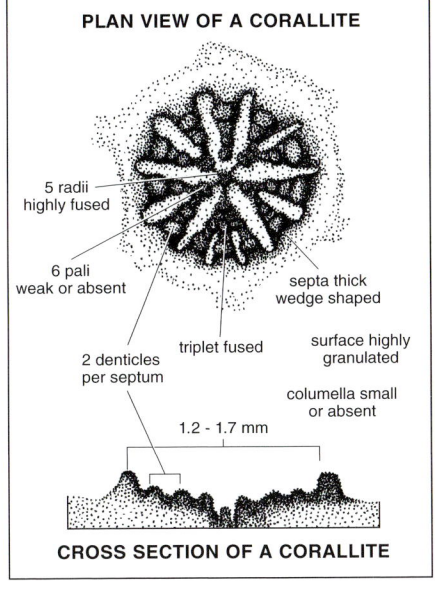

Family Poritidae   Genus *Porites*

Porites desilveri
Veron, this publication

Characters: Colonies are encrusting with a smooth or nodular surface forming short branches. Septa are so irregular that the *Porites* pattern is often unrecognisable. **Colour:** Grey with white tops to nodules. **Similar species:** *Porites heronensis*. **Habitat:** Shallow reef environments, especially lagoons. **Abundance:** Common.

Taxonomic note: See 'New species described in *Corals of the World*' (Veron, in preparation) for further information.

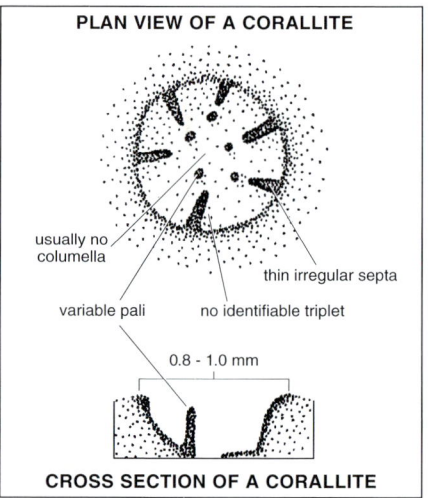

Porites vaughani
Crossland, 1952

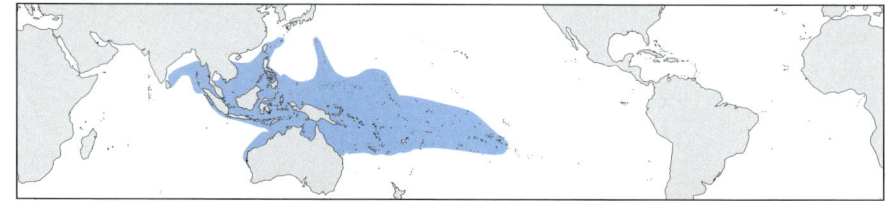

Characters: Colonies are encrusting, laminar or form columns. Corallites are widely spaced and separated by ridges. They are uniform in size within the same colony; those of colonies from exposed upper reef slopes are relatively small. **Colour:** Usually pale cream, pink or brown but may be bright green or purple. **Similar species:** *Porites annae* and *P. lichen*. **Habitat:** A wide range of environments. **Abundance:** Sometimes common.

Taxonomic reference: Veron and Pichon (1982). **Identification guides:** Veron (1986), Nishihira and Veron (1995).

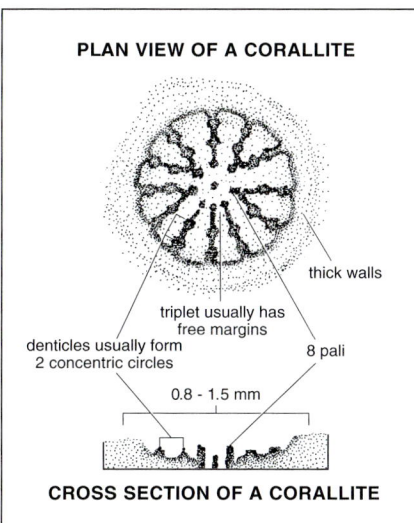

1 *Porites desilveri*. Typical appearance of colonies. SRI LANKA *Photograph: author*

2 *Porites desilveri*. Colony surface. SRI LANKA *Photograph: author*

3 *Porites vaughani*. Encrusting colonies in shallow water may have very thick ridges between corallites. GUAM *Photograph: Gustav Paulay*

4 *Porites vaughani*. Detail of *P. vaughani* (below) and *P. lichen* (above). PAPUA NEW GUINEA *Photograph: author*

5 *Porites vaughani*. Surface of an encrusting colony. GREAT BARRIER REEF, AUSTRALIA *Photograph: Ed Lovell*

Porites annae
Crossland, 1952

Characters: Colonies have nodular anastomosing branches or columns with encrusting or laminar bases. In turbid water colonies are predominantly laminar. **Colour:** Pale or dark green, yellow, purple or brown. **Similar species:** *Porites cocosensis*. See also *P. lichen*, which is best distinguished by its different growth-forms and larger corallites, although the size range overlaps. **Habitat:** May form single species stands on sloping reef faces in clear or turbid water. **Abundance:** Common.

Taxonomic reference: Veron and Pichon (1982). **Identification guides:** Veron (1986), Nishihira and Veron (1995).

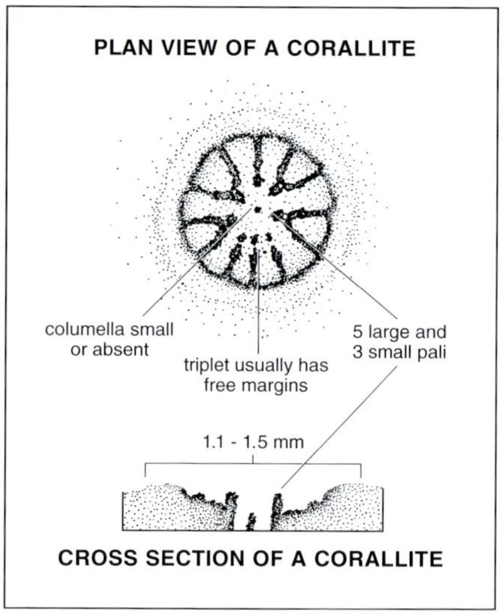

| Family Poritidae | Genus *Porites* |

1 *Porites annae*. Characteristically irregular branches. GREAT BARRIER REEF, AUSTRALIA Photograph: Ed Lovell

2 *Porites annae*. Side of irregularly fused columns. GUAM Photograph: Gustav Paulay

3, 4 *Porites annae*. Branch ends. **3** PUERTO GALERA, PHILIPPINES **4** SOLOMON ISLANDS Photographs: 3 author 4 Neville Coleman

5 *Porites annae*. Fused column tops on a subtidal reef flat. GUAM Photograph: Gustav Paulay

6 *Porites cocosensis*. Colonies have a nodular upper surface. Tentacles are partly extended. FLORES, INDONESIA Photograph: author

Porites cocosensis
Wells, 1950

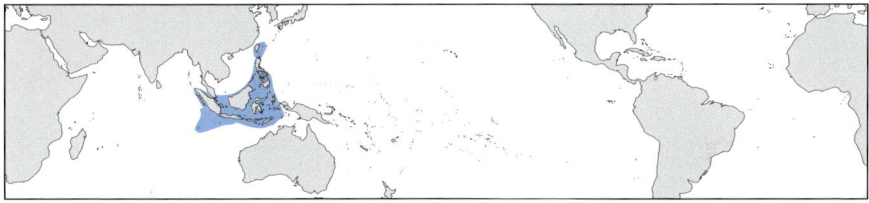

Characters: Colonies are encrusting with a nodular surface forming short branches. Corallites are deeply excavated giving a cellular appearance. Very tall pali are visible underwater where they resemble slightly extended tentacles. **Colour:** Usually grey, green or blue. **Similar species:** *Porites annae*, which has less cellular corallites and more irregularly fused branches. **Habitat:** Shallow reef environments. **Abundance:** Occasionally locally common.

Taxonomic reference: Wells (1950).

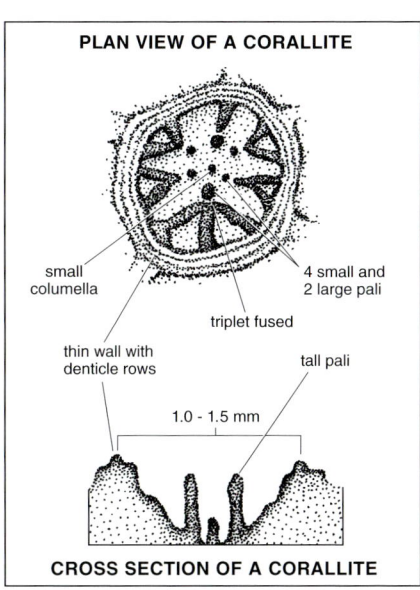

Family Poritidae | Genus *Porites*

Group 5: Species forming composites of laminae and branches

Porites rus
(Forskål, 1775)

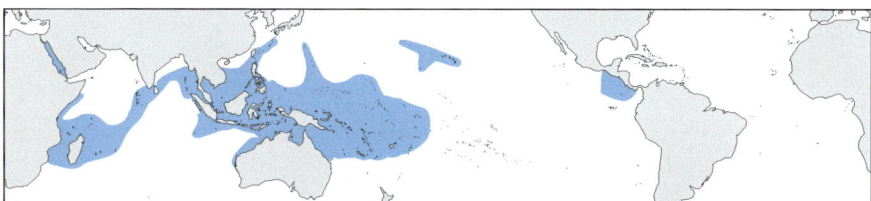

Characters: Colonies are submassive, laminar or contorted anastomosing branches and columns, commonly over 5 metres across. Corallites are separated into groups by ridges which characteristically converge towards each other forming flame-shaped patterns. **Colour:** Pale cream, yellow or dark bluish-brown, often with pale branch tips. Sometimes brightly coloured in shallow water. **Similar species:** *Porites rus* can be confused with *Montipora* (Acroporidae) underwater. Closest to *P. monticulosa*. See also *P. deformis*, *P. cumulatus* and *P. ornata*, all of which have small corallites. **Habitat:** Shallow reef environments. **Abundance:** Common and may be a dominant species in a wide range of habitats.

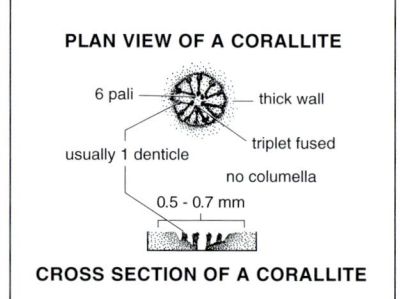

Taxonomic note: This species is divisible into several smaller semi-distinct taxonomic units, see 'What are species?' p425. **Taxonomic reference:** Veron and Pichon (1982). **Identification guides:** Randall and Myers (1983), Veron (1986), Sheppard and Sheppard (1991), Nishihira and Veron (1995).

Family Poritidae | Genus *Porites*

1 A large hemispherical colony. PEMBA ISLAND, TANZANIA *Photograph: author*

2 A small colony composed of branches on the upper part and plates on the lower part. PEMBA ISLAND, TANZANIA *Photograph: author*

3 Forming a cluster of branches. GREAT BARRIER REEF, AUSTRALIA *Photograph: Valerie Taylor*

4 In an intertidal habitat. SCOTT REEF, WESTERN AUSTRALIA *Photograph: author*

5 Detail of an irregular plate. VIETNAM *Photograph: author*

6 Detail of a branch. COCOS (KEELING) ATOLL, WESTERN AUSTRALIA *Photograph: author*

Family Poritidae — Genus *Porites*

Porites monticulosa
Dana, 1846

Characters: Colonies are massive, columnar, laminar, branching or encrusting and are usually less than one metre across. They are commonly mixtures of these growth-forms. Corallites are separated into groups by ridges. **Colour:** Usually brown or blue. **Similar species:** *Porites rus*, which tends to form branches whereas *P. monticulosa* tends to be massive or form plates. **Habitat:** Shallow reef environments. **Abundance:** Common in the western Indian Ocean.

Taxonomic reference: Dana's original description/specimens.

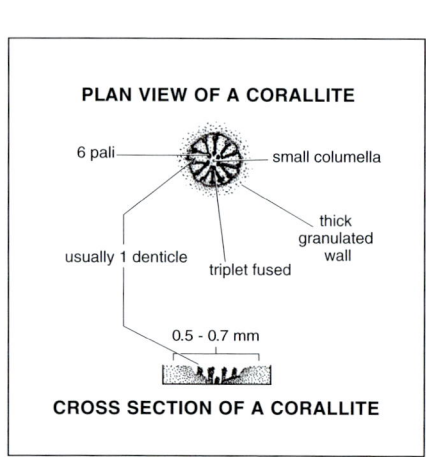

Family Poritidae Genus *Porites*

1 Comparison between *P. rus* (yellow and grey, centre and right) with a small colony of *P. cylindrica* (grey, lower centre) and *P. monticulosa* (left). PEMBA ISLAND, TANZANIA *Photograph: author*

2 Colonies composed primarily of plates with sparse branches. MADAGASCAR *Photograph: author*

3 Colony composed of plates and branches. MADAGASCAR *Photograph: author*

4 Comparison between a massive part of a *P. monticulosa* colony (left) and a branching *P. rus* colony (right). COCOS (KEELING) ATOLL, WESTERN AUSTRALIA *Photograph: author*

5 Colonies composed of compact branches. MADAGASCAR *Photographs: author*

6, 7 Colonies composed of plates. **6** HAWAII **7** SEYCHELLES *Photographs: 6 Doug Fenner 7 author*

Family Poritidae | Genus *Porites*

Porites horizontalata
Hoffmeister, 1925

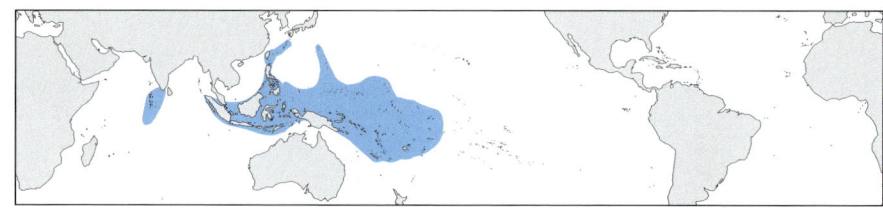

Characters: Colonies are composites of encrusting laminae and contorted anastomosing branches. Corallites are separated into groups by ridges. **Colour:** Pale brown with cream extremities of branches and plates. Sometimes brightly coloured in shallow water. **Similar species:** Laminar parts of colonies resemble those of *Porites vaughani*. Branching parts of colonies may have corallites arranged in a *P. rus*-like pattern. **Habitat:** Shallow reef environments. **Abundance:** Sometimes common.

Taxonomic references: Wells (1954), Veron and Pichon (1982). **Identification guide:** Nishihira and Veron (1995).

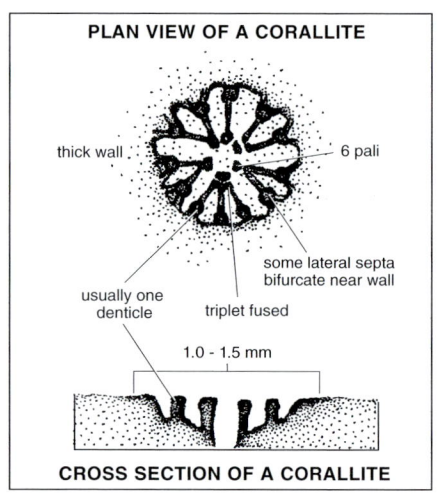

PLAN VIEW OF A CORALLITE

thick wall — 6 pali — some lateral septa bifurcate near wall — usually one denticle — triplet fused — 1.0 - 1.5 mm

CROSS SECTION OF A CORALLITE

1 An irregularly branching colony. RYUKYU ISLANDS, JAPAN *Photograph: Moritaka Nishihira*

2 Colony composed of flat plates. CAROLINE ISLANDS, MICRONESIA *Photograph: Pat Colin*

3 An encrusting colony. SOLOMON ISLANDS *Photograph: Neville Coleman*

4 Surface of a plate. PAPUA NEW GUINEA *Photograph: author*

5 Corallite detail. FLORES, INDONESIA *Photograph: author*

6 Forming irregular upgrowths. FLORES, INDONESIA *Photograph: author*

Porites napopora
Veron, this publication

Characters: Colonies are broad basal laminae with irregular clumps of tapered irregularly fused branches. Corallites are irregularly spaced and are in excavated pits. Those on branches are especially deeply excavated, giving branches a rough surface. Walls between corallites are thin. **Colour:** Brown with white corallite centres. **Similar species:** *Porites nigrescens*, and *P. negrosensis*, both of which have similarly excavated corallites. *Porites nigrescens* does not have basal laminae and *P. negrosensis* has corallites with rounded walls. *Porites horizontalata* has similar corallites but forms primarily explanate plates. See also *P. flavus* and *P. tuberculosa*. **Habitat:** Shallow reef environments. **Abundance:** Sometimes common.

Taxonomic note: See 'New species described in *Corals of the World*' (Veron, in preparation) for further information.

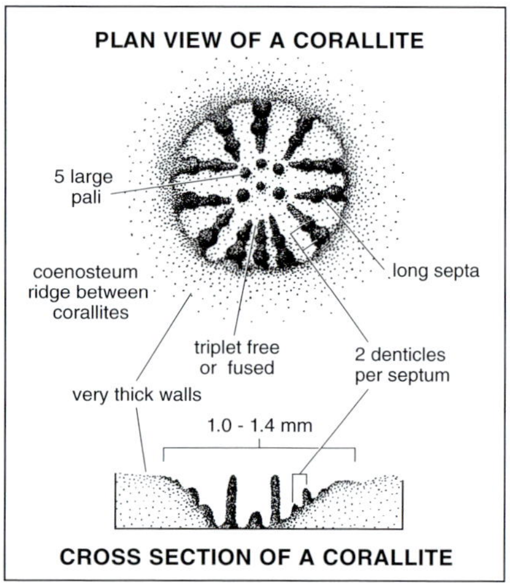

Porites sillimaniana
Nemenzo, 1976

Characters: Colonies are clumps of thin anastomosing branches or thin basal laminae with upright irregular branches arising from central parts. Corallites are often aligned in rows on the laminae. Tentacles are often extended during the day. **Colour:** Pale yellow or brown, usually with white corallite centres. **Similar species:** *Porites latistella*, which has similar growth-forms but more excavated corallites. Corallites on branches resemble those of *P. nigrescens*. Branches are like those of *P. flavus*. **Habitat:** Mostly on reef flats. **Abundance:** Uncommon.

Taxonomic references: Nemenzo (1976), Veron and Hodgson (1989). **Identification guide:** Nishihira and Veron (1995).

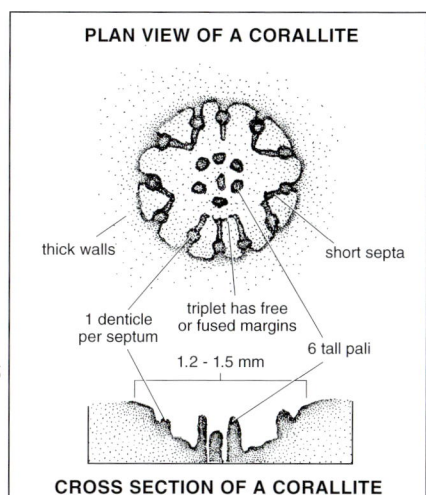

PLAN VIEW OF A CORALLITE

thick walls — short septa
1 denticle per septum — triplet has free or fused margins — 6 tall pali
1.2 - 1.5 mm

CROSS SECTION OF A CORALLITE

1 *Porites napopora*. Colony with well developed basal laminae. FLORES, INDONESIA *Photograph: author*

2 *Porites napopora*. Branches are sometimes highly fused. FLORES, INDONESIA *Photograph: author*

3 *Porites napopora*. Corallite detail. GUAM *Photograph: Gustav Paulay*

4 *Porites napopora*. Colony with compact branches. FLORES, INDONESIA *Photograph: author*

5 *Porites sillimaniana*. The characteristic appearance of branches growing on flat laminae. APO ISLAND, PHILIPPINES *Photograph: author*

6 *Porites sillimaniana*. Colony composed mostly of branches. SEYCHELLES *Photograph: author*

Porites latistella
Quelch, 1886

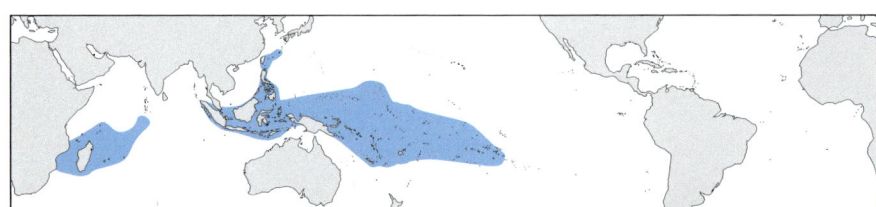

Characters: Colonies are thin basal laminae and twisted flattened branches. Corallites are aligned in irregular rows along the branches. **Colour:** Pale brown. **Similar species:** *Porites sillimaniana*. Corallites are similar to those of *P. cylindrica*. **Habitat:** Shallow protected reef environments. **Abundance:** Uncommon.

Taxonomic reference: Nemenzo (1971). **Identification guide:** Nishihira and Veron (1995).

1, 5 Colonies are usually composed of branches and flat plates. 1 PAPUA NEW GUINEA 5 FLORES, INDONESIA *Photographs: author*

2 Colony in shallow water composed of thick compact branches. SEYCHELLES *Photograph: author*

3 A colony composed mostly of twisted branches. CEBU, PHILIPPINES *Photograph: author*

4 Detail of branches and laminae. CEBU, PHILIPPINES *Photograph: author*

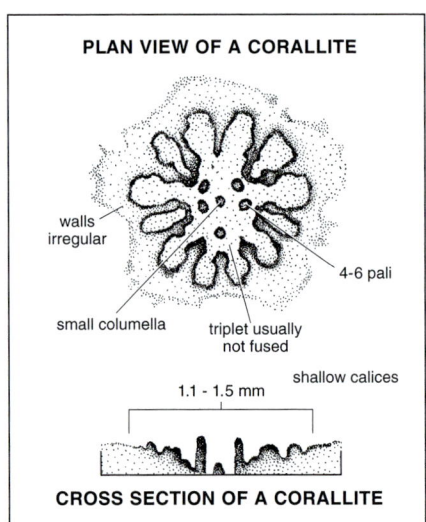

Family Poritidae — Genus *Porites*

Porites eridani
Umbgrove, 1940

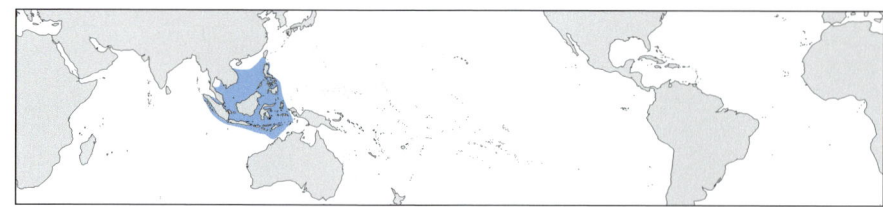

Characters: Colonies are large basal laminae, often with contorted branches. Corallites are irregularly distributed on both laminae and branches. **Colour:** Brown. **Similar species:** Corallites are similar to those of *Porites flavus*, which is primarily distinguished by lack of basal laminae. Growth-forms are similar to those of *P. sillimaniana* and *P. latistella*. **Habitat:** Shallow protected reef environments. **Abundance:** Usually uncommon.

Taxonomic reference: Nemenzo (1980).

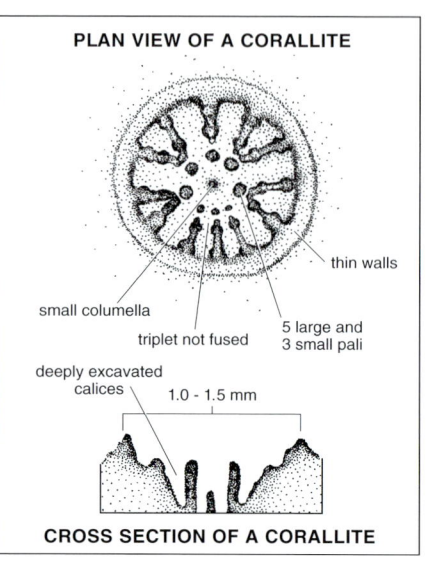

1 *Porites eridani*. Irregularly flattened branches on a laminate base. VIETNAM *Photograph: author*

2 *Porites eridani*. Colony with extensive basal plates. VIETNAM *Photograph: author*

Family Poritidae | Genus *Porites*

Porites deformis
Nemenzo, 1955

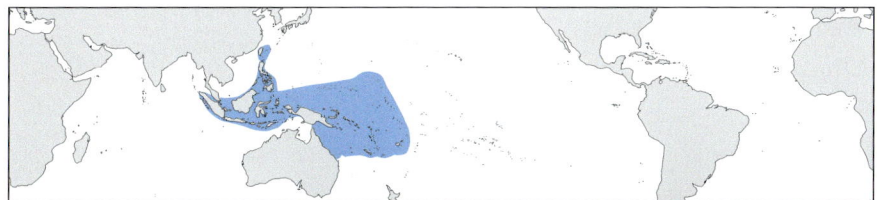

Characters: Colonies are thin basal laminae and nodular branches that fuse into clumps. Corallites are superficial and branch surfaces are smooth. Tentacles are sometimes extended during the day. **Colour:** Pale brown. **Similar species:** *Porites cumulatus*. Corallites are similar to those of *P. horizontalata*. **Habitat:** Shallow protected reef environments. **Abundance:** Uncommon.

Taxonomic references: Nemenzo (1955), Veron and Hodgson (1989). **Identification guide:** Nishihira and Veron (1995).

3 *Porites deformis*. Characteristic appearance on an upper reef slope. CALAMIAN ISLANDS, PHILIPPINES *Photograph: author*

4 *Porites deformis*. A large stand of contorted branches. CALAMIAN ISLANDS, PHILIPPINES *Photograph: author*

5 *Porites deformis*. Surface detail. SULAWESI, INDONESIA *Photograph: Doug Fenner*

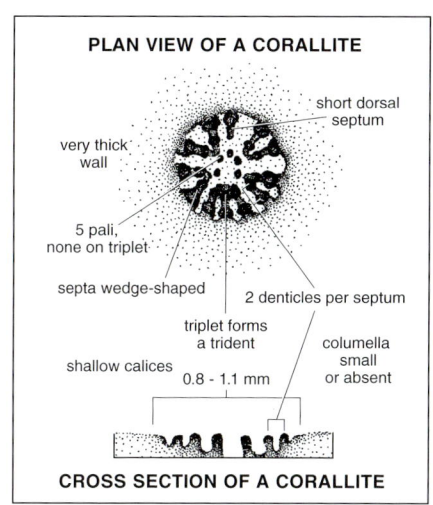

Family Poritidae   Genus *Porites*

Porites cumulatus
Nemenzo, 1955

Characters: Colonies are highly fused flattened branches. Corallites are angular and superficial. Branch surfaces are smooth. **Colour:** Cream or pale brown. **Similar species:** *Porites deformis*, which has more nodular branches, no conspicuous basal laminae and sometimes more excavated corallites. Corallites are similar to those of *P. rus*. **Habitat:** Shallow protected reef environments. **Abundance:** Uncommon.

Distribution note: Records from the Red Sea are doubtful. **Taxonomic reference:** Nemenzo (1955).

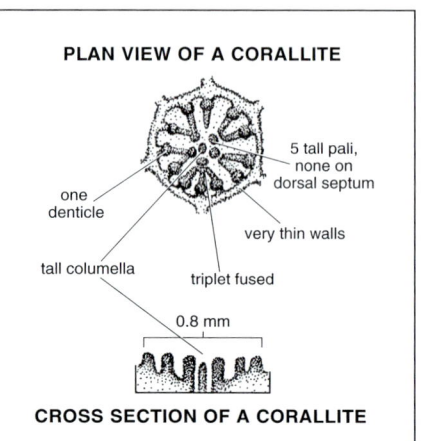

1 *Porites cumulatus*. Colony with well developed branches and a smooth surface. VIETNAM *Photograph: author*

2 *Porites cumulatus*. Surface of a columnar colony. SULAWESI, INDONESIA *Photograph: Doug Fenner*

3, 4 *Porites cumulatus*. Colony surface. **3** SINAI PENINSULA, EGYPT **4** SULAWESI, INDONESIA *Photographs:* **3** *author* **4** *Doug Fenner*

Family Poritidae | Genus *Porites*

Porites branneri
Rathbun, 1887

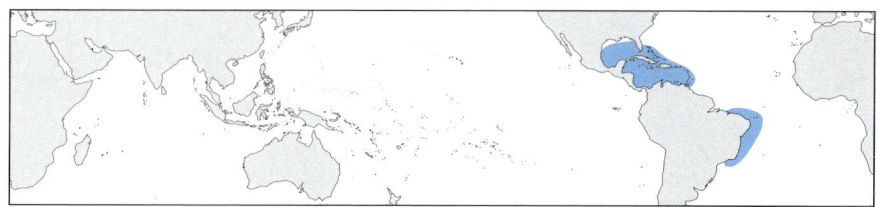

Characters: Colonies are thin encrusting plates with nodular upgrowths. Septa are uniform in spacing and length. **Colour:** Blue (which may photograph purple). **Similar species:** *Porites astreoides*, which is usually yellow and forms large massive colonies. **Habitat:** Protected shallow reef environments. **Abundance:** Uncommon.

Taxonomic note: Differences between Brazilian and Caribbean occurrences of this species are substantial. **Taxonomic reference:** Laborel (1969). **Identification guide:** Humann (1993).

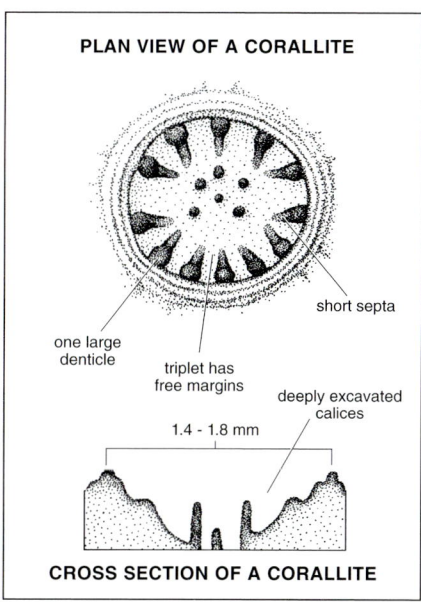

5 *Porites branneri*. Corallite detail. COZUMEL, MEXICO Photograph: Doug Fenner

6 *Porites branneri*. A small nodular colony. Tentacles are extended. BAHAMAS Photograph: Paul Humann

Family Poritidae   Genus *Porites*

Group 6: Species forming primarily branching colonies

Porites porites
(Pallas, 1766)

Characters: Colonies are sturdy fused branches with rounded tips. Corallites are deeply excavated and, with a coarse coenosteum, form a rough surface. Tentacles are usually extended during the day. **Colour:** Uniform grey-brown. **Similar species:** *Porites furcata*. See taxonomic note. **Habitat:** Shallow protected reef environments. **Abundance:** Sometimes common.

Taxonomic note: Several authors have considered *Porites porites*, *P. divaricata* and *P. furcata* to be different forms of the same species. See 'What are species?' p425. **Taxonomic references:** Squires (1958), Roos (1971). **Identification guides:** Colin (1978), Humann (1993).

1

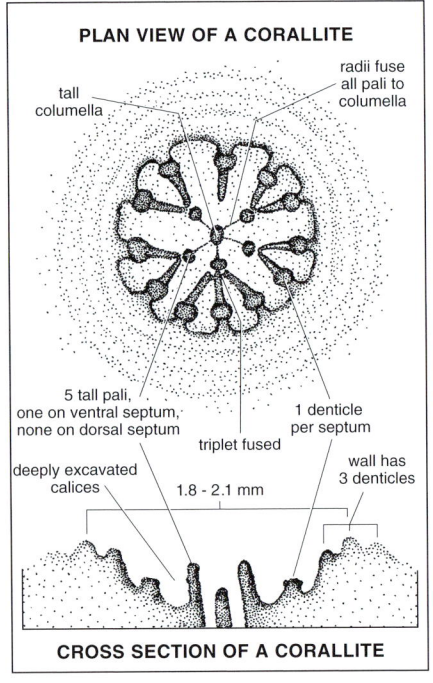

2

1 Colony composed of prostrate irregular branches, some of which have polyps extended. BAHAMAS *Photograph: Mary Stafford-Smith*

2 Colony composed of clumps of short thick branches. BAHAMAS *Photograph: author*

3 Branch tips on dead bases. JAMAICA *Photograph: author*

4 Branches with contracted polyps. CAYMAN ISLANDS *Photograph: Nancy Sefton*

5 Detail of partly extended polyps. BARBADOS *Photograph: author*

6 Branches with extended polyps. JAMAICA *Photograph: author*

Family Poritidae — Genus *Porites*

Porites furcata
Lamarck, 1816

Characters: Colonies are compact branches with flattened tops. Corallites are deeply excavated. Tentacles are usually extended during the day. **Colour:** Uniform grey-brown. **Similar species:** See taxonomic note. *Porites porites* has thicker corallite walls and smaller pali, but these distinctions are minor. **Habitat:** Shallow to mid-slope reef environments. **Abundance:** Common.

Taxonomic note: Several authors have considered *Porites porites*, *P. divaricata* and *P. furcata* to be different forms of the same species. See 'What are species?' p425. **Taxonomic references:** Squires (1958), Roos (1971). **Identification guides:** Colin (1978), Humann (1993).

1 *Porites furcata*. A compact colony in an intertidal rock pool. BELIZE *Photograph: author*

2 *Porites furcata*. A thicket on an upper reef slope. BELIZE *Photograph: author*

3 *Porites furcata*. Branch tips with polyps extended during the day. BELIZE *Photograph: Mary Stafford-Smith*

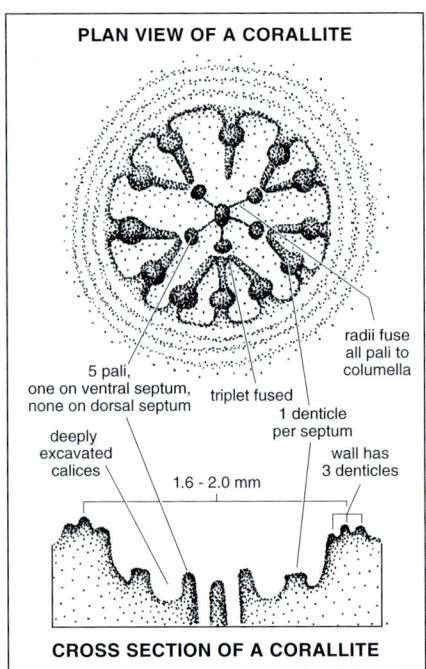

Porites divaricata
Lesueur, 1821

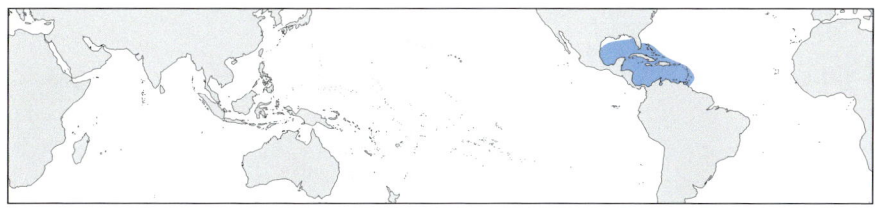

Characters: Colonies are thin fused branches with flattened tips. Corallites are deeply excavated and, with a coarse coenosteum, form a rough surface. Tentacles are usually extended during the day. **Colour:** Uniform grey-brown or greenish-yellow. **Similar species:** See taxonomic note. **Habitat:** Shallow protected reef environments, especially seagrass beds. **Abundance:** Usually common.

Taxonomic note: Several authors have considered *Porites porites*, *P. divaricata* and *P. furcata* to be different forms of the same species. See 'What are species?' p425. **Taxonomic references:** Squires (1958), Roos (1971). **Identification guides:** Colin (1978), Humann (1993).

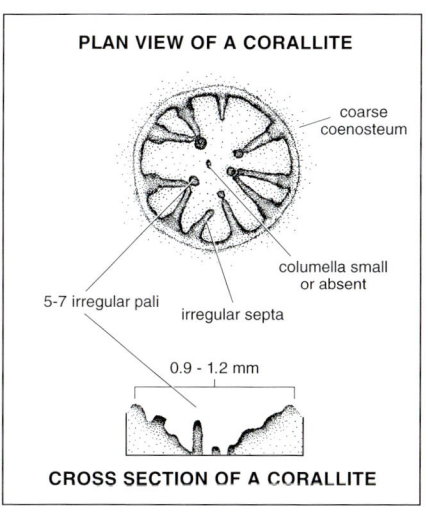

4 *Porites divaricata*. A clump of consolidated branches. BELIZE *Photograph: author*

5 *Porites divaricata*. A loosely aggregated bed of thin branches with dead bases and living tips. These beds only occur in shallow water. BELIZE *Photograph: author*

Family Poritidae · Genus *Porites*

Porites attenuata
Nemenzo, 1955

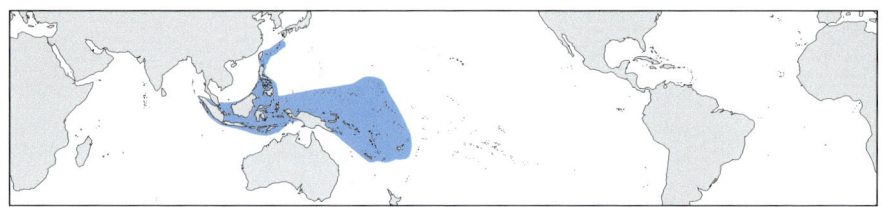

Characters: Colonies are sturdy fused branches with rounded tips. Corallites are moderately excavated. The coenosteum is coarse. **Colour:** Usually mustard or bright yellow-green, also pale brown. **Similar species:** *Porites cylindrica*, which has finer branches and less excavated corallites. See also *P. tuberculosa*. **Habitat:** Shallow protected reef environments. **Abundance:** Common.

Taxonomic references: Nemenzo (1955), Veron and Hodgson (1989). **Identification guide:** Nishihira and Veron (1995).

1 *Porites attenuata*. Branch detail. NEGROS ISLANDS, PHILIPPINES *Photograph: Doug Fenner*

2 *Porites attenuata* (left) with *P. cylindrica* (top right) and *P. nigrescens* (bottom centre). CEBU, PHILIPPINES *Photograph: author*

3 *Porites attenuata* (right) with *P. cylindrica* (left). BOLINAO, PHILIPPINES *Photograph: author*

4 *Porites attenuata* (right) with *P. cylindrica* (left). NEGROS, PHILIPPINES *Photograph: Doug Fenner*

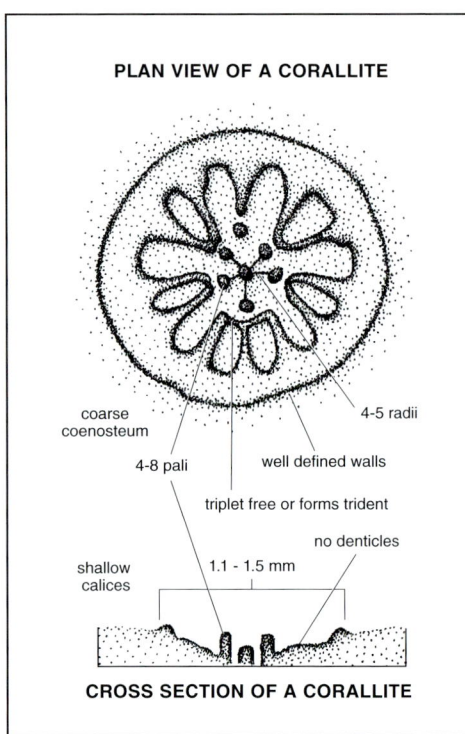

Porites tuberculosa
Veron, this publication

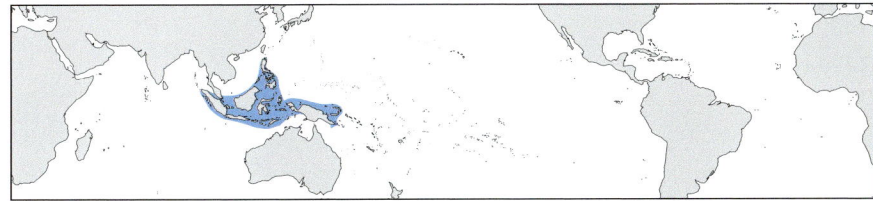

Characters: Colonies are sturdy fused branches, sometimes with basal plates. Branches usually have squared-off tips. Corallites are moderately excavated and connected by ridges of coenosteum. **Colour:** Grey or green. **Similar species:** *Porites attenuata*, which has larger corallites that are not connected by ridges of coenosteum. See also *P. napopora*, which has thinner branches. This species can readily be mistaken for a *Montipora* (Acroporidae) underwater. **Habitat:** Shallow protected reef environments. **Abundance:** Sometimes common in Indonesia.

Taxonomic note: See 'New species described in *Corals of the World*' (Veron, in preparation) for further information.

5 *Porites tuberculosa*. A large colony. BALI, INDONESIA *Photograph: Roger Steene*

6 *Porites tuberculosa*. A colony with both branches and basal plates. BALI, INDONESIA *Photograph: Gerry Allen*

7, 8 *Porites tuberculosa*. Variation in the surface detail of branches. FLORES, INDONESIA *Photographs: author*

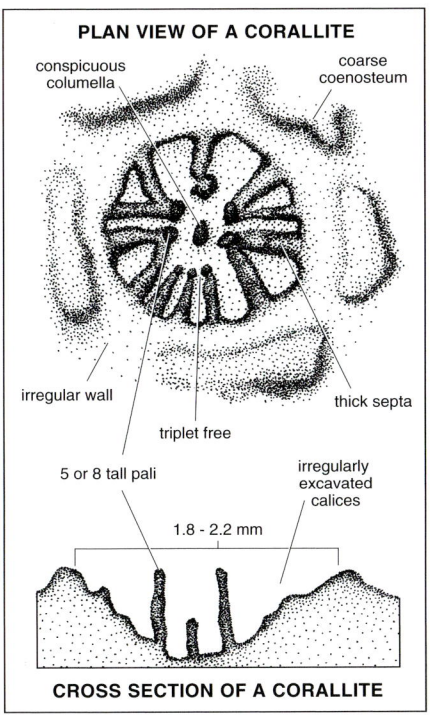

Family Poritidae Genus *Porites*

Porites cylindrica
Dana, 1846

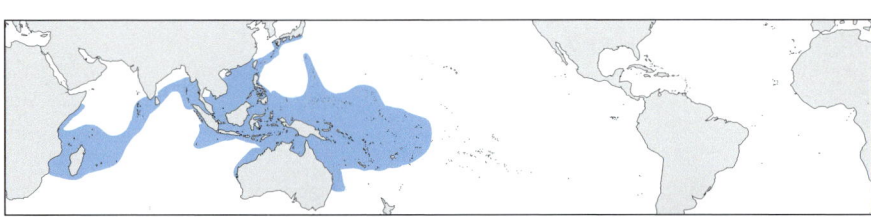

Characters: Colonies are branching, sometimes with an encrusting base. Corallites are shallow giving branches a smooth surface. **Colour:** Usually cream, yellow, blue or green. **Similar species:** See *Porites attenuata* and *P. compressa*. *Porites cylindrica* is readily confused with the astrocoeniid *Palauastrea ramosa* underwater. **Habitat:** May be a dominant species in lagoons or on back reef margins. **Abundance:** Common.

Taxonomic reference: Veron and Pichon (1982). **Identification guides:** Randall and Myers (1983), Veron (1986), Nishihira and Veron (1995).

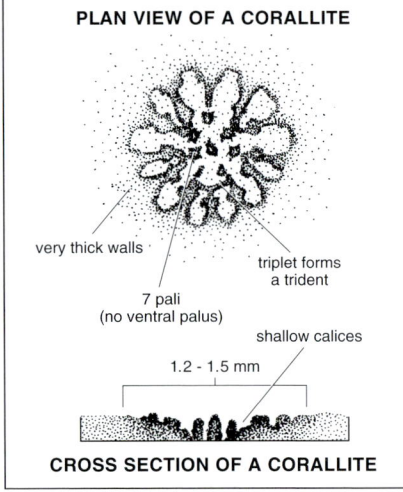

1 This species is sometimes a dominant one in sheltered lagoons. GREAT BARRIER REEF, AUSTRALIA *Photograph: author*

2 Colonies exposed to turbulence may have irregularly shaped branches. CALAMIAN ISLANDS, PHILIPPINES *Photograph: author*

3 Compact branches in a sheltered lagoon. GREAT BARRIER REEF, AUSTRALIA *Photograph: Ed Lovell*

4 An intertidal 'micro-atoll'. GREAT BARRIER REEF, AUSTRALIA *Photograph: Isobel Bennett*

5 A large colony. BALI, INDONESIA *Photograph: Roger Steene*

6 Surface of branches of a colony on a reef slope. PAPUA NEW GUINEA *Photograph: author*

7 Detail with polyps extended. RYUKYU ISLANDS, JAPAN *Photograph: author*

Family Poritidae — Genus *Porites*

Porites nigrescens
Dana, 1846

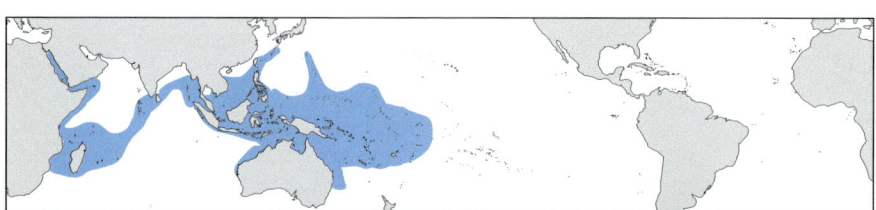

Characters: Colonies are branching, sometimes with an encrusting base. Concave calices give the surface a pitted appearance. Tentacles are frequently extended during the day. **Colour:** Brown or cream. **Similar species:** *Porites cylindrica*, which is usually found in the same habitat where it has less excavated corallites and thicker branches. See also *P. negrosensis*. **Habitat:** Common on lower reef slopes and lagoons protected from wave action. **Abundance:** Sometimes common.

Taxonomic reference: Veron and Pichon (1982). **Identification guides:** Veron (1986), Nishihira and Veron (1995).

1 *Porites nigrescens* (left, with yellow branch tips) with *P. cylindrica* (centre, mustard colour) and *P. attenuata* (top and centre right). CALAMIAN ISLANDS, PHILIPPINES Photograph: author

2 A large colony. CALAMIAN ISLANDS, PHILIPPINES Photograph: author

3 Common appearance of a small colony in shallow water. RYUKYU ISLANDS, JAPAN Photograph: author

4 Branch detail. ASHMORE REEF, WESTERN AUSTRALIA Photograph: author

5 *Porites nigrescens* (right) with *P. cylindrica* (left). VIETNAM Photograph: author

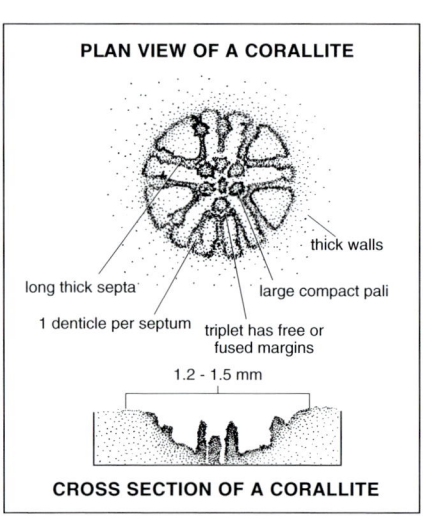

Family Poritidae   Genus *Porites*

Porites negrosensis
Veron, 1990

Characters: Colonies are clumps of anastomosing contorted branches. Large colonies may develop basal laminae. Corallites are widely spaced and conspicuously excavated. The coenosteum is fine. **Colour:** Pale brown or grey. **Similar species:** *Porites rugosa*. See also *P. nigrescens*, which has less excavated corallites. Resembles the acroporid *Montipora porites* underwater. **Habitat:** Shallow protected reef environments where diversity is high. **Abundance:** Uncommon.

Taxonomic reference: Veron (1990a). **Identification guide:** Nishihira and Veron (1995).

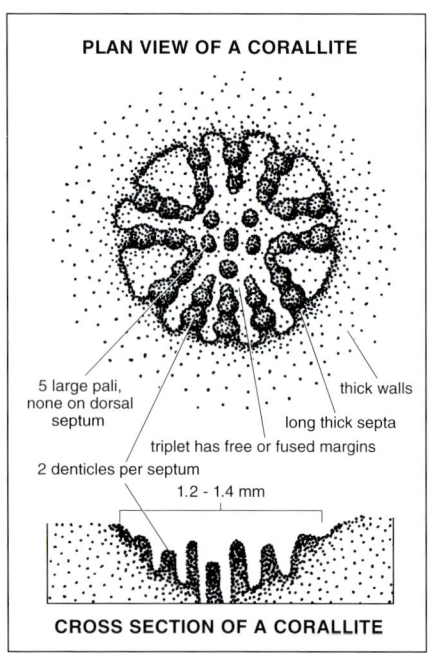

1, 2 Anastomosed branches and deeply excavated corallites. **1** FLORES, INDONESIA **2** PUERTO GALERA, PHILIPPINES *Photographs: author*

3 Colony on a shallow reef edge, with stunted branches. RYUKYU ISLANDS, JAPAN *Photograph: author*

4 A large colony of laminae and branches. RYUKYU ISLANDS, JAPAN *Photograph: author*

5 Surface detail of branches. RYUKYU ISLANDS, JAPAN *Photograph: author*

Family Poritidae | Genus *Porites*

Porites profundus
Rehberg, 1892

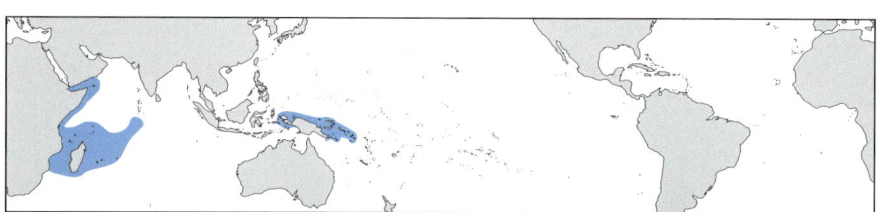

Characters: Colonies are composed of straight, slightly tapered branches. Corallites are deeply excavated and angular, with thin walls. **Colour:** Grey or brown, usually with pale walls and darker centres. **Similar species:** *Porites negrosensis*, which has corallites with thicker, more rounded walls and more twisted branches. **Habitat:** Shallow reef environments. **Abundance:** Uncommon or rare.

Taxonomic reference: Pillai and Scheer (1976).

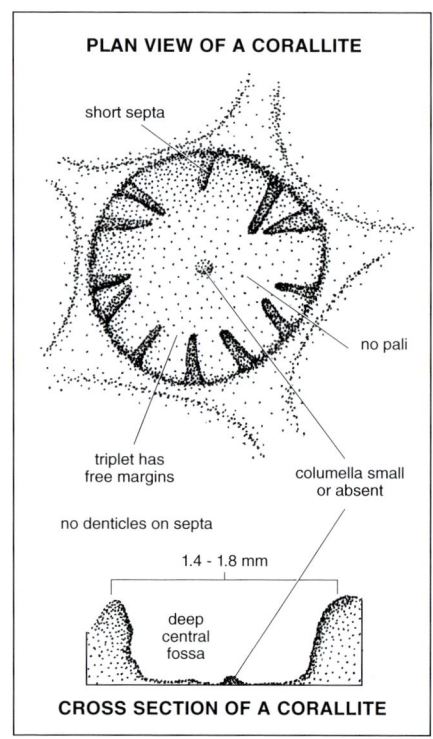

Family Poritidae   Genus *Porites*

1 Small colony with upright branches. ZANZIBAR, TANZANIA *Photograph: author*

2 A compact colony in shallow water. SOLOMON ISLANDS *Photograph: Neville Coleman*

3, 5 Detail of branches. **3** ZANZIBAR, TANZANIA **5** MADAGASCAR *Photographs: author*

4 Branches drooping over an overhang. PAPUA NEW GUINEA *Photograph: author*

Family Poritidae | Genus *Porites*

Porites ornata
Nemenzo, 1971

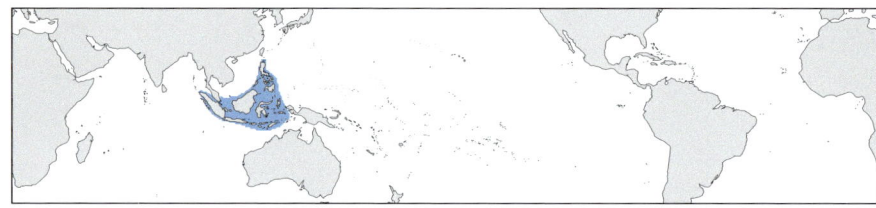

Characters: Colonies are clumps of fused branches which are usually tapered. Corallites are small and superficial. **Colour:** Pale cream. **Similar species:** *Porites cylindrica*, which has thicker, less fused branches. Corallites are similar to those of *P. rus*. **Habitat:** Shallow protected reef environments. **Abundance:** Uncommon or rare.

Taxonomic reference: Nemenzo (1971).

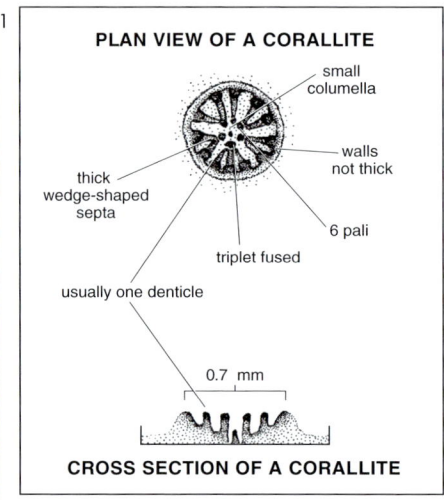

1 *Porites ornata*. Showing the characteristically rough (*Montipora*-like) surface of branches. PUERTO GALERA, PHILIPPINES *Photograph: author*

2 *Porites ornata*. An unusually large colony. BALI, INDONESIA *Photograph: Roger Steene*

Family Poritidae | Genus *Porites*

Porites flavus
Veron, this publication

Characters: Colonies consist of thin tapered branches which irregularly fuse. There are no basal laminae. Corallites are superficial giving branches a smooth appearance. **Colour:** Uniform pale grey with yellow tips to branches. **Similar species:** *Porites sillimaniana*, which has branches of similar size and appearance but has flat laminar bases. See also *P. ornata*. **Habitat:** Shallow protected fringing reefs. **Abundance:** Locally common.

Taxonomic note: See 'New species described in *Corals of the World*' (Veron, in preparation) for further information.

3 *Porites flavus*. Characteristic appearance of a large colony. PAPUA NEW GUINEA *Photograph: author*

4 *Porites flavus*. Branches are twisted and irregularly fuse. They characteristically have yellow tips. PAPUA NEW GUINEA *Photograph: author*

5 *Porites flavus*. Detail of branches. PAPUA NEW GUINEA *Photograph: author*

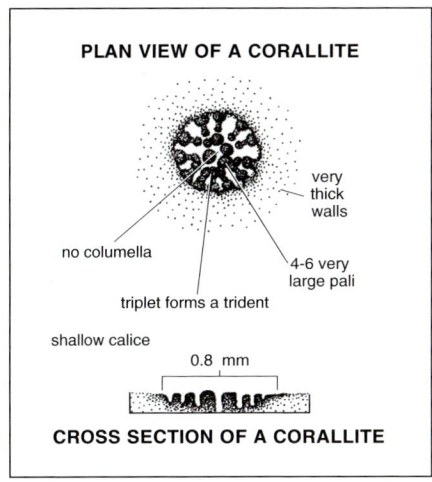

Family Poritidae · Genus *Porites*

Porites rugosa
Fenner and Veron, this publication

Characters: Colonies are branching, the branches being irregularly contorted and fused. Corallites are very small and are deeply embedded in the coenosteum. The coenosteum is very rough and seldom surrounds corallites, but rather forms irregular ridges. **Colour:** Pale brown or pinkish, with yellow tips to branches. **Similar species:** *Porites nigrescens*, which has larger corallites, less contorted branches and a more even coenosteum. **Habitat:** Shallow reef environments. **Abundance:** Uncommon.

Taxonomic note: See 'New species described in Corals of the World' (Veron, in preparation) for further information.

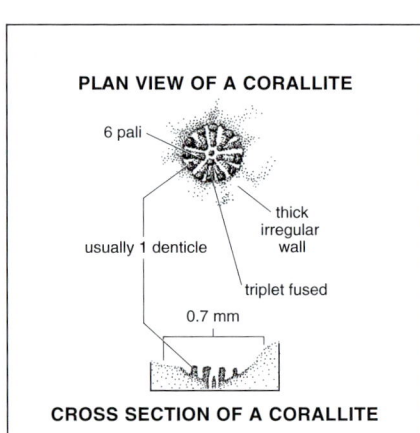

1 *Porites rugosa*. A small colony with straight branches. SULAWESI, INDONESIA. *Photograph: Doug Fenner*

2 *Porites rugosa*. Surface detail of branches showing the rough coenosteum and small corallites. SULAWESI, INDONESIA *Photograph: Doug Fenner*

3 *Porites harrisoni*. Side view of a columnar colony. KUWAIT *Photograph: Peter Harrison*

4 *Porites harrisoni*. Colonies have an irregular to nodular upper surface. KUWAIT *Photograph: Peter Harrison*

Family Poritidae | Genus *Porites*

Porites harrisoni
Veron, this publication

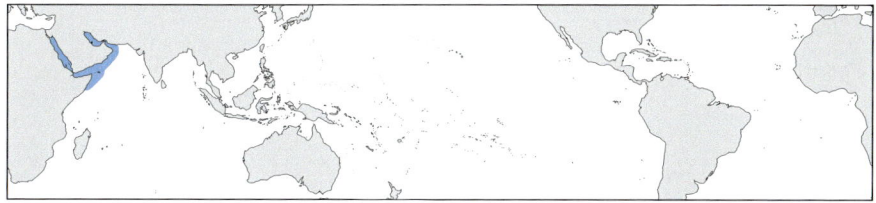

Characters: Colonies are usually less than one metre across. They have a wide range of submassive, nodular, columnar and branching growth-forms on a broad encrusting base. **Colour:** Commonly dark brown, also pink or blue. **Similar species:** *Porites compressa*, which has distinct corallite characters. See also *P. nodifera*, which has a similar growth-form. **Habitat:** Shallow fringing reefs. **Abundance:** Locally common and may be a dominant species.

Taxonomic note: This species in the western Indian Ocean has been called *Porites compressa*. See 'New species described in *Corals of the World*' (Veron, in preparation) for further information. **Distribution note:** This species may occur elsewhere in the western Indian Ocean (see *Porites compressa* in Scheer and Pillai, 1983). **Identification guide:** Carpenter et al. (1997, as *Porites compressa*).

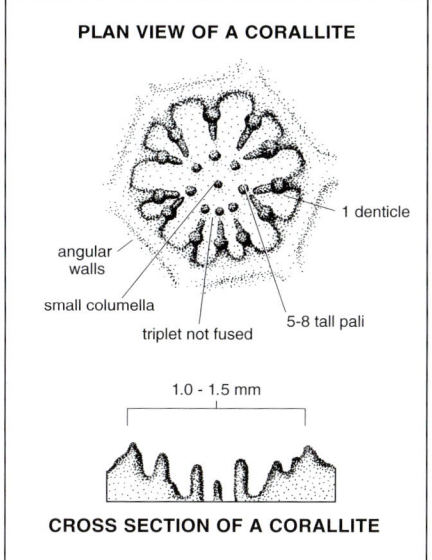

Porites compressa
Dana, 1846

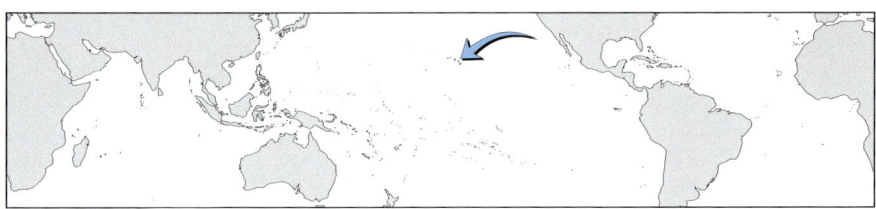

Characters: Colonies may form large patches of reefs. Branches are cylindrical and commonly fuse. Growth-forms and corallite characters are extremely variable so much so that single reef patches are composites of distinct races. **Colour:** Mostly dull greys and browns. **Similar species:** *Porites cylindrica* and *P. harrisoni*, which have a similar appearance underwater. **Habitat:** Shallow protected reef and lagoon environments. **Abundance:** The dominant coral in Kaneohe Bay, Hawaii.

Taxonomic note: *Porites duerdeni* (Vaughan, 1907) may be a valid Hawaiian endemic species like *P. compressa*. It is described as having thicker branches and has minor corallite distinctions.
Taxonomic reference: Vaughan (1907). **Identification guide:** Maragos (1977).

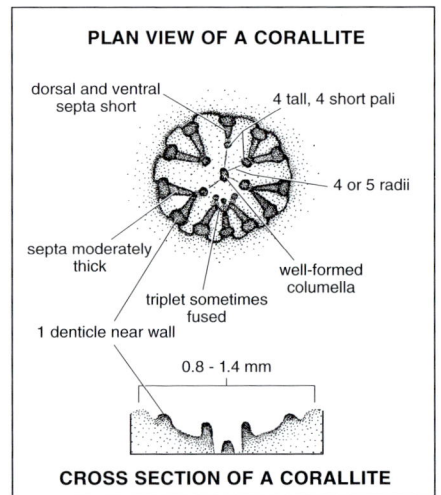

1 Forming large stands. HAWAII *Photograph: Jim Maragos*

2 Side view of branches. HAWAII *Photograph: Don Potts*

3 A small clump. HAWAII *Photograph: Cindy Hunter*

4 A colony of highly fused branches. HAWAII *Photograph: Don Potts*

Family Poritidae   Genus *Stylaraea*

GENUS *STYLARAEA*
Milne Edwards and Haime, 1851

Characters, abundance and distribution: This genus has only one species, see *Stylaraea punctata*.
Earliest fossil record: Plio-Pleistocene of Guam.

Stylaraea punctata
(Linneaus, 1758)

Characters: Colonies are circular, encrusting, less than 15 millimetres across with uniformly spaced corallites. Septa are in two cycles of 6 each. The columella is an irregular pinnule. **Colour:** Pale brown with white septa and columellae. **Similar species:** *Stylaraea* resembles *Porites* except that septa are short, are in two cycles and do not fuse. See also *Poritipora paliformis*. **Habitat:** Subtidal environments usually uninhabited by other corals. **Abundance:** Usually rare.

Taxonomic references: Crossland (1952), Veron and Pichon (1982).

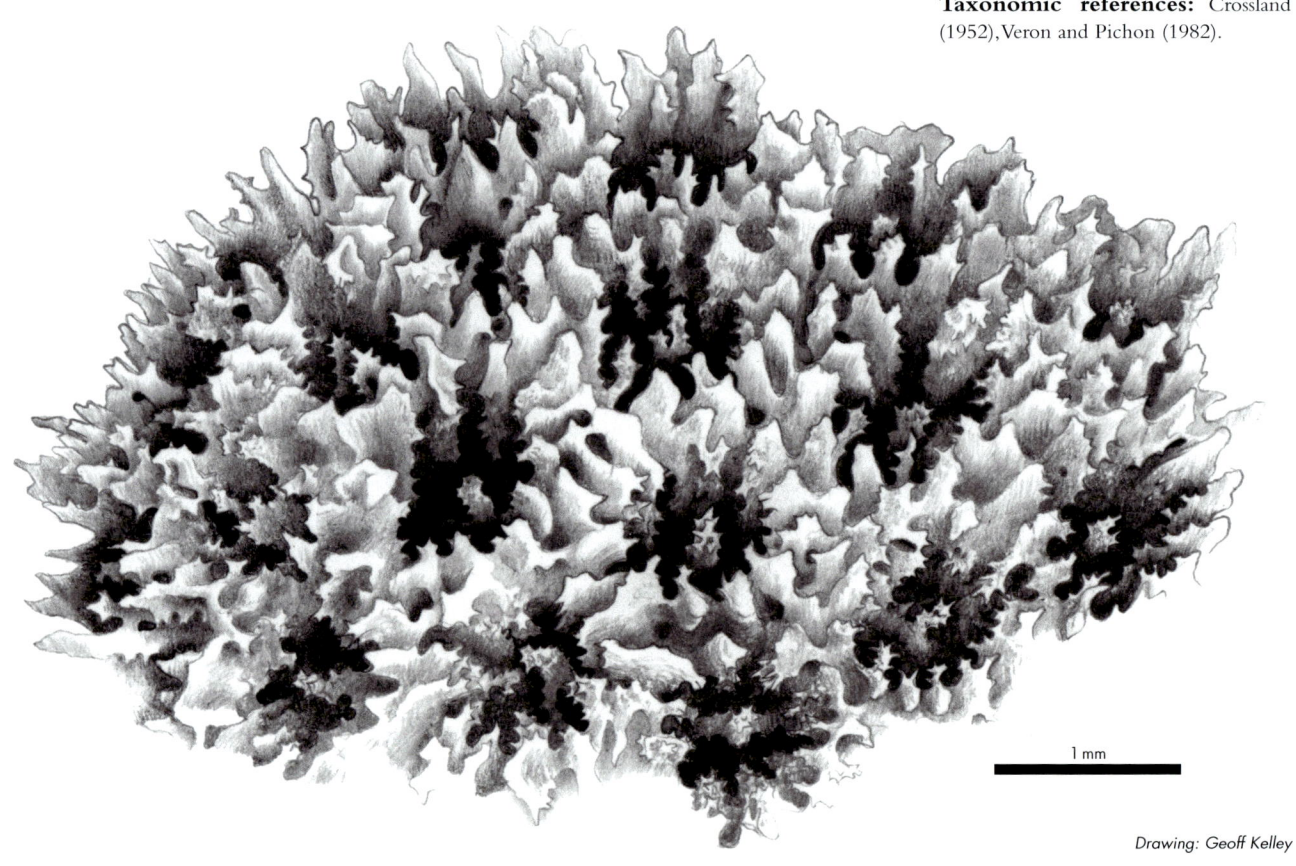

1 mm

Drawing: Geoff Kelley

Family Poritidae   Genus *Poritipora*

GENUS *PORITIPORA*
Veron, this publication

Characters, abundance and distribution: This genus has only one species, see *Poritipora paliformis*.
Fossil record: None.

Poritipora paliformis
Veron, this publication

Characters: Colonies are massive, usually hemispherical, and may be several metres across. The surface is smooth to undulating. Corallites are deeply excavated. There are two cycles of 12 septa each, 6 pali and no columella. Walls are thin. **Colour:** Dull brown. **Similar species:** Resembles *Porites densa*, which has slightly smaller corallites, but this species is easily recognised underwater. See also the astrocoeniid *Stylocoeniella armata*. **Habitat:** Shallow reef environments and lagoons. **Abundance:** Rare.

Taxonomic note: See 'New species described in *Corals of the World*' (Veron, in preparation) for further information. **Type species:** *Poritipora paliformis*.

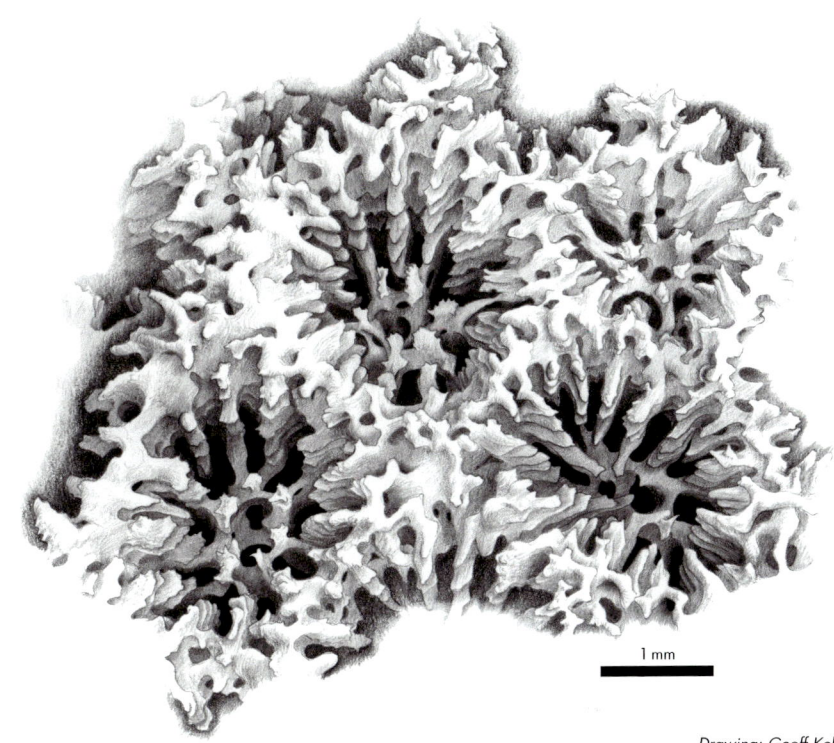

Drawing: Geoff Kelley

1 *Stylaraea punctata*. Colonies are massive but always inconspicuous and small. GUAM
Photograph: Gustav Paulay

2 *Stylaraea punctata*. Detail of corallites. GUAM Photograph: Gustav Paulay

3 *Poritipora paliformis*. Corallite detail. SRI LANKA Photograph: author

4 *Poritipora paliformis*. In a lagoon colonies have a cellular appearance. SRI LANKA Photograph: author

Family Poritidae | Genus Goniopora

GENUS *GONIOPORA*
Blainville, 1830

Characters: Colonies are usually branching, columnar or massive but may be encrusting. Corallites have thick but porous walls, and calices are filled with compacted septa and columellae. Polyps are long and fleshy and tentacles are normally extended day and night. Polyps have 24 tentacles. Different species have polyps of different shapes and colours, which allow them to be identified underwater. **Similar genus:** *Goniopora* is like *Alveopora* but skeletal structures are better developed. Polyps of these genera are similar except that *Goniopora* has 24 tentacles while *Alveopora* has 12. **Earliest fossil record:** Cretaceous of the Tethys (doubtful record), Eocene of the Caribbean.

Taxonomic reference: Veron and Pichon (1982). **Summary key:** See pp457-8.

Identification of *Goniopora* species.

Goniopora species are usually easier to identify underwater than they are from dried skeletons. They form six groups based on growth-form and corallite size:

Group 1: Massive species with large (over 5 mm diameter) corallites, page 350.

Group 2: Branching or columnar species with large (over 5 mm diameter) corallites, page 354.

Group 3: Encrusting species, page 358.

Group 4: Massive species with medium (3-5 mm diameter) corallites, page 362.

Family Poritidae | Genus *Goniopora*

Group 5: Branching or columnar species with medium (3–5 mm diameter) corallites, page 368.

Group 6: Species with small (less than 3 mm diameter) corallites, page 375.

1 *Goniopora* (left) is readily distinguished from *Alveopora* (right) by the number of tentacles (24 and 12 respectively). DAMPIER ARCHIPELAGO, WESTERN AUSTRALIA *Photograph: author*

2 Polyp size helps with underwater identification. Here, polyp size helps separate *Goniopora pandoraensis* (right) from *G. columna* (left). GREAT BARRIER REEF, AUSTRALIA *Photograph: author*

3 Colony shape is often hard to determine if polyps are extended. Different branches of this single *Goniopora lobata* colony have polyps retracted (left, revealing the colony shape) and extended (right, masking the colony shape). GREAT BARRIER REEF, AUSTRALIA *Photograph: author*

4 Different *Goniopora* species often occur clustered together. Here, *G. djiboutiensis* (left), *G. eclipsensis* (centre) and *G. pandoraensis* (right) form a continuous clump. GREAT BARRIER REEF, AUSTRALIA *Photograph: Ed Lovell*

5 *Goniopora* polyps of many species have distinctively coloured oral cones and tentacle tips. FLORES SEA, INDONESIA *Photograph: Valerie Taylor*

Family Poritidae | Genus *Goniopora*

> **Group 1: Massive species with large (over 5 mm diameter) corallites**

Goniopora pendulus
Veron, 1985

Characters: Colonies are hemispherical. Columellae are large. Polyps are large, with drooping tentacles. **Colour:** Brown or greenish-brown. **Similar species:** *Goniopora stokesi*. See also *G. djiboutiensis*, *G. ciliatus* and *G. lobata*, which have smaller polyps. *Alveopora gigas* has polyps of similar size and appearance except that polyps have twelve tentacles. **Habitat:** Protected turbid reef environments. **Abundance:** Common at the Houtman Abrolhos Islands, south-west Australia, uncommon elsewhere.

Taxonomic reference: Veron (1985). **Identification guide:** Veron (1986).

1 *Goniopora pendulus*. This species forms massive colonies which have a shaggy appearance due to long polyps and tentacles. HONSHU, JAPAN *Photograph: author*

2 *Goniopora pendulus*. Polyp detail. GREAT BARRIER REEF, AUSTRALIA *Photograph: author*

Goniopora djiboutiensis
Vaughan, 1907

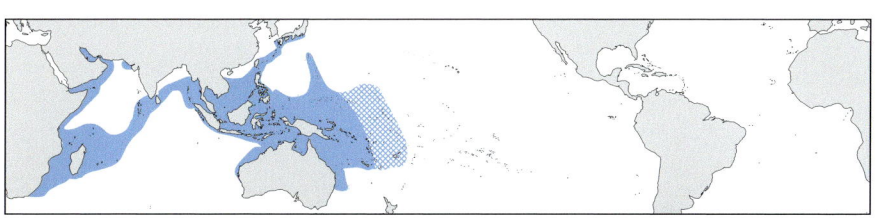

Characters: Colonies are submassive or are short thick columns. Columellae are prominent, dome-shaped, and commonly divided into six parts, each part having a deltaic pattern of four septa. Polyps have large oral cones. **Colour:** Pale or dark brown or green. Oral cones are usually white or blue (which may photograph pink). **Similar species:** *Goniopora pendulus* and *G. columna*, which also have large oral cones. The former has bigger polyps with much longer tentacles, the latter has a more columnar growth-form. **Habitat:** May form large single species stands in turbid water. **Abundance:** Common.

Distribution note: Records from the central west-Pacific are doubtful. **Taxonomic reference:** Veron and Pichon (1982). **Identification guides:** Veron (1986), Sheppard and Sheppard (1991), Nishihira and Veron (1995).

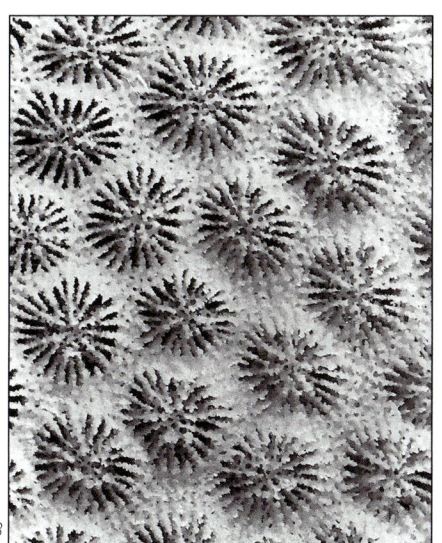

3 *Goniopora djiboutiensis* (centre) with *G. pandoraensis* (left and right). GREAT BARRIER REEF, AUSTRALIA *Photograph: Ed Lovell*

4 *Goniopora djiboutiensis*. A colony 20 cm across with polyps fully extended. PEMBA ISLAND, TANZANIA *Photograph: author*

5 *Goniopora djiboutiensis*. The typical appearance of polyps. GREAT BARRIER REEF, AUSTRALIA *Photograph: Ed Lovell*

6 *Goniopora djiboutiensis*. Polyp detail. GREAT BARRIER REEF, AUSTRALIA *Photograph: Ed Lovell*

Family Poritidae | Genus *Goniopora*

Goniopora stokesi
Milne Edwards and Haime, 1851

Characters: Colonies are free-living or attached, hemispherical or (rarely) have short thick columns. Calices have high walls which have a ragged appearance. Columellae are broad and irregular. Small satellite colonies often occur embedded in the living tissue of parent colonies. Polyps are of mixed sizes, the larger being elongate. **Colour:** Pale brown or green, usually with green tentacle tips. **Similar species:** *Goniopora pendulus*, which forms larger colonies with polyps of equal size and does not form satellite colonies. See also *G. lobata*. **Habitat:** Usually found free-living, on soft substrates. **Abundance:** Usually uncommon, but conspicuous.

Taxonomic reference: Veron and Pichon (1982). **Identification guides:** Veron (1986), Sheppard and Sheppard (1991), Nishihira and Veron (1995).

1 Colonies are commonly hemispherical clumps growing on soft substrates in turbid lagoons. CALAMIAN ISLANDS, PHILIPPINES *Photograph: author*

2 Showing satellite colonies still attached to the parent colony. When these become free-living they sometimes cover the substrate. MADAGASCAR *Photograph: author*

3 Clusters of satellite colonies intermixed with parent polyps at different stages of tentacle retraction. GREAT BARRIER REEF, AUSTRALIA *Photograph: Ed Lovell*

4 Colonies frequently have polyps retracted during the day. These look very unlike those with extended polyps. GREAT BARRIER REEF, AUSTRALIA *Photograph: Valerie Taylor*

5 When polyps are fully extended, tentacles resemble those of *G. pendulus*. HOUTMAN ABROLHOS ISLANDS, SOUTH-WEST AUSTRALIA *Photograph: author*

Family Poritidae | Genus *Goniopora*

Group 2: Branching or columnar species with large (over 5 mm diameter) corallites

Goniopora lobata
Milne Edwards and Haime, 1860

Characters: Colonies are hemispherical or, more usually, form short thick columns. Columellae and oral cones are small. Polyps are elongate when fully extended. **Colour:** Usually brown, yellow or green, often with contrasting oral cones and tentacle tips. **Similar species:** *Goniopora columna*, which has large columellae and oral cones. *Goniopora stokesi* has high ragged walls and broad columellae. **Habitat:** Forms large single species stands, especially in turbid water. **Abundance:** Common.

Taxonomic reference: Veron and Pichon (1982). **Identification guides:** Veron (1986), Nishihira and Veron (1995), Carpenter *et al.* (1997).

1 *Goniopora lobata.* The typical appearance of a large colony. PAPUA NEW GUINEA *Photograph: author*

2 *Goniopora lobata.* Polyps have small oral discs, a useful character in underwater identification. GREAT BARRIER REEF, AUSTRALIA *Photograph: author*

3 *Goniopora lobata.* The columnar structure of large colonies is masked if polyps are fully extended. RYUKYU ISLANDS, JAPAN *Photograph: author*

4 *Goniopora sultani.* Expanded polyps of a large colony. SINAI PENINSULA, EGYPT *Photograph: author*

x3

Goniopora sultani
Veron, DeVantier and Turak, this publication

Characters: Colonies are thick short columns, oval in transverse section. Corallites are similar on the tops and sides of columns. They are very large (7-8 mm diameter), compact, with thin walls and thus have have an angular shape. Columellae are broad and are mostly composed of six septal deltas. Septa are short and uniform in size and spacing. Polyps are elongate when fully extended. **Colour:** Pale cream. **Similar species:** *Goniopora lobata* and *G. columna*, which have similar columnar branches but smaller corallites. See also *G. stokesi* which has high, uneven septal walls. This species has the largest corallites of all the *Goniopora*. **Habitat:** Upper reef slopes. **Abundance:** Uncommon.

Taxonomic note: See 'New species described in *Corals of the World*' (Veron, in preparation) for further information.

Family Poritidae | Genus *Goniopora*

Goniopora columna
Dana, 1846

Characters: Colonies are short columns, oval in transverse section. Corallites near the tops of columns have fine irregular septa and diffuse columellae. Those on the sides of columns have broad compact columellae and short septa. Colonies have large polyps with large oral cones. **Colour:** Brown, green or yellow, usually with white oral cones. Contracted polyps usually have distinctly different colours. **Similar species:** *Goniopora stokesi*, which has larger polyps and a ragged wall structure. See also *G. lobata*. **Habitat:** Forms large single species stands, especially in turbid water. **Abundance:** Common.

Taxonomic reference: Veron and Pichon (1982). **Identification guides:** Veron (1986), Sheppard and Sheppard (1991), Nishihira and Veron (1995).

Family Poritidae Genus *Goniopora*

1 The columnar structure of large colonies is masked if polyps are fully extended. HOUTMAN ABROLHOS ISLANDS, SOUTH-WEST AUSTRALIA *Photograph: author*

2 Polyps at varying stages of extension. GREAT BARRIER REEF, AUSTRALIA *Photograph: author*

3, 4 This species has a wide range of colours, commonly with matching oral discs and tentacle tips or just white oral cones. **3** GREAT BARRIER REEF, AUSTRALIA **4** CALAMIAN ISLANDS, PHILIPPINES *Photographs: 3 Ed Lovell 4 author*

5 Polyps retracted after being disturbed. CALAMIAN ISLANDS, PHILIPPINES *Photograph: author*

6 The large oral discs are a useful character in underwater identification. GREAT BARRIER REEF, AUSTRALIA *Photograph: Mary Stafford-Smith*

Family Poritidae	Genus *Goniopora*

Group 3: Encrusting species

Goniopora somaliensis
Vaughan, 1907

Characters: Colonies are thick or thin encrusting plates, sometimes over 2 metres across, with shallow calices forming a smooth surface. Corallites are 3-5 millimetres diameter. Polyps are short, giving colonies a carpet-like appearance, but are often retracted during the day. **Colour:** Usually grey or brown, often with polyps and surrounding tissue of contrasting colour. **Similar species:** *Goniopora polyformis*, which does not form big colonies and has irregular upgrowths. *Alveopora spongiosa* may have a similar growth-form and colouration. **Habitat:** Sheltered reef environments, especially beneath overhangs. **Abundance:** Uncommon but conspicuous.

Taxonomic references: Wells (1954), Veron and Pichon (1982). **Identification guides:** Randall and Myers (1983), Veron (1986), Sheppard and Sheppard (1991), Nishihira and Veron (1995).

Family Poritidae | Genus *Goniopora*

1, 2 Large lobed colonies are distinctive. When polyps are extended, these resemble colonies of *Alveopora spongiosa* more than other *Goniopora* species. **1** RYUKYU ISLANDS, JAPAN **2** MADAGASCAR *Photographs: author*

3 Colonies frequently have polyps partly retracted and are pale brown in colour. PUERTO GALERA, PHILIPPINES *Photograph: author*

4 Showing the progressive retraction of polyps after touching the lower part of the colony. This causes a major colour change. SINAI PENINSULA, EGYPT *Photograph: author*

5 Small lobed colonies such as this are distinctive. VIETNAM *Photograph: author*

6 Detail of extended polyps. ZANZIBAR, TANZANIA *Photograph: author*

Family Poritidae | Genus *Goniopora*

Goniopora tenella
(Quelch, 1886)

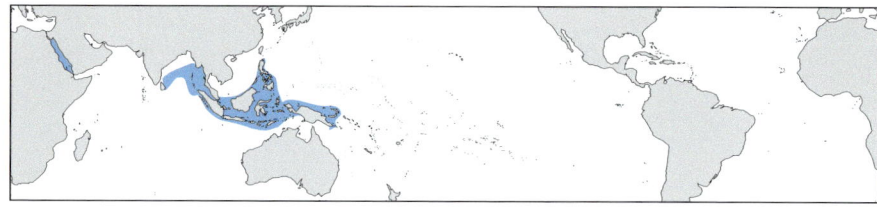

Characters: Colonies are encrusting or explanate plates. Corallites are shallow and have thin walls. Septa are numerous and even. Corallites are 5-7 millimetres diameter. Polyps are short, without tubular trunks. **Colour:** Brown, grey or mottled. **Similar species:** *Goniopora somaliensis*, which has smaller corallites and a distinctive colouration. **Habitat:** Shallow reef environments. **Abundance:** Rare.

Taxonomic reference: Scheer and Pillai (1983).

Goniopora polyformis
Zou, 1980

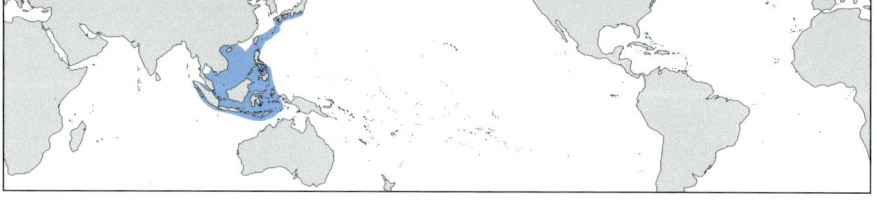

Characters: Colonies are encrusting with contorted upgrowths. Corallites are irregular in shape and have thin walls. Septa are numerous and even. **Colour:** Brown or grey, usually with white oral cones. **Similar species:** *Goniopora fruticosa*, which has smaller corallites and a distinctive colouration. **Habitat:** Shallow reef environments. **Abundance:** Rare.

Taxonomic references: Zou (1980), Veron (1992). **Identification guide:** Nishihira and Veron (1995).

Goniopora albiconus
Veron, this publication

Characters: Colonies are encrusting, forming thin irregular laminae. Corallites are shallow, polygonal, and have thin walls. Corallites vary greatly in size. Septa are irregularly fused but do not form deltas. Columellae are very small. Polyps are short and even. Oral cones are exceptionally large while tentacles are short and thin. Polyps retract rapidly if disturbed. **Colour:** White oral cones are conspicuous. **Similar species:** Resembles zoanthids underwater. *Goniopora columna* also has large white oral cones but polyps are elongate and tentacles are thicker. **Habitat:** Shallow reef environments. **Abundance:** Sometimes common.

Taxonomic note: See 'New species described in *Corals of the World*' (Veron, in preparation) for further information.

1 *Goniopora tenella*. An encrusting colony. PAPUA NEW GUINEA *Photograph: author*

2 *Goniopora polyformis*. The growth-form is typically irregularly encrusting. RYUKYU ISLANDS, JAPAN *Photograph: author*

3 *Goniopora albiconus*. A large massive colony. MADAGASCAR *Photograph: author*

4 *Goniopora albiconus*. Colonies form only a thin veneer, resembling zoanthids. The large white coral cones are very conspicuous. MALDIVE ISLANDS *Photograph: Neville Coleman*

5 *Goniopora albiconus*. With polyps fully extended. KOMODO, INDONESIA *Photograph: Valerie Taylor*

6 *Goniopora albiconus*. Colony detail with polyps retracted. MALDIVE ISLANDS *Photograph: Neville Coleman*

Family Poritidae   Genus *Goniopora*

> Group 4: Massive species with medium (3–5 mm diameter) corallites

Goniopora norfolkensis
Veron and Pichon, 1982

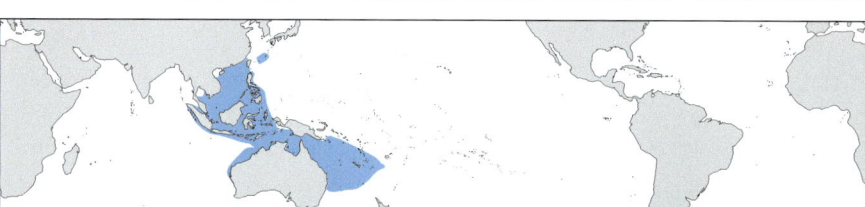

Characters: Colonies are hemispherical to encrusting. Calices have small columellae and long, regular, steeply plunging septa. Paliform lobes are absent. Columellae are small or absent. **Colour:** Usually greenish brown with distinctively coloured oral discs and pale tips to the tentacles. **Similar species:** *Goniopora pendulus*, which has larger polyps and dissimilar skeletal characters. See also *G. tenuidens* and *G. minor*, which are both distinguished by the presence of paliform lobes and/or broad columellae. **Habitat:** Shallow reef environments. **Abundance:** Uncommon except at Norfolk Island, western Pacific.

Taxonomic reference: Veron and Pichon (1982). **Identification guide:** Veron (1986).

1 *Goniopora norfolkensis*. A large massive colony. NORFOLK ISLAND, WESTERN PACIFIC Photograph: author

2 *Goniopora norfolkensis*. Polyps in the process of retracting after being disturbed. NORFOLK ISLAND, WESTERN PACIFIC Photograph: author

Family Poritidae | Genus *Goniopora*

3 *Goniopora norfolkensis*. Polyps have long shaggy tentacles. NORFOLK ISLAND, WESTERN PACIFIC Photograph: author

4 *Goniopora norfolkensis*. Polyp detail. NORFOLK ISLAND, WESTERN PACIFIC Photograph: author

5 *Goniopora norfolkensis*. Encrusting colonies with some polyps very extended. NORFOLK ISLAND, WESTERN PACIFIC Photograph: author

6 *Goniopora cellulosa*. This species forms small submassive colonies. KYUSHU, JAPAN Photograph: Moritaka Nishihira

Goniopora cellulosa
Veron, 1990

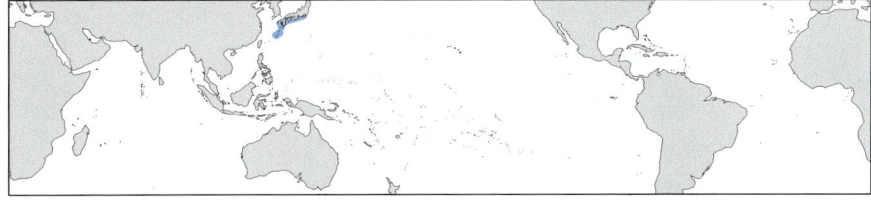

Characters: Colonies are massive. Walls are thin and perforated. Calices vary greatly in depth. Septa are short and irregularly fused. Columellae are irregular. **Colour:** Brown or tan. **Similar species:** *Goniopora minor*. **Habitat:** Shallow reef environments. **Abundance:** Uncommon.

Taxonomic reference: Veron (1990a). **Identification guide:** Nishihira and Veron (1995).

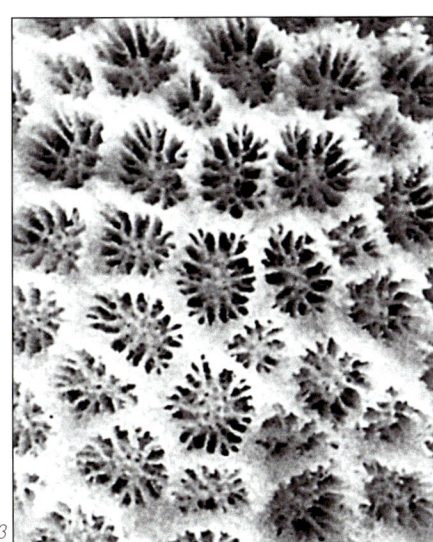

Goniopora tenuidens
(Quelch, 1886)

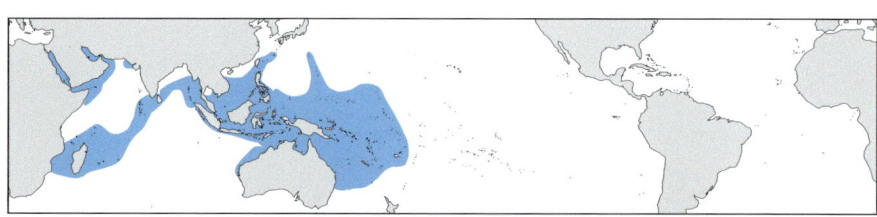

Characters: Colonies are massive, hemispherical or irregular. Corallites are rounded with thin walls and have six prominent paliform lobes. Polyps are closely compacted, uniform in length and have tentacles of uniform length. **Colour:** Uniform blue (which may photograph pink), green or brown, sometimes with white tips to the tentacles. **Similar species:** *Goniopora minor*, which has thick pali forming a crown. See also *G. pearsoni*; and *G. norfolkensis* which lacks paliform lobes. **Habitat:** Subtidal reef environments, especially lagoons. **Abundance:** Common.

Taxonomic reference: Veron and Pichon (1982). **Identification guides:** Veron (1986), Sheppard and Sheppard (1991), Nishihira and Veron (1995).

1 *Goniopora tenuidens*. Colonies are commonly hemispherical, especially in shallow water. GREAT BARRIER REEF, AUSTRALIA *Photograph: Ed Lovell*

2 *Goniopora tenuidens*. The species is readily recognised by the uniform length of the polyp tentacles. GREAT BARRIER REEF, AUSTRALIA *Photograph: author*

3 *Goniopora tenuidens*. Surface view of the polyps. GREAT BARRIER REEF, AUSTRALIA *Photograph: author*

4 *Goniopora pearsoni*. Large colonies on a shallow reef front. SINAI PENINSULA, EGYPT *Photograph: author*

5 *Goniopora pearsoni*. Polyps are long, with tubular trunks. SINAI PENINSULA, EGYPT *Photograph: author*

6 *Goniopora pearsoni*. Polyp detail. SINAI PENINSULA, EGYPT *Photograph: author*

Goniopora pearsoni
Veron, this publication

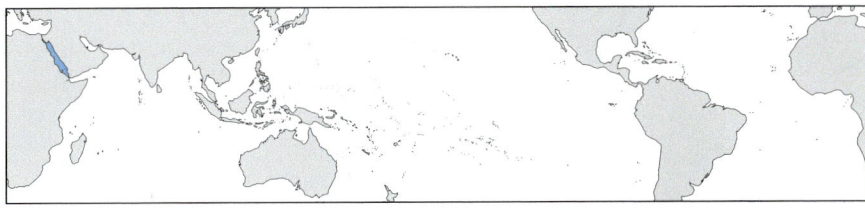

Characters: Colonies are massive, hemispherical or irregular. Corallites are rounded with thick walls and have six prominent paliform lobes. Polyps are elongate. **Colour:** Uniform grey with blue mouths (which may photograph pink). **Similar species:** *Goniopora tenuidens*, which has closely packed uniform polyps and corallites with thin walls. See also *G. djiboutiensis*, which has larger polyps. **Habitat:** Shallow reef environments. **Abundance:** Common.

Taxonomic note: See 'New species described in *Corals of the World*' (Veron, in preparation) for further information.

Family Poritidae | Genus *Goniopora*

Goniopora minor
Crossland, 1952

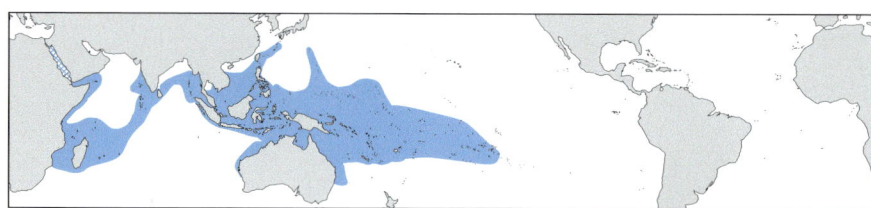

Characters: Colonies are hemispherical or encrusting. Calices are circular in outline, with thick walls. There are usually six thick pali which are in contact, forming a crown. All septal structures are heavily granulated. **Colour:** Brown or green, usually with distinctively coloured oral discs and pale tips to the tentacles. **Similar species:** *Goniopora tenuidens*, which has blunt tentacles of uniform length. **Habitat:** Subtidal reef environments, especially lagoons. **Abundance:** Common.

Distribution note: Records from the Red Sea are doubtful. **Taxonomic reference:** Veron and Pichon (1982). **Identification guides:** Randall and Myers (1983), Veron (1986), Sheppard and Sheppard (1991), Nishihira and Veron (1995).

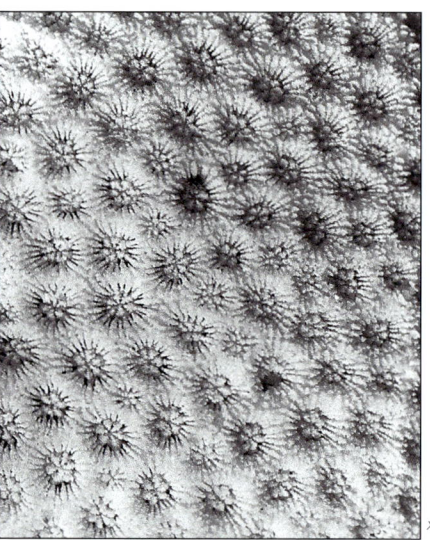

x3

Family Poritidae — Genus *Goniopora*

1 Colonies are massive or lobed. NINGALOO REEFS, WESTERN AUSTRALIA *Photograph: author*

2 A typically even cover of fully extended polyps. HOUTMAN ABROLHOS ISLANDS, SOUTH-WEST AUSTRALIA *Photograph: author*

3 Colonies commonly have partly retracted polyps. RYUKYU ISLANDS, JAPAN *Photograph: author*

4 A carpet of tentacles. CALAMIAN ISLANDS, PHILIPPINES *Photograph: author*

5 Polyps have pointed tentacle tips. GREAT BARRIER REEF, AUSTRALIA *Photograph: Ed Lovell*

> Group 5: Branching or columnar species with medium (3-5 mm diameter) corallites

Goniopora planulata
(Ehrenberg, 1834)

Characters: Colonies are submassive with small compacted columns or mounds. Corallites have thin walls. Septa are thin and irregular and do not form deltas except in colonies in very shallow water. Paliform lobes form a diffuse crown. Polyps are short with tentacles of uniform length. **Colour:** Dark grey-brown, usually with white mouths. **Similar species:** *Goniopora columna*, which forms thicker columns and has larger corallites. **Habitat:** Shallow reef environments. **Abundance:** Usually uncommon.

Taxonomic reference: Scheer and Pillai (1983).

Family Poritidae   Genus *Goniopora*

1 A large submassive colony. SEYCHELLES *Photograph: author*

2, 3 Colonies forming short compact columns. **2** SEYCHELLES **3** CALAMIAN ISLANDS, PHILIPPINES *Photographs: author*

4 Detail of columns. SEYCHELLES *Photograph: author*

5 Polyp detail. SEYCHELLES *Photograph: author*

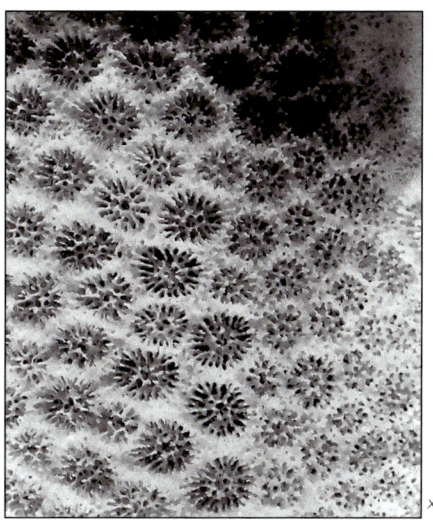

369

Family Poritidae | Genus *Goniopora*

Goniopora pandoraensis
Veron and Pichon, 1982

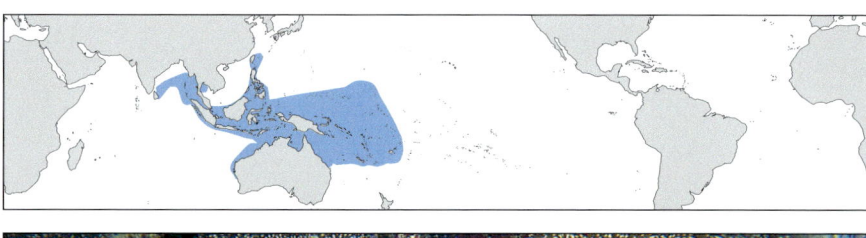

Characters: Colonies consist of small branching columns, usually oval in transverse section. Corallites have thick walls and septa. Six thick paliform lobes form a crown. **Colour:** Dark grey-brown with white mouths and tentacle tips. **Similar species:** *Goniopora pandoraensis* and *G. eclipsensis* can only reliably be distinguished if both species occur together or by the colour of the polyps. *Goniopora columna* has larger columns and larger corallites. **Habitat:** Shallow turbid reef environments. **Abundance:** Usually uncommon.

Taxonomic reference: Veron and Pichon (1982). **Identification guides:** Veron (1986), Nishihira and Veron (1995).

1 *Goniopora pandoraensis.* Colonies are irregularly branched. GREAT BARRIER REEF, AUSTRALIA *Photograph: author*

2 *Goniopora pandoraensis.* A colony with some bleached branches that have no symbiotic algae. GREAT BARRIER REEF, AUSTRALIA *Photograph: Ed Lovell*

3 *Goniopora pandoraensis.* Adjacent colonies with polyps of different colours. GREAT BARRIER REEF, AUSTRALIA *Photograph: author*

Family Poritidae | Genus *Goniopora*

Goniopora eclipsensis
Veron and Pichon, 1982

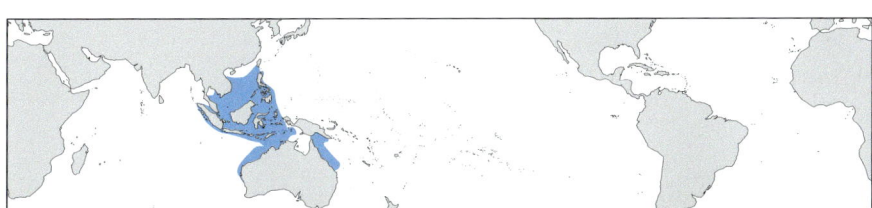

Characters: Colonies are small branching cylindrical columns. Corallites are circular in outline, with thick walls and septa. Six paliform lobes form a crown. **Colour:** Uniform brown, with or without paler tentacle tips. **Similar species:** *Goniopora palmensis* and *G. pandoraensis*, which are best distinguished by minor differences in growth-form. **Habitat:** Shallow reef environments. **Abundance:** Uncommon.

Taxonomic reference: Veron and Pichon (1982). **Identification guide:** Veron (1986).

4 *Goniopora eclipsensis* (right) with *G. pandoraensis* (left), showing differences in polyp structure and colour. GREAT BARRIER REEF, AUSTRALIA *Photograph: Ed Lovell*

5 Polyps of *Goniopora eclipsensis* (right) compared with the much larger *G. lobata* (left). GREAT BARRIER REEF, AUSTRALIA *Photograph: Ed Lovell*

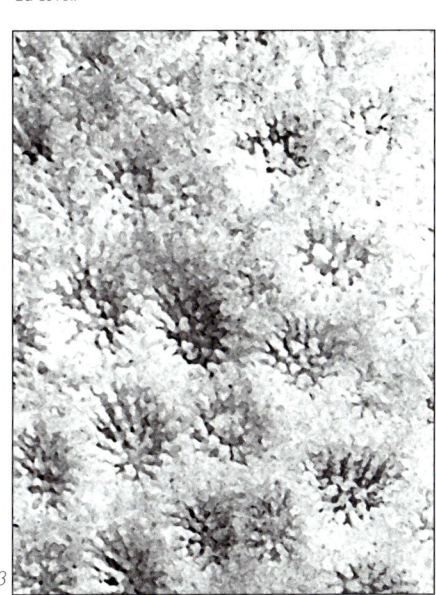

Goniopora ciliatus
Veron, this publication

Characters: Colonies are submassive or columnar. Columns are short and usually form clumps. Corallites are uniform, and circular in outline. Septa are widely spaced and irregular. Columellae are small. Polyps have long thin tentacles. **Colour:** Pale brown. **Similar species:** *Goniopora eclipsensis* which has short tentacles and well developed columellae. See also *G. pendulus*. No other columnar species has medium sized corallites with such long tentacles. **Habitat:** Shallow reef environments exposed to turbulence. **Abundance:** Sometimes common.

Taxonomic note: See 'New species described in *Corals of the World*' (Veron, in preparation) for further information.

1 Colonies may form extensive stands of large rounded clumps. Sinai Peninsula, Egypt *Photograph: author*

2 *Goniopora ciliatus* (below) next to *G. pearsoni*. Sinai Peninsula, Egypt *Photograph: author*

3 Rounded clumps are each composed of a cluster of short branches. Sinai Peninsula, Egypt *Photograph: Mary Stafford-Smith*

4 A colony surface with polyps at different stages of extension. Sinai Peninsula, Egypt *Photograph: author*

5 Polyps have long fine tentacles making this species distinctive underwater. Sinai Peninsula, Egypt *Photograph: author*

Goniopora palmensis
Veron and Pichon, 1982

Characters: Colonies are submassive or branching. Branches seldom fuse. Corallites are uniform, and circular in outline. Polyps have large oral cones and short pointed tentacles. **Colour:** Brown, green or cream, often with white tentacle tips. **Similar species:** *Goniopora eclipsensis*. See also *G. fruticosa*, which has interlocking branches and is always dark brown with white oral discs. **Habitat:** Shallow reef environments. **Abundance:** Common.

Taxonomic reference: Veron and Pichon (1982). **Identification guide:** Veron (1986).

1, 2 *Goniopora palmensis*. Polyps have pointed tentacles. GREAT BARRIER REEF, AUSTRALIA *Photographs: Ed Lovell*

3 *Goniopora burgosi*. A large lobed colony. MADAGASCAR *Photograph: author*

4 *Goniopora burgosi*. Surface of a colony. Polyps have fully extended tentacles partly masking the underlying growth-form. MADAGASCAR *Photograph: author*

5 *Goniopora burgosi*. Polyps are retracted on the lower parts of the colony and fully extended on the upper part. RYUKYU ISLANDS, JAPAN *Photograph: author*

6 *Goniopora burgosi*. Surface detail of a massive colony. SINAI PENINSULA, EGYPT *Photograph: author*

| Group 6: Species with small (less than 3 mm diameter) corallites |

Goniopora burgosi
Nemenzo, 1955

Characters: Colonies are submassive, usually lobed, and may be large. Corallites are deep, with short septa. **Colour:** Polyps are a mass of grey or brown tentacles when extended, the surrounding tissue is brown. **Similar species:** *Goniopora stutchburyi*, which has a similar growth-form and colouration but polyps are much smaller. See also *G. planulata*, which forms upright columns. **Habitat:** Shallow reef environments. **Abundance:** Uncommon but distinctive.

Taxonomic references: Nemenzo (1955), Veron and Hodgson (1989). **Identification guide:** Nishihira and Veron (1995).

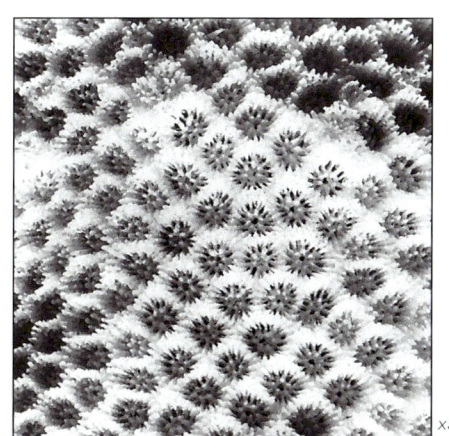

Goniopora savignyi
Dana, 1846

Characters: Colonies are columnar, usually with thin explanate basal plates. They are usually lobed, and may be large. Corallites are deep, polygonal to rounded, with short septa. Paliform lobes are prominent. Columellae are poorly developed. **Colour:** Polyps are a mass of grey tentacles when extended. **Similar species:** *Goniopora fruticosa*, which has fused branches rather than columns and is distinctively coloured. See also *G. burgosi*, which has similar polyps but is submassive and *G. planulata*, which has larger polyps. **Habitat:** Shallow reef environments. **Abundance:** Uncommon but distinctive.

Taxonomic references: Dana's original description/specimens, Scheer and Pillai (1983). **Identification guide:** Sheppard and Sheppard (1991).

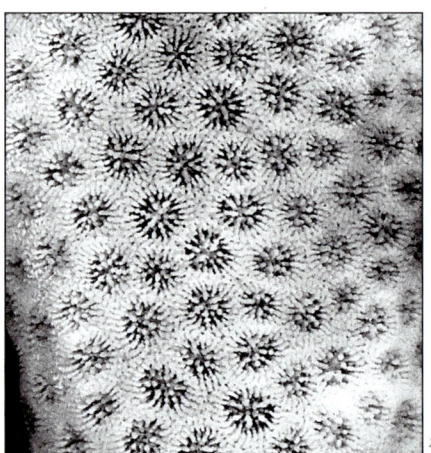

x3

1 *Goniopora savignyi*. Colonies are usually irregular clusters of columns with explanate bases. SINAI PENINSULA, EGYPT *Photograph: author*

2 *Goniopora savignyi*. Branch ends. SINAI PENINSULA, EGYPT *Photograph: Mary Stafford-Smith*

3 *Goniopora savignyi*. Surface of a colony with some polyps extended. SINAI PENINSULA, EGYPT *Photograph: author*

4 *Goniopora savignyi*. Surface of a colony with polyps contracted. SINAI PENINSULA, EGYPT *Photograph: author*

Goniopora stutchburyi
Wells, 1955

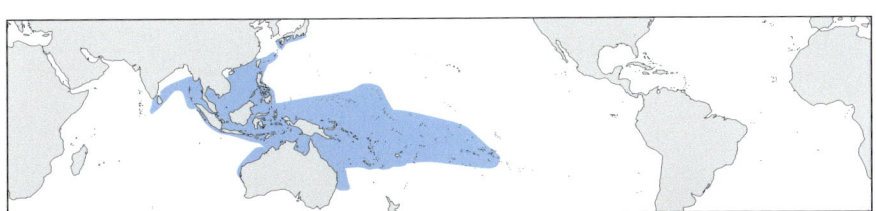

Characters: Colonies are submassive to encrusting. Calices are small and shallow, giving colonies a smooth surface. Polyps have short tapered tentacles which may not be extended during the day. **Colour:** Usually pale brown or cream, sometimes with pale blue mouths. **Similar species:** None, but could be confused with *Porites*. **Habitat:** Shallow reef environments. **Abundance:** Uncommon.

Taxonomic reference: Veron and Pichon (1982). **Identification guides:** Veron (1986), Nishihira and Veron (1995).

5 *Goniopora stutchburyi*. A large lobed colony. HOUTMAN ABROLHOS ISLANDS, SOUTH-WEST AUSTRALIA *Photograph: author*

6 *Goniopora stutchburyi*. Typical appearance with polyps fully extended. BALI, INDONESIA *Photograph: author*

7 *Goniopora stutchburyi*. Polyp detail, with polyps partly retracted. GREAT BARRIER REEF, AUSTRALIA *Photograph: author*

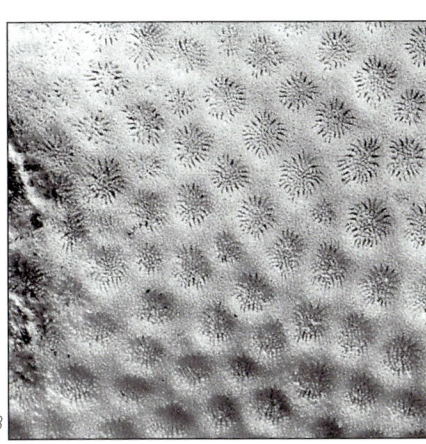

Goniopora fruticosa
Saville-Kent, 1893

Characters: Colonies are encrusting or branching, with highly fused branches. Corallites are rounded or polygonal. Columellae are broad and conspicuous. Polyps have fine, tapered tentacles. **Colour:** Dark brown with white oral discs. **Similar species:** *Goniopora savignyi*, which has larger corallites, elongate polyps and lacks well developed columellae. See also *G. palmensis*. **Habitat:** Upper reef slopes protected from strong wave action. **Abundance:** Usually uncommon.

Taxonomic reference: Veron and Pichon (1982). **Identification guides:** Veron (1986), Nishihira and Veron (1995).

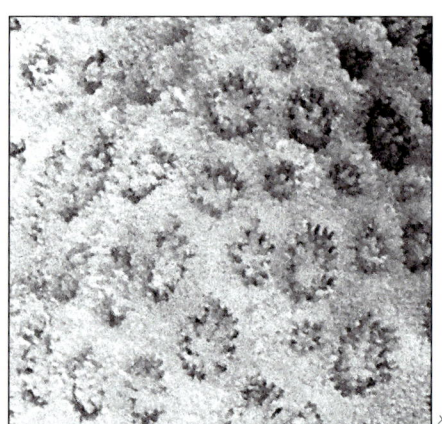

Family Poritidae | Genus *Goniopora*

1, 2 Colonies have irregular shapes. **1** FLORES, INDONESIA **2** BOLINAO, PHILIPPINES *Photographs: author*

3 A small colony forming branches. GREAT BARRIER REEF, AUSTRALIA *Photograph: Ed Lovell*

4 Colony surface. GREAT BARRIER REEF, AUSTRALIA *Photograph: Ed Lovell*

5 Polyps of this species are distinctive. GREAT BARRIER REEF, AUSTRALIA *Photograph: author*

GENUS ALVEOPORA
Blainville, 1830

Characters: Colonies are massive or branching, often with irregular shapes. The skeletal structure is light, consisting of interconnecting rods and spines. Corallites have lattice-like walls and septa that are mostly composed of fine spines which may connect in the centre to form a columella tangle. Polyps are large and fleshy and are normally extended day and night. They have 12 tentacles, often with swollen knob-like tips. **Similar genus:** All *Goniopora* have 24 tentacles and much greater skeletal development. **Earliest fossil record:** Eocene of the Caribbean and Tethys.

Taxonomic reference: Veron and Pichon (1982).
Summary key: See p457.

Identification of *Alveopora* species

It is rare to see many *Alveopora* species together in the same place as habitats of individual species are very different, more so than for any other genus. These habitats include protected turbid environments (the majority of species), exposed upper reef slopes (for example, *A. marionensis*) and rocky foreshores of subtropical locations (for example, *A. japonica*).

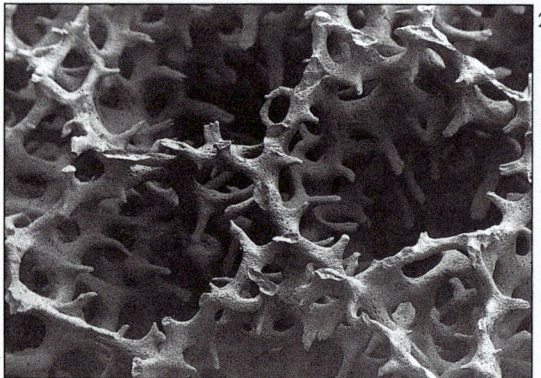

Alveopora gigas
Veron, 1985

Characters: Colonies are composed of blunt-ended irregular columns. Corallites have thin, highly perforated walls of interconnected rods and spines. Polyps are up to 100 millimetres long and 20 millimetres diameter when fully extended. **Colour:** Oral discs and tentacle tips are white, the rest of the polyps are brown or greenish-brown. **Similar species:** This is the largest *Alveopora*. *Alveopora allingi* has similar skeletal structures and *A. catalai* has similar polyps. **Habitat:** Protected turbid environments. **Abundance:** Common and conspicuous at the Houtman Abrolhos Islands, south-west Australia, uncommon elsewhere.

Taxonomic reference: Veron (1985). **Identification guide:** Veron (1986).

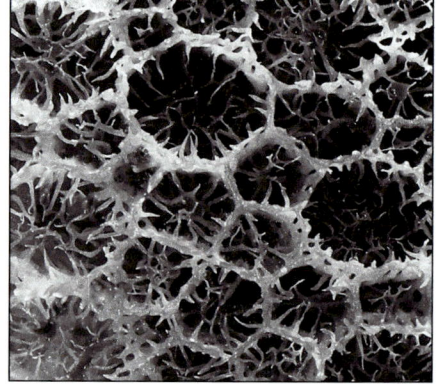

1 All *Alveopora* have tubular polyps with 12 tentacles. The trunk of the polyps often have ribs which correspond to the tentacles. Tentacles have tips of varying shapes and colours, which can aid identification. AMBON, INDONESIA *Photograph: Valerie Taylor*

2 Electron micrograph of *Alveopora* showing detail of skeletal structure. All elements are loosely interconnecting rods and spines.

3 *Alveopora gigas*. A large colony. HOUTMAN ABROLHOS ISLANDS, SOUTH-WEST AUSTRALIA *Photograph: author*

4 *Alveopora gigas*. With its long flower-like polyps, this species is one of the most beautiful of all corals. HOUTMAN ABROLHOS ISLANDS, SOUTH-WEST AUSTRALIA *Photograph: Ed Lovell*

5 *Alveopora gigas*. Fully extended polyps: the species shows little variation. HOUTMAN ABROLHOS ISLANDS, SOUTH-WEST AUSTRALIA *Photograph: author*

6 *Alveopora gigas*. Polyps in the process of retracting. PAPUA NEW GUINEA *Photograph: Neville Coleman*

Alveopora catalai
Wells, 1968

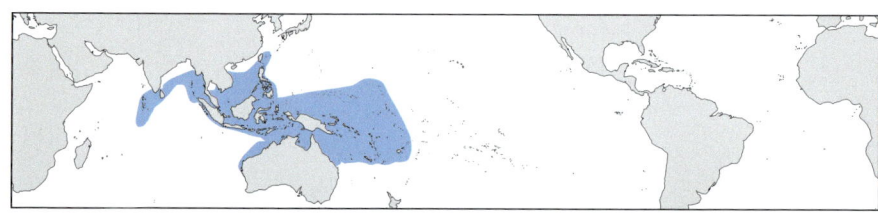

Characters: Colonies are composed of gnarled branches that divide irregularly. They may form stands over 5 metres across. Corallites are large, composed of an interlocking network of rods and spines. Polyps are large with knob-like tentacle tips. **Colour:** Pale tan when polyps are retracted. Extended polyps are amber or yellowish with white oral discs, sometimes with white tentacle tips. **Similar species:** *Alveopora gigas* and *A. allingi*. **Habitat:** Occurs only on soft substrates in deep water or in shallow turbid water protected from wave action and currents. **Abundance:** Uncommon but conspicuous.

Taxonomic reference: Veron and Pichon (1982). **Identification guides:** Veron (1986), Nishihira and Veron (1995).

Family Poritidae   Genus *Alveopora*

1 This species commonly forms extensive beds of flower-like polyps. CALAMIAN ISLANDS, PHILIPPINES *Photograph: author*

2 Characteristic appearance of fully extended polyps. PAPUA NEW GUINEA *Photograph: author*

3 The colony shape is revealed when polyps are retracted. PAPUA NEW GUINEA *Photograph: author*

4 When polyps are fully extended they are long tubes. APO ISLAND, PHILIPPINES *Photograph: author*

5 Detail of retracting polyps. GREAT BARRIER REEF, AUSTRALIA *Photograph: author*

383

Family Poritidae   Genus *Alveopora*

Alveopora allingi
Hoffmeister, 1925

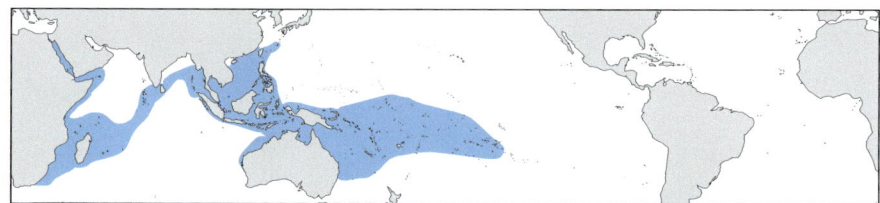

Characters: Colonies are encrusting or have short irregular lobes with rounded surfaces, or are columnar. Corallites have walls composed of interconnected rods and spines and long spine-like septa. Columellae are usually present and are sometimes well developed. Polyps are tightly compacted and are long, usually with slightly expanded tentacle tips. **Colour:** Usually yellow, green or brown with white oral cones. **Similar species:** *Alveopora catalai*, which has larger corallites and a branching growth-form. *Alveopora marionensis* has smaller corallites but a similar growth-form. **Habitat:** Protected reef environments. **Abundance:** Usually uncommon.

Taxonomic references: Wells (1954), Veron and Pichon (1982). **Identification guides:** Veron (1986), Sheppard and Sheppard (1991), Nishihira and Veron (1995).

x3

Family Poritidae | Genus *Alveopora*

Alveopora marionensis
Veron and Pichon, 1982

Characters: Colonies are composed of short irregular lobes. Corallites have walls formed from compacted rods and spines. Septa are composed of tapered spines which are only connected low down in the corallite. Polyps have short straight tentacles. **Colour:** Uniform grey, brown or pinkish, sometimes with white tentacle tips. **Similar species:** *Alveopora fenestrata*, which has a growth-form intermediate between *A. marionensis* and *A. verrilliana*, but only larger colonies can be distinguished by differences in growth-form. Polyp shapes and colours are always distinctive. **Habitat:** Found only in environments with moderate wave action. **Abundance:** Rare.

Taxonomic reference: Veron and Pichon (1982). **Identification guide:** Veron (1986).

×3

5

1 *Alveopora allingi*. A carpet of long tentacles. NORFOLK ISLAND, WESTERN PACIFIC *Photograph: author*

2 *Alveopora allingi*. Polyps in the process of contraction. NORFOLK ISLAND, WESTERN PACIFIC *Photograph: author*

3 *Alveopora allingi*. A large colony. The lobed structure of the underlying skeleton is masked by the long polyps. HOUTMAN ABROLHOS ISLANDS, SOUTH-WEST AUSTRALIA *Photograph: author*

4 *Alveopora allingi*. Polyp detail. HOUTMAN ABROLHOS ISLANDS, SOUTH-WEST AUSTRALIA *Photograph: author*

5 *Alveopora marionensis*. A small colony. MARION REEF, CORAL SEA *Photograph: Ed Lovell*

6 *Alveopora marionensis*. Polyp detail. BALI, INDONESIA *Photograph: author*

Family Poritidae | Genus *Alveopora*

Alveopora fenestrata
(Lamarck, 1816)

Characters: Colonies are generally hemispherical with the surface divided into lobes. Corallites have walls composed of compacted rods and spines. Septa are composed of tapered spines which are connected low in the corallite. Polyps are long, with long tentacles giving a ragged appearance. **Colour:** Grey or greenish-brown, sometimes with white oral cones. **Similar species:** *Alveopora marionensis* and *A. verrilliana*. **Habitat:** Shallow reef environments. **Abundance:** Uncommon.

Taxonomic reference: Veron and Pichon (1982). **Identification guide:** Veron (1986).

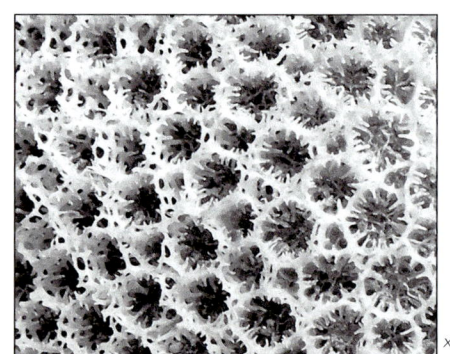
x3

Alveopora verrilliana
Dana, 1872

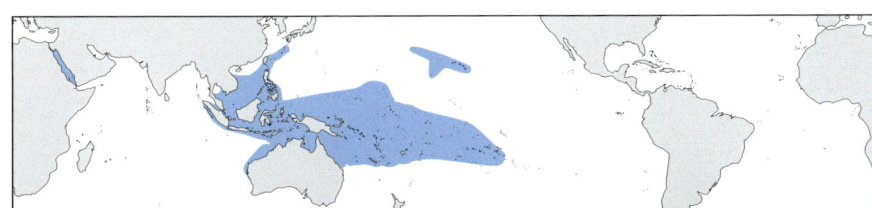

Characters: Colonies are composed of short irregularly dividing knob-like branches. Corallites have short blunt septal spines and a palisade of vertical spines above the wall. Polyps are long when extended. **Colour:** Dark greenish-brown, grey or chocolate, sometimes with white oral cones and/or tentacle tips. **Similar species:** *Alveopora fenestrata*, which is distinguished by growth-form, lack of spines around the corallites and larger corallites. **Habitat:** Shallow reef environments. **Abundance:** Uncommon.

Taxonomic reference: Veron and Pichon (1982). **Identification guides:** Veron (1986), Nishihira and Veron (1995).

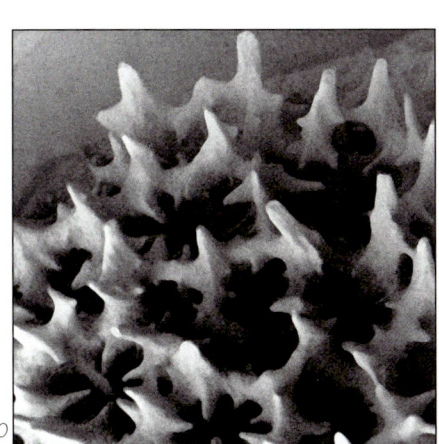

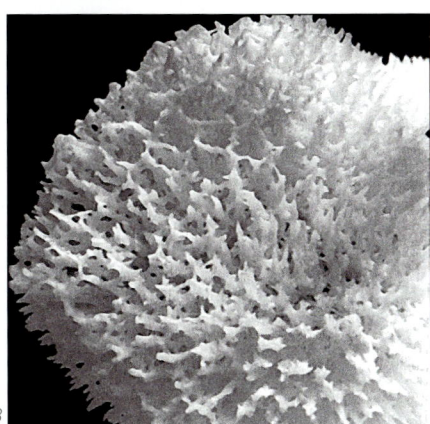

1 *Alveopora fenestrata*. A small hemispherical colony in a wave washed habitat. Polyps are partly retracted. GREAT BARRIER REEF, AUSTRALIA *Photograph: Ed Lovell*

2 *Alveopora fenestrata*. The typically mop-like appearance of colonies with extended polyps of varying sizes. PAPUA NEW GUINEA *Photograph: author*

3 *Alveopora fenestrata*. Polyp detail. CEBU, PHILIPPINES *Photograph: author*

4 Adjacent colonies of *Alveopora verrilliana* (with polyps extended) and *A. fenestrata* (with polyps retracted). MARION REEF, CORAL SEA *Photograph: Ed Lovell*

5 *Alveopora verrilliana*. The characteristic appearance of a lobed colony. MARION REEF, CORAL SEA *Photograph: Ed Lovell*

6 *Alveopora verrilliana*. Electron micrograph showing vertical spines surrounding the corallite. *Electron micrograph: author*

Alveopora spongiosa
Dana, 1846

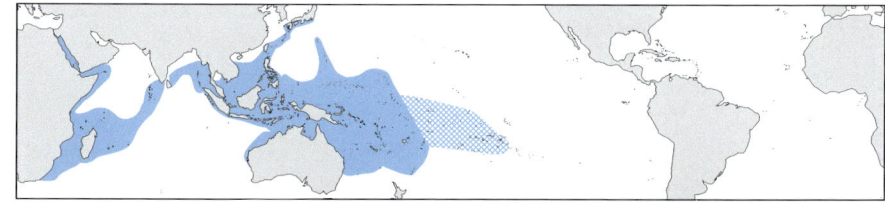

Characters: Colonies are encrusting, submassive thick plates or cushions, or columnar. They have a flat or irregular surface and may be over 2 metres across. Corallites have long or short fine septal spines which seldom connect. Tips of polyp tentacles may be pointed or knob-like. Sometimes six large tentacles alternate with six small ones. **Colour:** Usually uniform pale or dark brown, rarely green or other colours. Polyps sometimes have white tentacle tips. **Similar species:** *Alveopora daedalea*. See also *A. tizardi*, which has smaller corallites, a more solid skeletal structure and distinctive knobs on tentacle tips. **Habitat:** Protected upper reef slopes. **Abundance:** Usually uncommon.

Distribution note: Records from the eastern central Pacific are doubtful. **Taxonomic reference:** Veron and Pichon (1982). **Identification guides:** Veron (1986), Sheppard and Sheppard (1991), Nishihira and Veron (1995).

Family Poritidae | Genus *Alveopora*

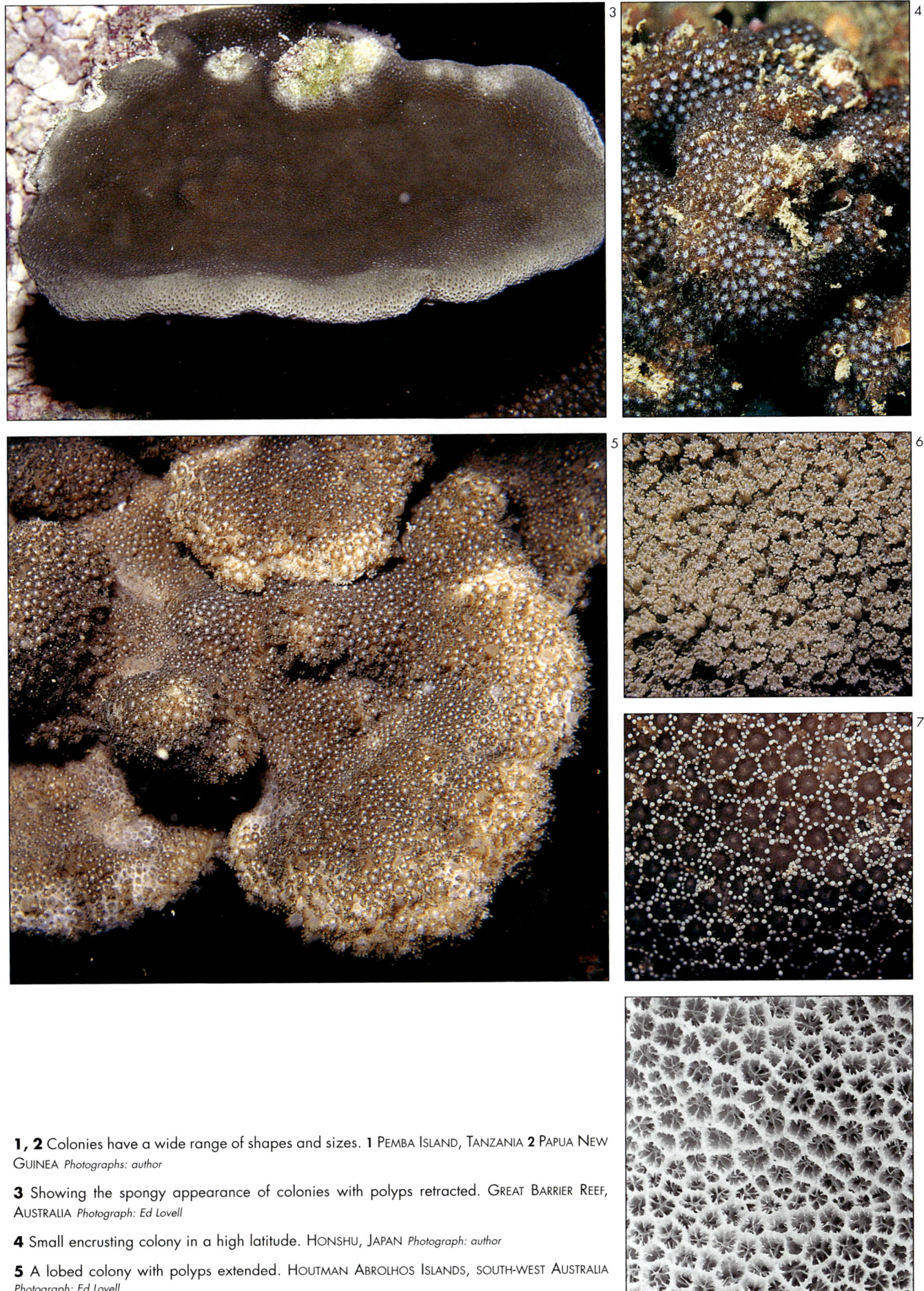

1, 2 Colonies have a wide range of shapes and sizes. **1** PEMBA ISLAND, TANZANIA **2** PAPUA NEW GUINEA *Photographs: author*

3 Showing the spongy appearance of colonies with polyps retracted. GREAT BARRIER REEF, AUSTRALIA *Photograph: Ed Lovell*

4 Small encrusting colony in a high latitude. HONSHU, JAPAN *Photograph: author*

5 A lobed colony with polyps extended. HOUTMAN ABROLHOS ISLANDS, SOUTH-WEST AUSTRALIA *Photograph: Ed Lovell*

6 Colony with extended polyps. GREAT BARRIER REEF, AUSTRALIA *Photograph: author*

7 Polyp detail. GREAT BARRIER REEF, AUSTRALIA *Photograph: Ed Lovell*

Alveopora daedalea
(Forskål, 1775)

Characters: Colonies are encrusting, thick plates, or columnar. They have a smooth surface and may be large, with columns up to one metre high. Corallites have alternating long and short fine septal spines. Tips of polyp tentacles are truncated giving a squared appearance. Sometimes six large tentacles alternate with six small tentacles. **Colour:** Uniform pale or dark brown. **Similar species:** *Alveopora spongiosa*, which is less columnar, does not have truncated tentacles, and has less clearly alternating septa. **Habitat:** Protected upper reef slopes. **Abundance:** Relatively common in the western Indian Ocean.

Taxonomic reference: Scheer and Pillai (1983).

1, 3 Colonies forming tall columns. **1** CALAMIAN ISLANDS, PHILIPPINES **3** SEYCHELLES *Photographs: author*

2 An irregular colony without tall columns. PAPUA NEW GUINEA *Photograph: author*

4 Polyp detail. SEYCHELLES *Photograph: author*

Family Poritidae　　　　　　　　　　　　　　　　　　　　　　　　　　　Genus *Alveopora*

391

Alveopora japonica
Eguchi, 1968

Characters: Colonies are small (less than 40 mm across) and hemispherical or encrusting. Septa have long and short fine spines which seldom connect. Tips of tentacles are knob-like. **Colour:** Distinctive dark green with a white stripe on the top of tentacle tips. **Similar species:** *Alveopora tizardi*, which forms larger colonies with a different colouration. **Habitat:** Found only in shallow, partly wave washed rocky foreshores nested among algae and soft corals. **Abundance:** Uncommon.

Taxonomic reference: Eguchi (1968). **Identification guide:** Nishihira and Veron (1995).

Alveopora tizardi
Bassett-Smith, 1890

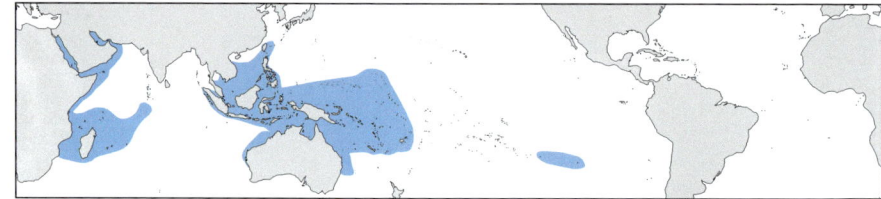

Characters: Colonies are flat or undulating plates. Corallites usually have regularly tapered septal spines. Polyps are short with knob-like tentacle tips. **Colour:** Pale pinkish-brown to bright pink, sometimes with grey oral discs and white tentacle tips. **Similar species:** *Alveopora japonica*. See also *A. spongiosa*, which has larger corallites and distinctive polyps. **Habitat:** Shallow reef environments. **Abundance:** Usually uncommon.

Taxonomic reference: Veron and Pichon (1982). **Identification guides:** Veron (1986), Sheppard and Sheppard (1991), Nishihira and Veron (1995).

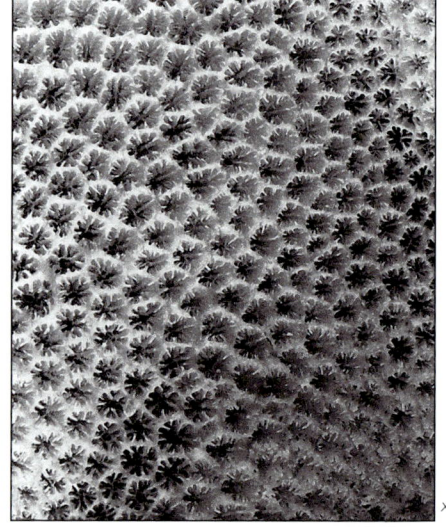

Family Poritidae | Genus *Alveopora*

1 *Alveopora japonica.* This species occurs only as small colonies in shallow non-reef habitats. KYUSHU, JAPAN *Photograph: author*

2 *Alveopora japonica.* Typical colour and appearance of polyps. KYUSHU, JAPAN *Photograph: author*

3, 5 *Alveopora tizardi.* Polyp detail. **3** AMBON, INDONESIA **5** GREAT BARRIER REEF, AUSTRALIA *Photographs: 3 Valerie Taylor 5 Roger Steene*

4 *Alveopora tizardi.* This species commonly forms small colonies growing from crevices. Tentacles retract quickly if disturbed. PEMBA ISLAND, TANZANIA *Photograph: author*

6 *Alveopora tizardi.* A large lobed colony. HOUTMAN ABROLHOS ISLANDS, SOUTH-WEST AUSTRALIA *Photograph: Ed Lovell*

Alveopora excelsa
Verrill, 1863

Characters: Colonies are submassive and usually lobed. They may be over 2 metres across. Corallites have irregular spiny walls. Polyps are often retracted during the day, but are elongate when extended giving colonies a mop-like appearance. **Colour:** Grey or pinkish brown when polyps are retracted, brown when polyps are extended. **Similar species:** *Alveopora verrilliana*, which can readily be distinguished by the palisade of spines around the corallites. **Habitat:** Exposed reef slopes. **Abundance:** Uncommon.

Taxonomic reference: Veron and Hodgson (1989). **Identification guide:** Nishihira and Veron (1995).

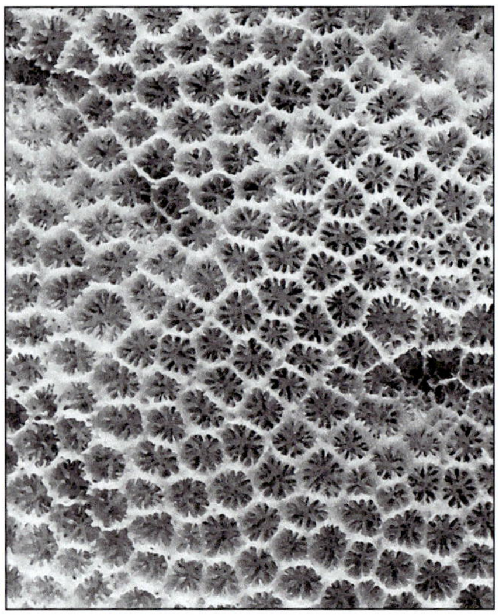

Alveopora viridis
Quoy and Gaimard, 1833

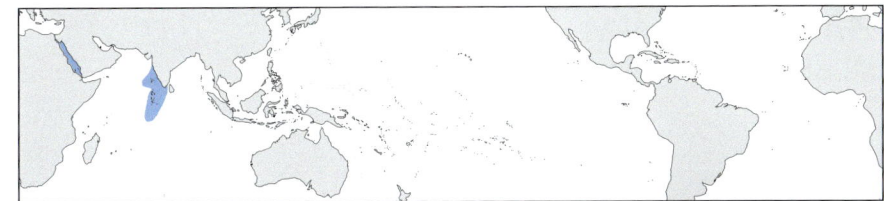

Characters: Colonies are submassive, sometimes forming columns. Corallites are as small as many species of *Porites*. Septal spines are of two lengths. **Colour:** Grey-brown or greenish-brown. **Similar species:** *Alveopora minuta*. Readily mistaken for a *Porites* underwater. **Habitat:** Lower reef slopes protected from wave action. **Abundance:** Rare.

Taxonomic reference: Scheer and Pillai (1983). **Identification guide:** Sheppard and Sheppard (1991).

x3

1 *Goniopora excelsa*. The characteristically lobed appearance of colonies with polyps retracted. VIETNAM *Photograph: author*

2 *Goniopora excelsa*. A colony with partly retracted polyps. RYUKYU ISLANDS, JAPAN *Photograph: author*

3 *Goniopora viridis*. A columnar colony with polyps retracted. SINAI PENINSULA, EGYPT *Photograph: author*

4 *Goniopora viridis*. Surface detail with polyps partly extended. SINAI PENINSULA, EGYPT *Photograph: author*

Family Poritidae Genus *Alveopora*

Alveopora minuta
Veron, this publication

Characters: Colonies are composed of short irregularly dividing knob-like branches. Corallites have no septa or one or two septa reduced to a single spine. A palisade of vertical spines occurs above the wall. Corallites are as small as many species of *Porites* (approximately 1 mm diameter) although this species always looks *Alveopora*-like underwater. **Colour:** Greenish-brown. **Similar species:** *Alveopora viridis*, which has coarser skeletal structures. See also *A. verrilliana*, which has larger corallites. **Habitat:** Rocky surfaces exposed to currents. **Abundance:** Rare.

Taxonomic note: See 'New species described in Corals of the World' (Veron, in preparation) for further information.

1 *Alveopora minuta*. A small colony. BALI, INDONESIA *Photograph: author*

2 *Alveopora minuta*. Surface detail. BALI, INDONESIA *Photograph: author*

Alveopora ocellata
Wells, 1954

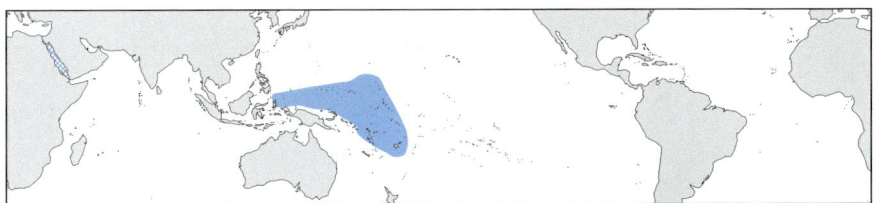

Characters: Colonies are flat plates. Corallites are circular, with neat well defined walls. Septa are in two unequal cycles. **Colour:** Dull brown. **Similar species:** *Alveopora viridis*, which does not have neat circular corallites. **Habitat:** Lower reef slopes protected from wave action. **Abundance:** Rare.

Taxonomic references: Wells (1954), Scheer and Pillai (1983). **Identification guide:** Sheppard and Sheppard (1991).

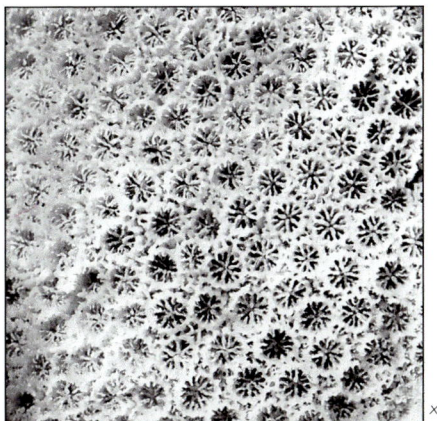

x3

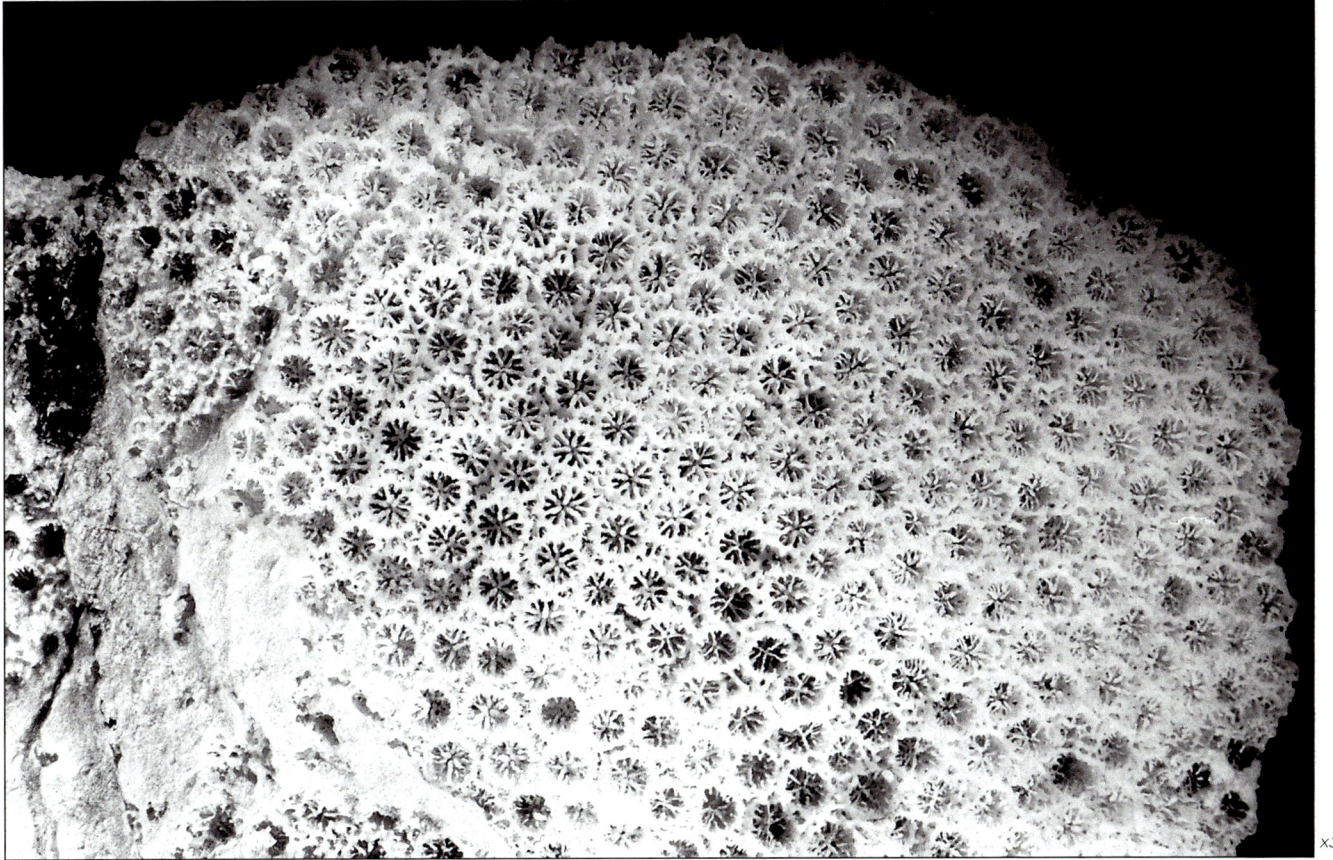

x5

Non-Scleractinian Corals

NON-SCLERACTINIAN CORALS

This book is about zooxanthellate Scleractinia – stony corals with symbiotic algae (variously called hermatypic or reef-building corals) that are primarily responsible for the construction of modern coral reefs. The azooxanthellate Scleractinia, amounting to 117 extant genera, mostly inhabit the deep ocean although some are commonly found on reefs and other shallow water habitats. The latter (Families Astrocoeniidae, Oculinidae, Rhizangiidae and Dendrophylliidae) are briefly illustrated in Volume 2 of this book. Some azooxanthellate species within these families belong to the same genus as zooxanthellate relatives (Genera *Madracis*, *Oculina*, *Astrangia* and *Balanophyllia*), but mostly they are distinct non-colonial forms dominated by the Family Caryophylliidae.

There are also groups of stony corals which are not scleractinians. The relationships of these groups with each other and with other groups of Phylum Cnidaria (Coelenterata) are summarised in the taxonomic hierarchy below. Note that these are only very distantly related groups which have very little in common except a body wall consisting of two tissue layers separated by a non-tissue layer (see 'Structure', v1, p47). Note also that there are several recently published variations of this classification.

Class Hydrozoa
    Order Hydroidea (hydroids)
    Order Milleporina (including Genus *Millepora*)
    Order Stylasterina (including Genera *Distichopora* and *Stylaster*)
Class Scyphozoa (jellyfishes)
Class Cubozoa (sea wasps)
Class Anthozoa
    Subclass Octocorallia
        Order Helioporacea (Genus *Heliopora*)
        Order Alcyonacea (soft corals, *Tubipora*, sea fans and relatives)
        Order Pennatulacea (sea pens)
    Subclass Hexacorallia
        Order Actiniaria (simple sea anemones)
        Order Zoanthidia (colonial anemones)
        Order Corallimorpharia (corallimorpharians)
        Order Scleractinia (true stony corals)
        Order Rugosa (extinct corals)
        Order Tabulata (extinct corals)
    Subclass Ceriantipatharia
        Order Antipatharia (black corals)
        Order Ceriantharia (tube anemones)

Groups having some or all species with stony skeletons are indicated in bold.

Opposite: *Distichopora violacea* is a common sight under wave-washed overhangs of Indo-Pacific reefs. GREAT BARRIER REEF, AUSTRALIA
Photograph: Valerie Taylor

Non-Scleractinian Corals

Class Hydrozoa
Two groups of Hydrozoa form stony colonies. Hydrozoan corals, unlike anthozoan corals, have different types of polyps that have different functions. These polyps are of near microscopic size.

Order Milleporina

Millepora are common on reefs. They have growth-forms ranging from arborescent to submassive and encrusting. Approximately 50 species have been described. All are zooxanthellate. Polyps are mostly embedded in the skeleton where they are linked by a network of minute canals, the cyclosystem. All that can be seen on the smooth surface are pores of two sizes, gastropores and dactylopores. The gastropores house the retractile gastrozooids, each of which is surrounded by up to nine minute dactylopores that house the dactylozooids. The latter have long fine hairs that protrude from the skeleton, visible underwater. The hairs contain batteries of nematocysts capable of inflicting mild stings on soft human skin. These capture prey, which is engulfed by the gastrozooid. Reproductive ampullae, which produce medusae, can also be seen on the colony surface. Generations of sexual medusae and asexual polyps alternate.

Non-Scleractinian Corals

1 *Millepora*, commonly known as 'fire coral' because of its stings to divers, may form extensive outcrops where currents are strong. Branches of colonies such as these are easily broken and have hollow cores which contain oxygen. GREAT BARRIER REEF, AUSTRALIA *Photograph: Katharina Fabricius*

2 *Millepora*, has many growth-forms, some of which are columnar. GREAT BARRIER REEF, AUSTRALIA *Photograph: Ed Lovell*

3, 6 Branching *Millepora*. These branches are brittle and are easily damaged by divers. GREAT BARRIER REEF, AUSTRALIA *Photographs: author*

4 A very large branching colony. RYUKYU ISLANDS, JAPAN *Photograph: author*

5 Surface detail of a finely branched *Millepora* showing the long hair-like dactylozooids. GREAT BARRIER REEF, AUSTRALIA *Photograph: author*

Non-Scleractinian Corals

Order Stylasterina

Distichopora and *Stylaster*, being azooxanthellate, are widely distributed in temperate as well as tropical localities and also occur at abyssal depths. *Distichopora* occurs throughout the Indo-west Pacific and the Galapagos Islands and is restricted to deep water in some regions, including Hawaii. *Stylaster* occurs throughout the Indo-Pacific and Atlantic Oceans.

The second group of hydrozoans are divisible into three subfamilies, most of which are found only in deep water. All are azooxanthellate. Two genera, *Distichopora* and *Stylaster* are commonly found in caves and under overhangs in shallow reef environments. *Distichopora* are ornate corals that branch in one plane. They have no cyclosystem; instead gastropores are aligned along the edge of branches and these are flanked on each side by a row of dactylopores. These pores are visible underwater. Reproductive ampullae are clustered towards the ends of side branches. *Stylaster* also branches in one plane but branches are fine, tapered and delicate. Gastropores are linked by individual cyclosystems and surrounded by dactylopores. These alternate on the sides of branches, giving the latter a zig-zag pattern. Wart-like reproductive ampullae occur on the sides of older branches.

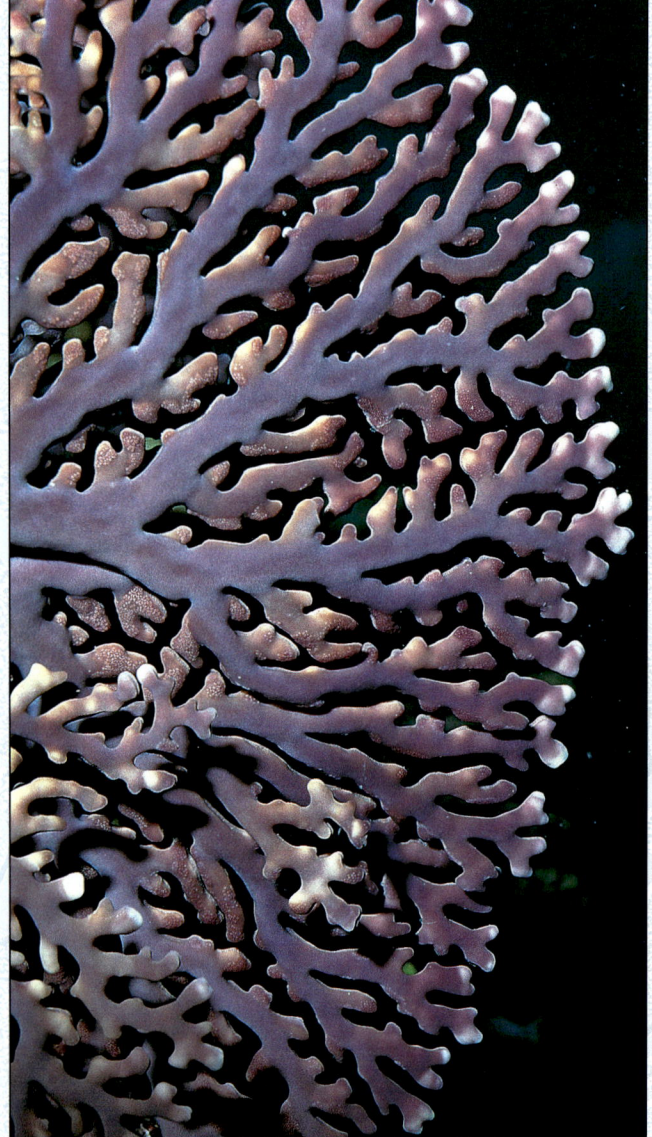

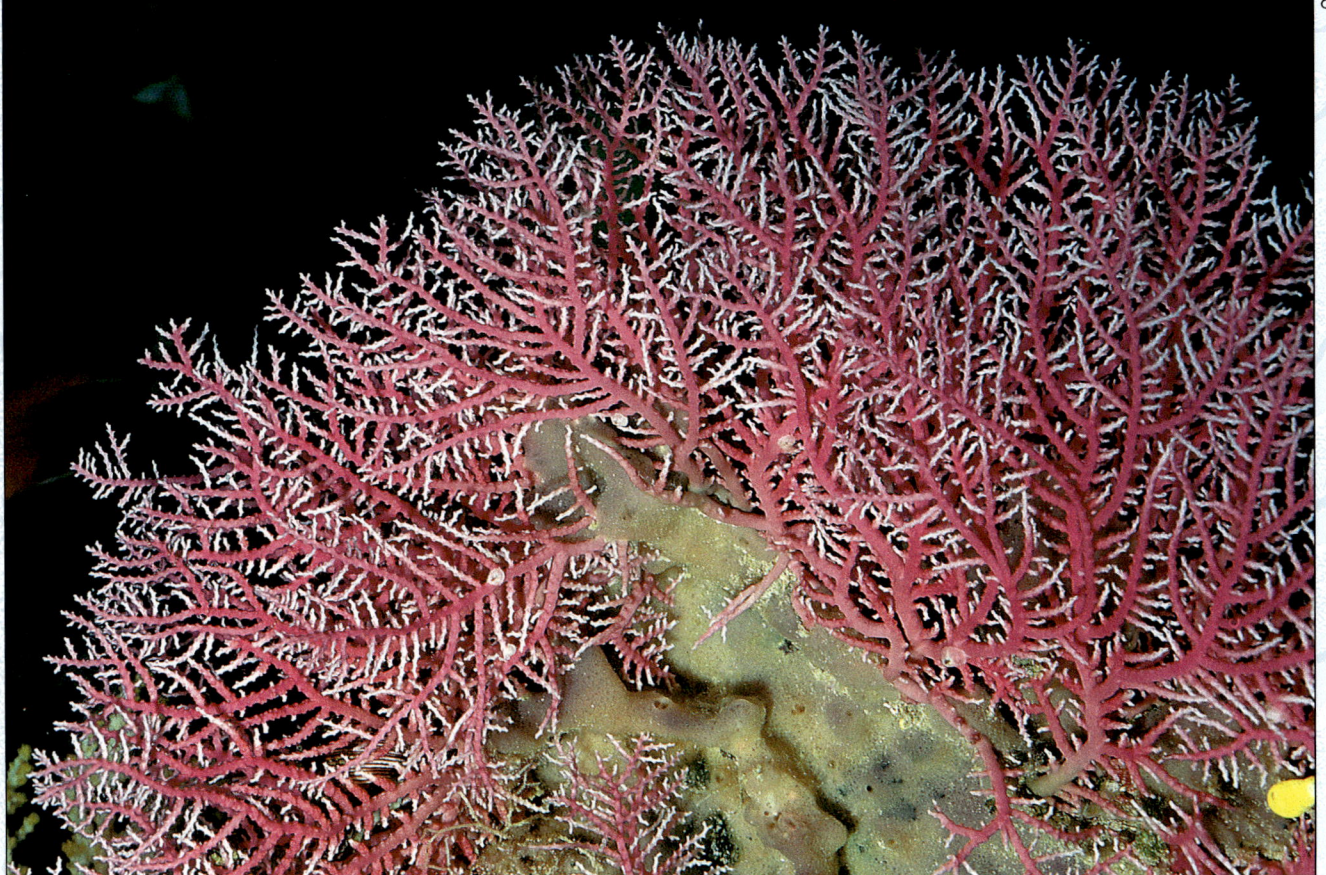

1-3, 5 *Distichopora* is commonly found in crevices, under ledges and in caverns. They come in a wide variety of colours, but most appear to be one species. **1, 2, 5** Great Barrier Reef, Australia **3** Sulawesi, Indonesia *Photographs: 1, 2, 5 Valerie Taylor 3 Doug Fenner*

4, 6 *Stylaster* is commonly found in caverns where it may occur as clumps (6). On close inspection, the long spines of a few dactylozooids can be seen with each gastrozooid (4). Great Barrier Reef, Australia *Photographs: Valerie Taylor*

Non-Scleractinian Corals

> **Class Anthozoa: Subclass Octocorallia**
> The Octocorallia are a large and diverse group. Most do not develop skeletons: *Heliopora* and *Tubipora* are exceptions. Polyps have eight tentacles.

Order Helioporacea

Heliopora coerulea or 'blue coral' is the sole member of Order Helioporacea. It is blue or greenish underwater, but the skeleton, composed of fibrocrystalline aragonite, is always permanently blue. It is easily recognised in fossil outcrops and can be traced back to the Cretaceous Period. Throughout this time it appears to have remained a single species, in which case it would have by far the greatest geological longevity of any coral.

1 *Heliopora coerulea* commonly occurs on intertidal reef flats. Colonies show little resemblance to other colonies in nearby deeper water, a variability of growth-form similar to that found in many Scleractinia. SCOTT REEF, NORTH-WEST AUSTRALIA *Photograph: author*

2 Surface detail of branches with minute transparent polyps extended. Unlike those of Scleractinia, the polyps have eight tentacles. GREAT BARRIER REEF, AUSTRALIA *Photograph: author*

3 At Ishigaki Island, the site of the world's largest known outcrop of this living fossil. RYUKYU ISLANDS, JAPAN *Photograph: author*

4 Large colonies of *Heliopora coerulea* can be a beautiful sight. SCOTT REEF, NORTH-WEST AUSTRALIA *Photograph: author*

Non-Scleractinian Corals

Order Alcyonacea

The second group of octocorals to form skeletons is part of the very large Order Alcyonacea and comprises the genus *Tubipora*, the 'organ-pipe' corals. There are at least two species, only one of which, *T. musica*, is named. Both are zooxanthellate. Like *Heliopora*, the skeleton is permanently coloured – a dark red. Unless the colony has been damaged, the mass of brown-green tentacles obscures the skeleton underwater and thus this common coral is not easily recognised.

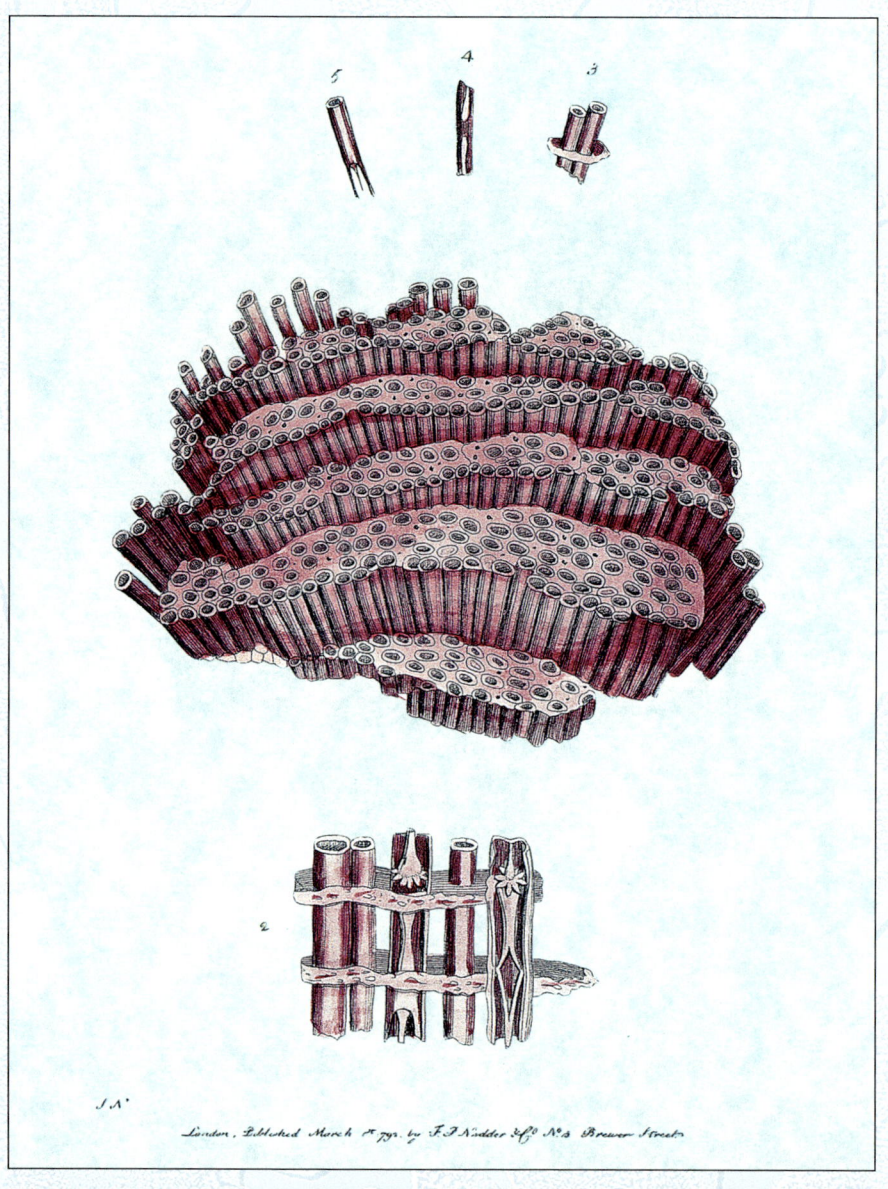

1 Drawing of the skeleton of *Tubipora musica* from the Joseph Banks expedition, published in 1792. GREAT BARRIER REEF, AUSTRALIA

2 The red skeleton is sometimes seen beneath extended polyps. GREAT BARRIER REEF, AUSTRALIA Photograph: author

3 *Tubipora musica* may form extensive clumps in shallow water. The thick mass of tentacles usually obscures the underlying red organ-like tubes of the corallites. PAPUA NEW GUINEA Photograph: author

4, 5 Detail of polyps of *Tubipora musica*. The tentacle structure of different colonies varies greatly. Some have broad (feather-like) pinnae while others have no pinnae at all. **4** PAPUA NEW GUINEA **5** NEGROS, PHILIPPINES Photograph: 4 Jim Maragos 5 Doug Fenner

Class Anthozoa: Subclass Hexacorallia
The Hexacorallia are divided into three skeletal and three non-skeletal Orders. Of the former, the Scleractinia are extant, while the Rugosa and Tabulata were extinct by the close of the Palaeozoic Era.

Order Rugosa

Order Tabulata

Two of the above groups, Order Rugosa and Order Tabulata, were major reef builders of the Palaeozoic Era. Neither of these groups survived the end-Permian extinctions (see 'Geological history of corals', v1, p33). They do not have well defined relationships with each other, or with the Scleractinia.

Corals of the World

Biogeography

Biogeography

Patterns of diversity – endemism – regional comparisons – latitudinal changes – dispersal – ocean currents – reproduction – hybridisation

'What are they?, where are they?, why?' These are the central themes of this book. The what? (taxonomy) is the oldest and most traditional of biological sciences. The where? (biogeography) confronts the practical and theoretical issues that arise when geographic variation is added to traditional taxonomy. The why? leads to reticulate evolution, a highly explanatory concept that arises when biogeographic patterns are seen in evolutionary time. These subjects, and themes, are all interlinked.

Turbinaria frondens

This chapter summarises the information in the maps that have been presented throughout the book. Global contours of diversity are given for three taxonomic levels: family, genus and species (next page). Species differences among biogeographic regions are then examined in terms of various measures of regional similarity and/or difference (pp412-4). It is important to note that geographic patterns of diversity at these different taxonomic levels (see box, left) involve somewhat separate explanations, ranging from issues of ancient geological history (following on from 'Geological History', v1, p33) to those of ocean currents and reproductive mechanisms (pp414-21). When all explanations are combined, the concept of reticulate evolution, as presented in the last chapter (pp437-43) emerges.

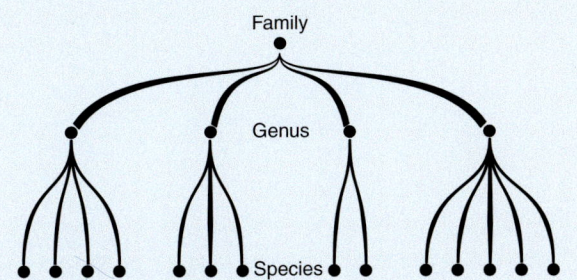

Taxonomic hierarchies and evolutionary time

The distribution patterns of extant corals at these different taxonomic levels have largely separate explanations. At the family level, they are mostly explained by continental drift and mass extinctions: events that occurred many tens of millions years ago. At the generic level, they are mostly explained by closure of the Tethys Sea and the Central American Seaway: events that occurred over the past ten million years or so. At the species level, they are mostly explained by changing patterns of ocean currents and reproduction.

This is a progression of level of detail from the family to the species and from the most distant times to the most recent. Like the taxonomic hierarchy itself, there is a progression of inheritance in explanations of distributions.

Previous page: A platform reef. The reef front (left) faces the prevailing wind. GREAT BARRIER REEF, AUSTRALIA *Photograph: Roger Steene*

Opposite: A large colony of the azooxanthellate *Tubastrea micrantha* in a shallow reef environment. Very little is known of the reproductive behaviour of azooxanthellate corals. PAPUA NEW GUINEA *Photograph: author*

Distribution patterns

Accounts of every species in this book are accompanied by a map showing the species' global distribution. When added together, these maps are used below to derive broad-scale patterns of diversity, to determine patterns of endemism, to examine similarities and differences among geographic regions and to show how species 'drop out' with increasing distance from the equator (so-called 'latitudinal attenuation').

Patterns of diversity

The three maps (next page) indicate morphological (as opposed to genetic) diversity at the three taxonomic levels – family, genus and species – referred to above. As indicated in the box (left), each map is the outcome of an inheritance from an evolutionary past together with a history of subsequent change. The lower the taxonomic level, the more recent are the events that are responsible for creating the distribution.

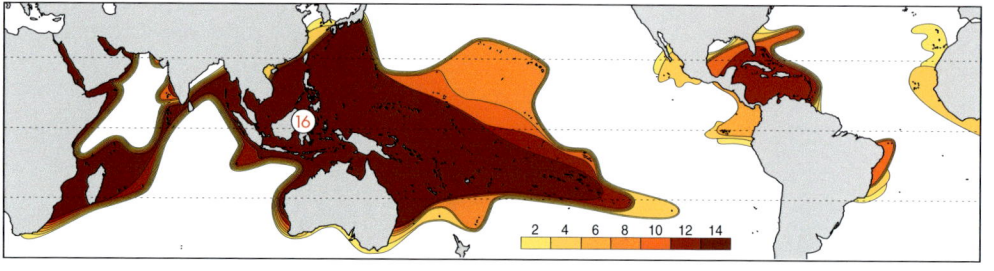

Family diversity. Contours of family diversity. This map has been produced by combining the 18 family distribution maps (zooxanthellate species only) in this book.

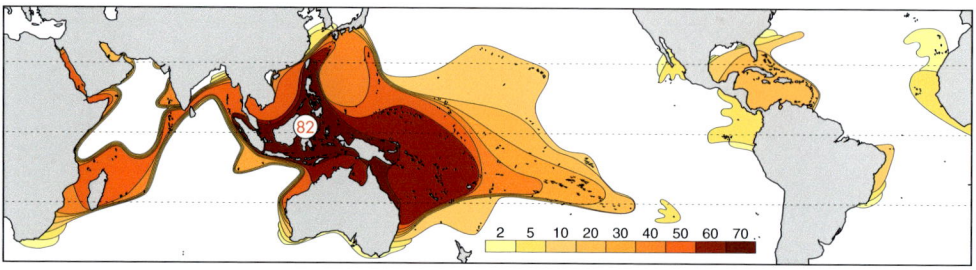

Generic diversity. Contours of generic diversity. This map has been produced by combining the 111 generic distribution maps (zooxanthellate species only) in this book.

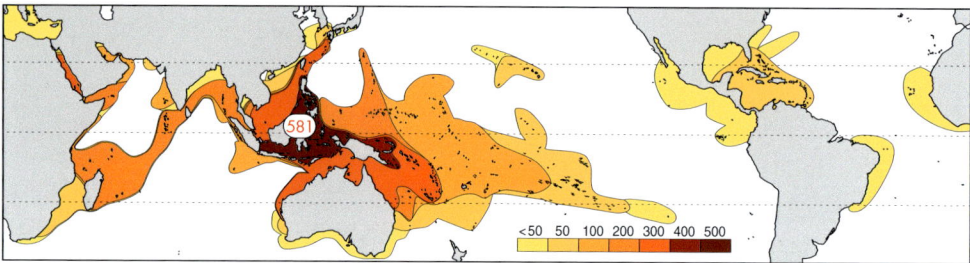

Species diversity. Contours of species diversity. This map has been produced by combining the 793 species distribution maps in this book.

Family diversity. The pattern of diversity at family level is mostly a matter of distant geological history as seen in the fossil record. That is, it reflects a time in the distant geological past when the presence of the Tethys Sea and the Central American Seaway led to a uniform circumglobal fauna (v1, p37). At this taxonomic level, there is no well defined Indo-Pacific centre of diversity and the Caribbean is almost as diverse as the Indo-Pacific.

Generic diversity. The pattern of diversity at generic level is primarily created by the many taxa that evolved in the Tethys Sea and which now survive only in the Indo-Pacific (v1, pp37-41), together with genera that evolved in, and remained in, the Indo-Pacific. At this taxonomic level there is a well defined Indo-Pacific centre of diversity and the Caribbean has a substantially lower diversity than the Indo-Pacific. From the central Indo-Pacific there is little attenuation of diversity west across the Indian Ocean, but a regular attenuation occurs east across the Pacific.

Species diversity. The pattern of diversity at species level is primarily the outcome of ocean circulation patterns and resulting geographic and evolutionary changes from the Plio-Pleistocene to the present time. There is a well defined Indo-Pacific centre of diversity in the Indonesian-Philippines archipelago. This centre attenuates sequentially in the north along the Ryukyu Island chain and in the east along the northern Papua New Guinea coast. It stops west of Indonesia. Within the tropical Indian Ocean, species diversity remains mostly uniform; the Red Sea is a secondary centre. The diversity of the Caribbean is no more than that of a depauperate outlying location of the Indo-Pacific.

Endemism

Indo-Pacific corals are mostly widely dispersed, with few species being restricted to one country or to one geographic region. That is, few species are regional endemics. At species level, the maps of this book show two types of endemism. The first is where species are restricted to centres of diversity. This is best seen in the Indonesian-Philippines centre of diversity, although even here the high diversity (581 species) is primarily the outcome of a confluence of distributions of widespread species rather than the presence of regional endemics (of which there are 31 species or 5% of the total). The Red Sea has 18 endemic species (6% of the total), a lower number than might be expected because many Red Sea species also occur at Socotra, east of the Gulf of Aden. Higher proportions of regional endemics are found in the Atlantic, simply because of geographic limitations. The Caribbean and Gulf of Mexico region has 21 endemics (37% of the total species) and Brazil has 4 endemics (25% of the total species, partly depending on arbitrary taxonomic decisions).

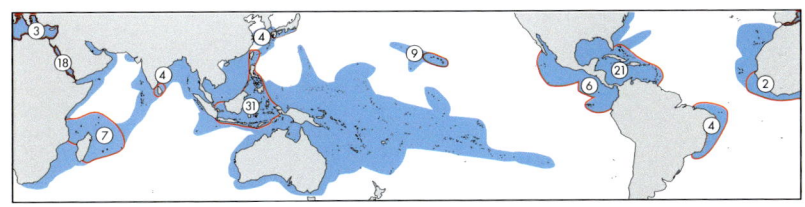

Endemics. Numbers of regional endemics.

The second type of endemic is largely due to isolation and is best seen in the northern Pacific (the Hawaiian islands and Johnston Atoll) and the far-eastern Pacific. In these regions, the number of endemics is high in relation to the total number of species (14% and 17% respectively, the latter number also being partly dependent on arbitrary taxonomic decisions). In most such regions, the percentage of endemics increases with remoteness and may be much higher in restricted geographic areas or particular island groups.

There is actually a third type of endemic: where the taxonomic level is below that of species. These are below the level of detail presented in this book. They may be called geographic subspecies, of which many thousands are readily recognisable. There are only arbitrary distinctions between species and subspecies, involving the complex of issues raised in 'What are species?' (p425). Most regions that have a high level of subspecies endemism are high latitude locations where reefs are not developed, but also large areas which are geographically isolated, including the Red Sea and the south Pacific islands.

Regional similarity

The distribution maps of this book allow measures of regional similarity to be calculated which take in to account not only the number of species, but also the identity of those species. The result, shown below, is intuitive: the broad pattern of similarity among regions primarily reflects their degree of geographic proximity. Atlantic corals are very distinct from Indo-Pacific corals. The Caribbean/Gulf of Mexico is a distinctive region, with the corals of Bermuda and the east American coast most closely allied to it. In the Indo-Pacific, the far eastern fauna is also very distinctive. To a lesser extent so also are the corals of the Red Sea. The least distinctive are the regions of the central equatorial Indo-Pacific.

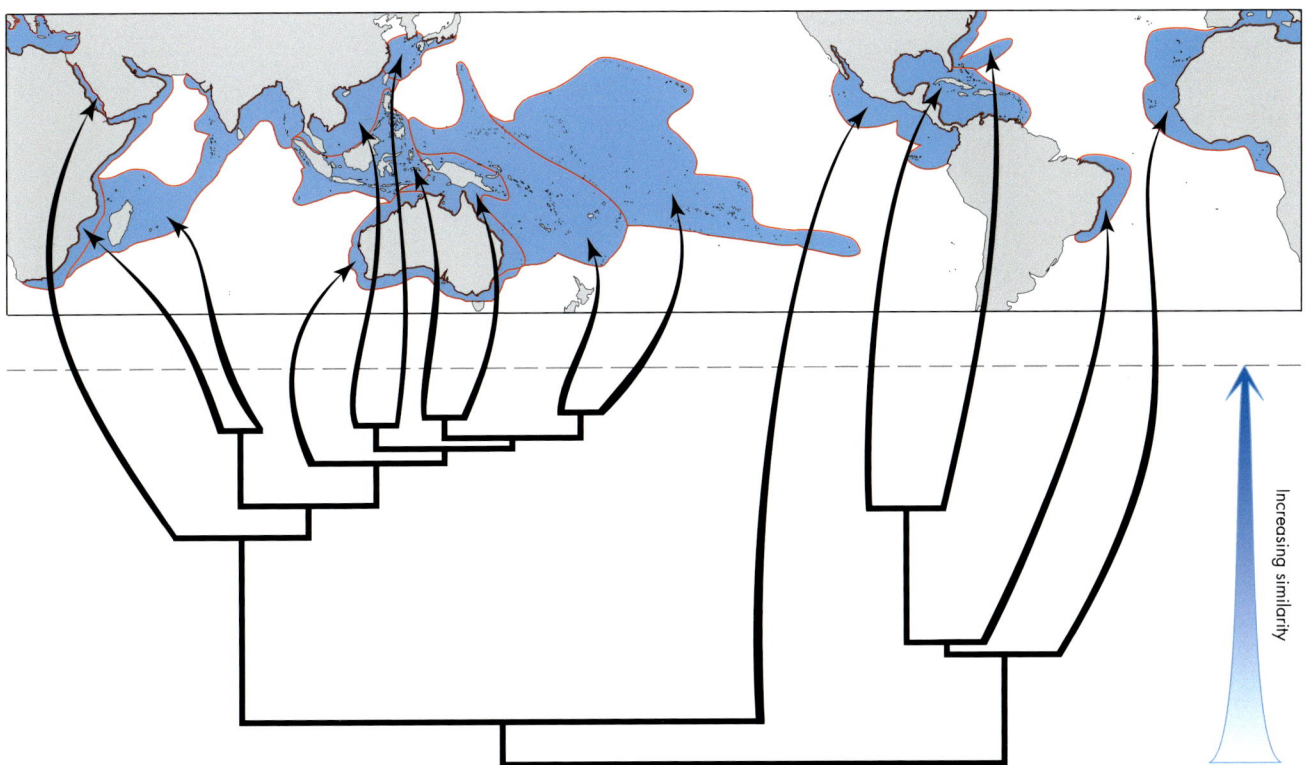

Global coral species similarity. Similarity between the main coral biogeographic regions of the world, calculated from all the species distribution maps in this book. The closer two linked regions on the dendrogram are to the broken line, the greater the similarity of their coral faunas.

Latitudinal attenuation

The diagrams on this page are based on detailed information about the corals along the island arc from the Philippines to Japan, and along the eastern and western Australian coastlines. Tropical species of corals are not replaced with subtropical species as latitude increases, but rather there is a gradual attenuation (thinning out) of tropical species. This is correlated with decreasing temperature, but it is also correlated with the poleward direction of the main surface currents. What the currents do is perpetually move planulae away from the equatorial centre of diversity to higher latitudes. This creates a ratchet effect: the corals can go in one direction only as there is no return. Subtropical locations are thus genetically connected to tropical locations, but not the reverse. This makes it possible to 'lock up' corals in high latitude regions and this is the probable reason why there are a large number of geographic subspecies found at high latitudes.

These patterns also show that there is usually a substantial difference in species diversity between reef and non-reef coral communities. This is not directly related to temperature. The majority of corals can tolerate water temperatures as low as 14°C for protracted periods of time, but corals can only build large consolidated reefs if the temperature normally stays above 18°C. Rather, the difference between reef and non-reef communities appears to be related to the capacity of corals to outcompete macroalgae. In effect, coral species act cooperatively to create habitats which will support a diverse coral community (v1, p28).

Explanations of distribution patterns

Corals, more than any other group of marine invertebrates, have attracted the attention of biogeographers over the past thirty years. This is partly due to the interest created by Darwin's (1842) book about coral reefs and subsequent debates about some of the issues it raised, but also because of the obvious concentration of both corals and reefs in tropical locations. The result has been a progression of publications which contain contour maps giving various revisions of the generic level map presented earlier (p412). A wealth of hypotheses has been put forward to give evolutionary explanations for these maps.

The explanation of distributions is best understood as a progression of layers of detail. The family level

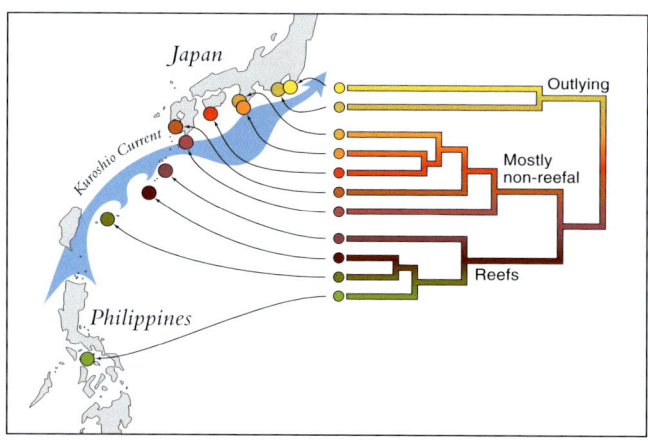

Coral species similarity in Japan and the Philippines. Geographic patterns of similarity of coral species of the Philippines and Japan. There are three main biogeographic zones: tropical reefs, subtropical non-reef communities and temperate (high latitude) depauperate communities. The surface current shown is the core of the Kuroshio.

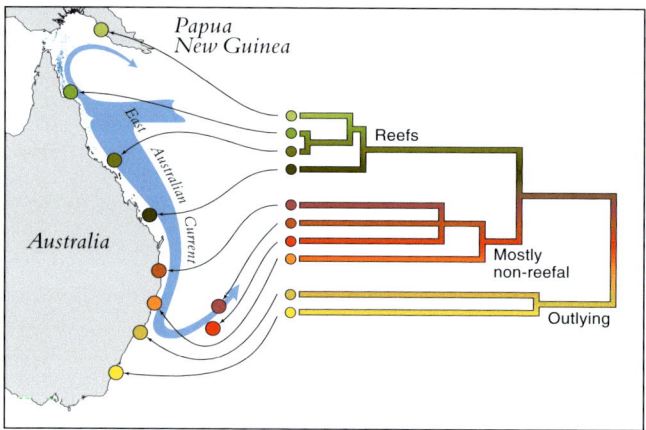

Coral species similarity in Papua New Guinea and Eastern Australia. Geographic patterns of similarity of coral species of southern Papua New Guinea and eastern Australia. There are three main biogeographic zones: tropical reefs, subtropical communities and high latitude depauperate communities. The surface current shown is the core of the East Australia Current.

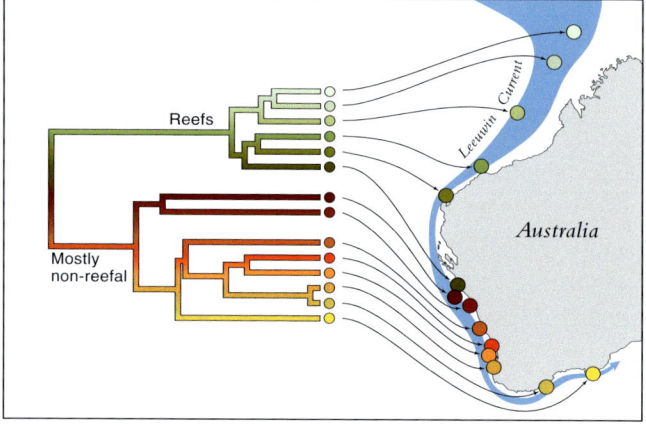

Coral species similarity in Western Australia. Geographic patterns of similarity of coral species of Western Australia. There are two main biogeographic zones: reefs and mostly non-reef communities. The surface current shown is the core of the Leeuwin Current.

distribution map (p412) provides the ancient basic template. This pattern is the present day remnant of major global geological and climatological changes, especially continental movements and major extinction events. Superimposed on this is the generic level template. This 'inherits' the family level background but is largely the outcome of more recent geological events, especially the obliteration of the Tethys Sea and the closure of the Central American Seaway. Superimposed on this again is the species level pattern. This 'inherits' the generic pattern but is largely the outcome of Pliocene to modern climates and their effect on ocean currents.

Ocean currents and diversity

Outlying areas of the Pacific have, to a large extent, subsets of the species of the centre of diversity: most peripheral patterns are therefore the result of outward dispersal from that centre, they are not the result of evolution by regional isolation.

There are essentially two reasons why there is an Indo-Pacific centre of diversity. The first is due to currents. The main currents of the equatorial Pacific flow in a westward direction to the centre (see map below), which therefore acts as a catch-all for planula larvae. The second is due to habitat diversity. The centre of diversity is the world's largest tropical archipelago of islands and contains 37% of the world's coral reefs. These islands have convoluted coastlines offering a wide range of habitats, all in close proximity. These habitats are closely linked by variable currents. Species diversity is maintained by habitat diversity and habitat accessibility: again, this diversity is not an outcome of regional evolution.

What, then, is the role of evolution in present day species diversity? There is a balance between species diversity and dispersal capability, a balance maintained or changed by ocean circulation patterns. If currents weaken, diversity will appear to increase because genetic isolation increases, creating pockets of semi-isolated taxa which are not clearly distinct from their relatives. If currents strengthen, diversity will appear to decrease because genetic isolation decreases, causing taxa to become widespread through the genetic intermixing that occurs with dispersal. These taxa will be relatively well defined.

In effect, changing currents cause genetic diversity to be endlessly 'repackaged' from large numbers of relatively isolated taxa into smaller numbers of relatively widespread taxa, then back again. This occurs in cycles that have variable impacts and frequencies. Substantial biogeographic change may occur at intervals of thousands of years or less and these may be associated with sufficient repackaging to form geographic races or varieties. Major cycles, or the

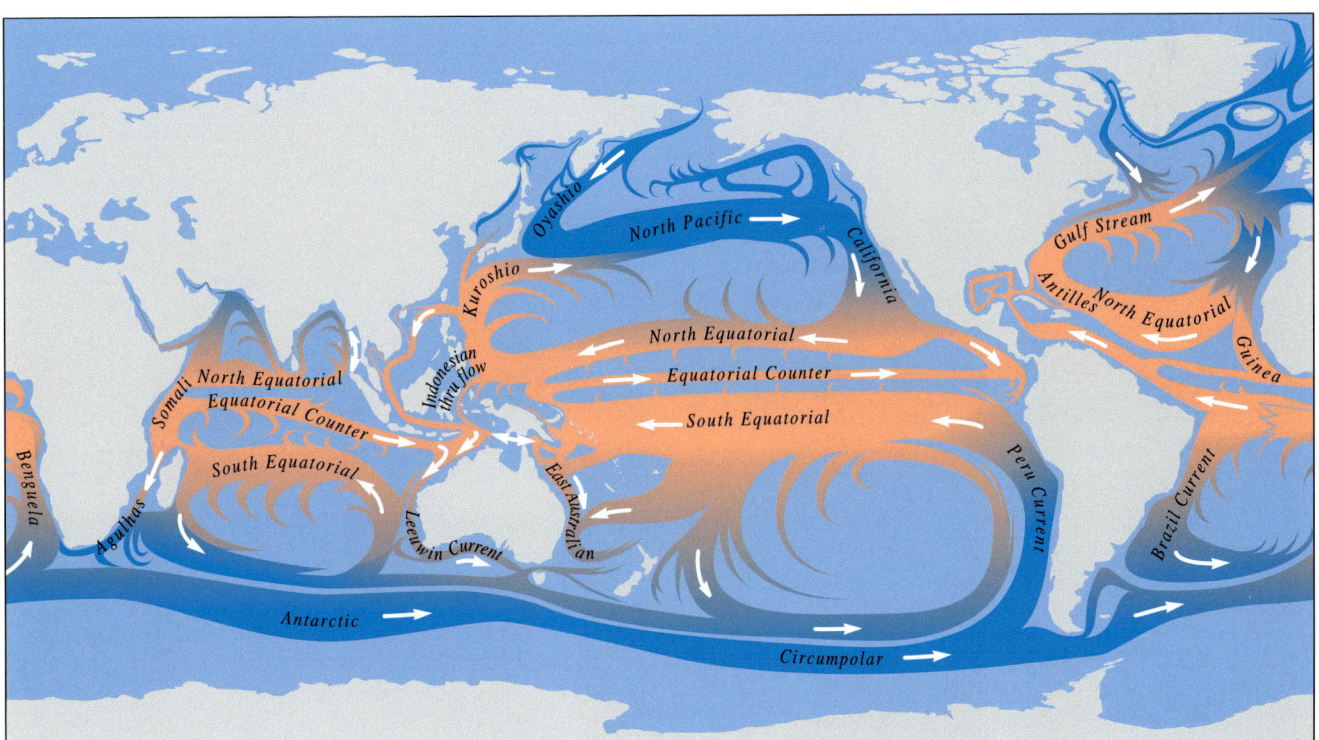

Global ocean currents. Principal global circulation patterns today. The main currents illustrated are the highways of dispersal, but there are many other pathways created by regional detail and seasonal, annual and long-term variations. All but the greatest of these currents are subject to changes associated with climate, tectonic alteration of coastlines, and fluctuations in sea level.

additive effects of many cycles, may create species level distinctions. This is further explained in 'Evolution of species' (p437); suffice to note here that none of the many biogeographic hypotheses that have been put forward to explain the mechanism of coral species formation have in fact done so. The subjects of distribution change and that of species level evolutionary change have only very tenuous connections because the former is much more frequent than the latter.

Ocean currents and dispersal

Dispersal of most terrestrial animals is an active undertaking by the organism itself. It requires a capacity for mobility, it allows choice of direction, and there is a high probability of surviving the journey. Dispersal of most marine animals is passive: it is controlled by ocean currents. It requires little effort, there is no choice of direction, and there is a low probability of survival. For these reasons, most dispersal on land is undertaken by adult animals, whilst in the ocean dispersal is usually undertaken by the most abundant life stage, the larvae. Reproduction and dispersal, for most marine animals, are therefore closely linked subjects. Only some larger vertebrates – oceanic fish, reptiles and mammals – are able to defy the currents and undertake active dispersal. For the rest of marine life, the pathways of dispersal cannot be controlled, or (except for timing) even favoured by natural selection. As outlined in the remaining chapters of this volume, this has enormous significance for all aspects of coral biology, including taxonomy and evolution.

Because most marine organisms (including corals) are dispersed by larvae, the paths of the currents are the paths of gene flow – the paths of genetic connections. Currents are largely responsible for creating, and breaking, the distribution ranges of species. For most species, these ranges are very large, so large that the genetic composition of a species in one part of the range may be very different from that in another. For corals, as with many other major groups of plants and animals, this variation in genetic composition reaches a point where variation within species becomes indistinguishable from variation between species. When this happens, geographic patterns are created where the morphological boundaries of species (the limits to what a taxonomist might call a species) become arbitrary.

Importantly, distribution patterns are the outcomes of dispersal by currents: not just the currents of today, but all the currents that existed during the evolutionary history of the species. This subject is developed in 'Evolution of species' (p437).

Ocean currents and reproductive mechanisms

Most marine invertebrates, including corals, have only minor control of dispersal; they can regulate the time of release of larvae and they can vary the time the larvae remain afloat. For corals, being sessile, timing is also important because males and females cannot

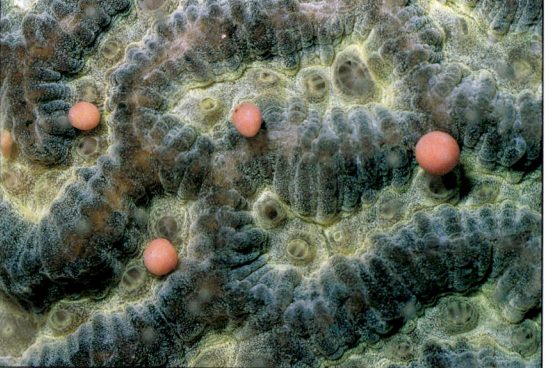

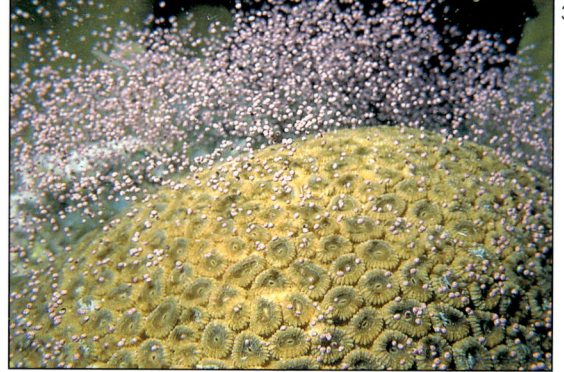

1-3 Spawning in hermaphrodite corals. (1) A *Platygyra* with egg and sperm bundles that have just been released. (2) An *Acropora* showing egg and sperm bundles that have moved to the mouths of the polyps just prior to spawning. (3) A *Favia* showing the upward moving shower of egg and sperm bundles that can come from a single colony. **1** KUWAIT **2** GREAT BARRIER REEF, AUSTRALIA **3** LORD HOWE ISLAND, SOUTH-EASTERN AUSTRALIA *Photographs:* **1, 3** *Peter Harrison* **2** *Valerie Taylor*

move into reproductive contact: species which spawn must release their gametes into the water simultaneously. As colonies may be a long way apart, this release must be both precisely and broadly synchronised. This is done in response to environmental cues, responses which have a wide array of options.

Sexual reproduction. About three-quarters of all zooxanthellate corals are hermaphrodite; the remainder have separate male and female colonies or (in solitary species) separately sexed individuals. The sexuality of corals – whether they are hermaphrodite or separately sexed – tends to be generally consistent within species and genera, although there are exceptions. There is sometimes also geographic variation within species.

About three-quarters of all zooxanthellate corals spawn eggs and sperm for external fertilisation rather than brood planulae within their bodies after internal fertilisation. Spawning is associated with high numbers of eggs and planulae, while brooding results in fewer, larger and better developed planulae. This tends to be variable, even within genera. Because of numbers, release of gametes into the ocean facilitates long distance dispersal and thereby the creation of genetic links between one reef region and another. In contrast, corals that brood planulae can readily concentrate planulae close to the parent colony. Reproductive mode may therefore involve finding balances. These include balancing local abundance (by having planulae that settle rapidly) and long distance dispersal, self fertilisation (internally or within egg and sperm bundles) versus outcrossing, and within species fertilisation versus 'hybridisation'. When transferred to biogeographic scales, these balances affect the genetic composition of species and how they form geographic patterns. As a result, a single species of coral may have different reproductive modes in different geographic regions. Some species combine brooding with long distance dispersal by rafting (see below). Others have planulae that have both rapid settling and long distance dispersal capability. Others release gametes as well as brood planulae.

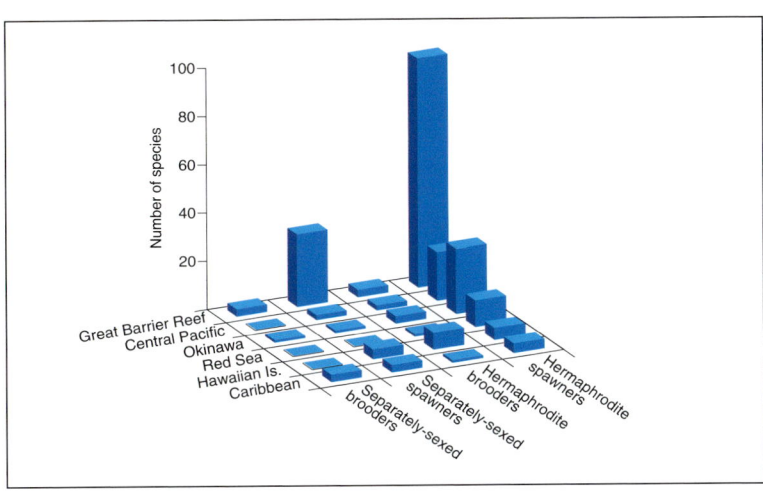

4-7 Spawning in separately sexed corals. (4), (5) Male and female *Galaxea*. The male (4) has a white globule of sperm and undeveloped eggs, the female (5) has only eggs. (6), (7) Male and female *Fungia*. The male (6) is releasing a smoke-like cloud of sperm, the female (7) is releasing a stream of eggs which are not clustered into bundles. GREAT BARRIER REEF, AUSTRALIA
Photographs: 4, 6, 7 Peter Harrison 5 Zollie Florian

Geographic variation in modes of reproduction. The Great Barrier Reef of Australia has an unusually high number of species which spawn eggs and sperm.

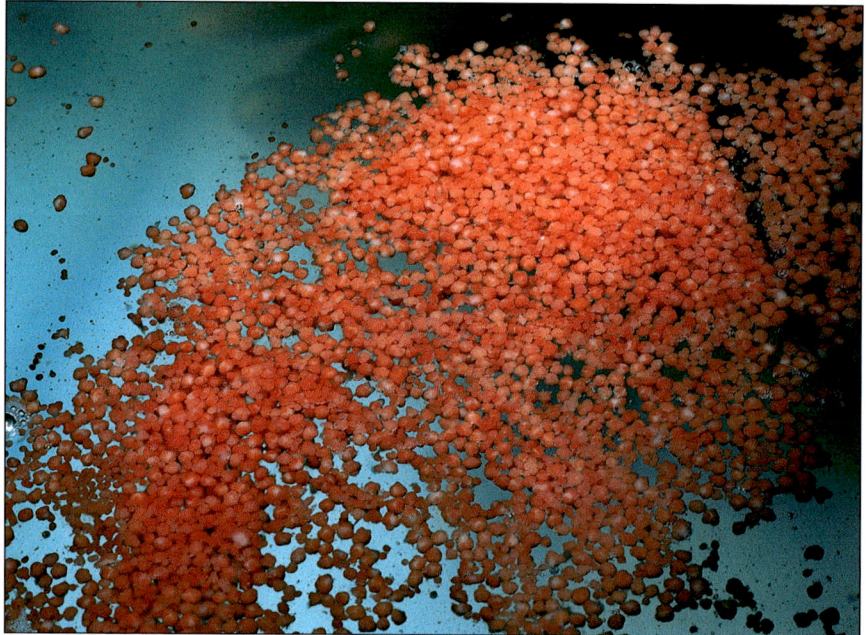

1, 2 Egg and sperm masses on the ocean surface. (1) Reproductive material forming slicks floating over reefs. These huge slicks gradually break up into countless millions of planulae. In most reef situations, this leaves little time for the development and settlement of planulae before they are transported away from the reef of origin by currents. (2) A concentration of egg and sperm bundles on the surface. Some are breaking apart and some are starting to develop into planulae. GREAT BARRIER REEF, AUSTRALIA *Photographs:* **1** *Bette Willis* **2** *Valerie Taylor.*

Synchrony. The long-term control of spawning (control of the maturation of gonads) may be temperature, day length and/or rate of temperature change (either increasing or decreasing). The short-term (getting ready to spawn) control is usually lunar. The final cue (release of spawn) is usually the time of sunset. There are many variations on these controls, probably because synchrony is usually linked to whatever environmental cues are most effective within a given region. Cues may also be biological (involving chemical messengers) as well as physical. Synchrony by chemical messengers may not only involve corals, but a host of other groups of marine organisms as well.

Brooding species, which release planulae and not gametes, can store unfertilised ova for weeks and thus require less synchrony for fertilisation. Spawning species require synchrony within a time frame of hours. This regional synchrony varies geographically: for example, on the Great Barrier Reef, spawning occurs in October/November, in Western Australia it occurs in March and in Japan in June/July. These

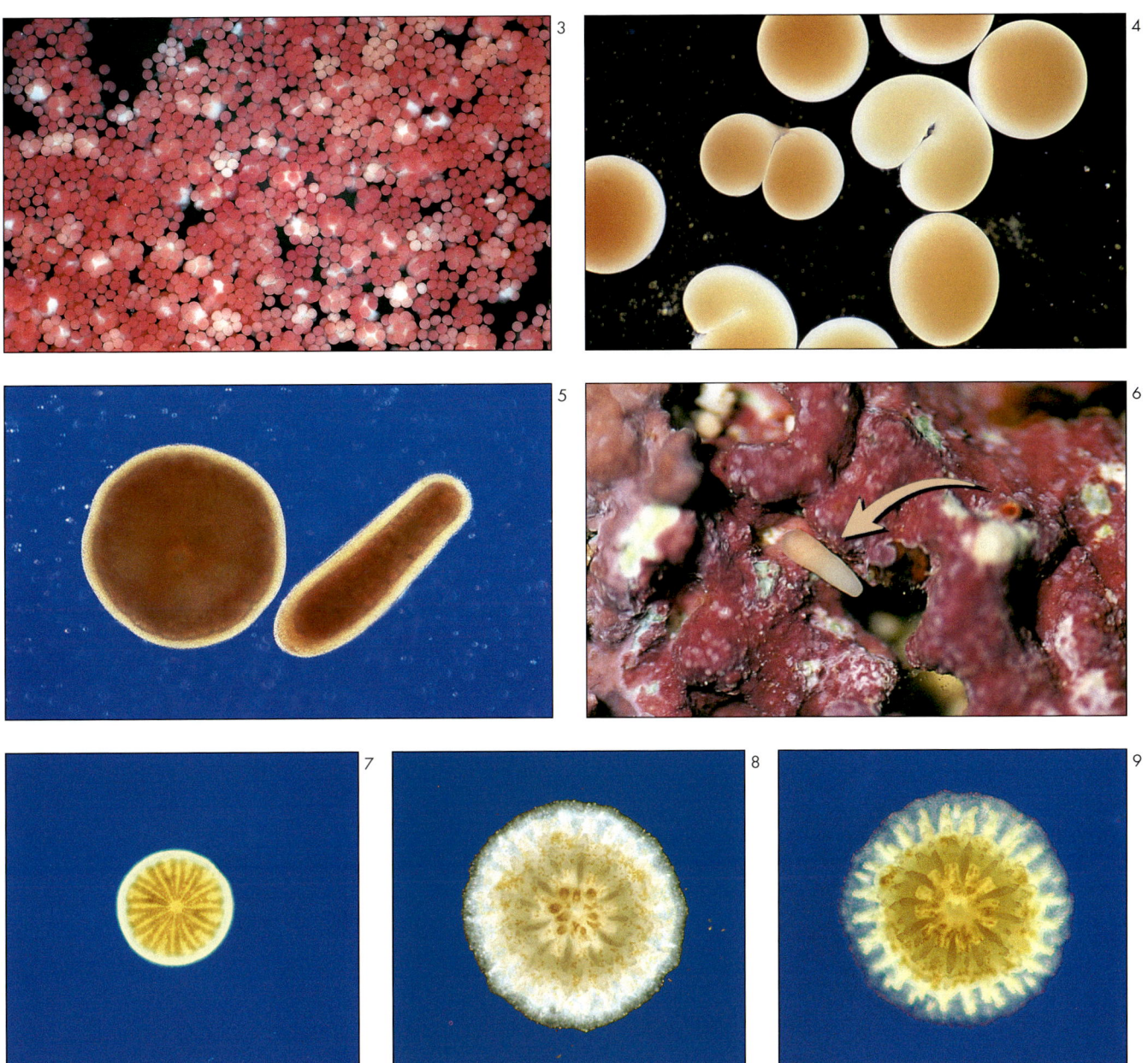

3-6 Development of planulae from fertilised eggs. (3) A mixture of eggs and developing embryos, still part of a slick on the ocean surface. (4) Different stages of embryo development. (5) Planulae of *Acropora*. Planula larvae typically change shape as they develop. The outer surface is covered with cilia allowing some motility, the interior is darkened with zooxanthellae. (6) A planula searching for a surface on which to settle. **3-5** GREAT BARRIER REEF, AUSTRALIA **6** LORD HOWE ISLAND, SOUTH-EASTERN AUSTRALIA *Photographs: Peter Harrison*

7-9 The early growth of 'spat'. (7) On day 1, after settling onto a glass slide, a planula larva becomes radially symmetrical and develops mesenteries packed with zooxanthellae. (8) By day 5 skeletal elements (the basal plate and septo-costae) are visible. (9) By day 10 all the main skeletal elements have developed. GREAT BARRIER REEF, AUSTRALIA *Photographs: Roger Steene*

differences suggest that synchrony, if under a genetic control that is not modified by local environment, could be a barrier to effective long distance dispersal: a planula larva could successfully make a long journey, only to develop into a colony which is reproductively isolated (spawns asynchronously) from other colonies in the region.

Hybridisation. In most major coral regions, not only do different colonies of the same species synchronise their spawning, but colonies of different species have the same synchrony. The outcome, in many regions, is 'mass' spawning. When mass spawning occurs, the ocean surface becomes a soup of genetic material creating endless possibilities for cross fertilisation. The extent to which hybridisation occurs — that eggs of one species are fertilised by the sperm of another — is not known. What is known is that different species within the same genus (and rarely between species of different genera) can readily hybridise and that the progeny can be normal looking corals. This is not what might be expected because if all species within a genus were to hybridise in this way, distinctions between the hybridising species would disappear and there would only be one species. In such a case, maintenance of species distinctions could only be by geographic separation. As described in the next chapter, geographic separation greatly affects the genetic composition of species, but clearly there must be other mechanisms for maintaining species distinctions within a single region. In many groups of plants and animals, there are behavioural, genetic, or anatomical barriers to hybridisation. Under conditions of mass spawning, such barriers might include reduced fertility or reduced capacity for hybrid survival. They may also include mechanisms that mask genetic change, such as a small number of genes being responsible for major morphological outcomes. Whatever the reality, it must always be remembered that in very large amounts of geographic space and evolutionary time, rare events can be the events that matter when it comes to creating change. With corals, genetic exchange between one species and another clearly occurs. Over vast amounts of space and time, this creates the interacting dynamic geographic patterns within and between species that we observe today.

Distance. An important aspect of coral reproduction, and one that underpins biogeographic patterns, concerns the capacity of corals to undertake extended ocean voyages. It is now known that planulae can spend months being transported by currents and still be competent to settle. It is likely that the planulae of most species of coral make long distance journeys and probably do so frequently. The likelihood of survival once a distant destination is reached is extremely small, but again, rare events are likely to be the events that matter. Planulae can live on the ocean surface for long periods of time because they have zooxanthellae to provide nutrients. They may even undergo partial metamorphosis while afloat. In some instances, adult colonies may also disperse on floating objects including pumice. This 'rafting' of colonies is probably an important mechanism of dispersal in some species, especially of Pocilloporidae.

Asexual reproduction. Asexual reproduction also affects the distribution and abundance of many species. Parts of branching corals are commonly scattered by storms only to regrow as a multitude of new colonies. This does not in itself aid long distance dispersal, but it may greatly enhance the reproductive output, hence the genetic impact, from the settlement of a single planula larva. There are many other mechanisms of asexual reproduction that may play the same role. These include the formation of satellite colonies in *Goniopora*, the 'bail-out' of polyps as a result of stress, the expulsion of fully developed polyps in place of planulae, the budding of acanthocauli in fungiids and other corals, and autotomy in *Diaseris*.

1, 2 Field and aquarium studies are used in conjunction to study coral reproduction. Here, a polyp of *Oculina* with a clearly visible calcareous stalk is being released from the parent colony (1). Polyps settle (2) and grow into new colonies. MEDITERRANEAN ISRAEL *Photographs: Amikam Shoob*

3 An acanthocaulus of *Fungia*. These juveniles develop long stalks. The stalks are broken by boring sponges, releasing the tiny discs as free-living corals. GREAT BARRIER REEF, AUSTRALIA *Photograph: Valerie Taylor*

4 A colony of *Platygyra* with a neoplasm or cancerous growth attached to it. Neoplasms are found in most coral species and they can be reproductive, just like the rest of the colony. Although neoplasms originate from somatic (non sex cell) mutations, they may well be a source of instant new species. BALI, INDONESIA *Photograph: author*

Other options. Other aspects of reproduction and colony formation may have biogeographic and evolutionary impacts. One is the possibility that polyploidy (the possession of more than two entire chromosome complements) occurs in corals just as it does in many groups of plants and some groups of animals. Another is the possibility that mutations in body cells can be transferred to sex cells by the formation of cancerous growths or 'neoplasms'. Neoplasms, which typically consist of many polyps that are presumed to have grown from a single mutant-containing cell, are common in corals and, in theory at least, raise the possibility of instant new species following sexual reproduction by the mutant polyps. Another type of colony formation occurs when spat or planulae of different genetic origin fuse to form postfertilisation hybrids or 'chimeras'. It is common for colonies to form from two or more original spat that fuse as they grow. Similarly, planulae can grow from two or more original eggs that fuse in early development. The outcome, in both cases, is a chimera (a colony that is not what it seems) and these may be postfertilisation 'hybrids' if developed from eggs of different species. Chimeras of many different types (reflecting differing levels of cell integration) are known in many groups of plants and in some groups of animals. However, unless there has been some sort of nuclear DNA fusion or polyploid formation, their reproductive output, if any, will be one or other of the parent species.

The Scleractinia are an ancient group of organisms and, as with many groups of plants, do not have a close genetic control over morphological characteristics. Like many groups of plants, this allows them to have a wide range of lifecycle options in reproduction and dispersal and hence biogeographic and evolutionary change. Both reproduction and dispersal are closely linked to concepts of what species are. Corals, like many major groups of plants, have fuzzy species boundaries: species are not single reproductively isolated units and, because of biogeographic variation, there are no essential distinctions between species, subspecies and hybrids. All such taxonomic units are parts of interconnected genetic patterns, created by ocean currents, and forever changing.

Further reading: This account is based on the author's book *Corals in Space and Time* (Veron, 1995) except for the information derived from maps, which have not been previously published. *Corals in Space and Time* gives sources of original information and references for further reading. For further detail about reproduction see Harrison and Wallace (1990) and Richmond and Hunter (1990).

What are species?

Traditional concept of species – variation within species merges with variation among species – 'syngameons' – genetic linkages – taxonomic uncertainty – continua in Nature – human need for units – species as arbitrary units

'What are species?'. The answer to this question goes to the very heart of the way Nature is organised. It is endlessly complex and is probably a little different for each and every species. Both practical and theoretical issues arise from it that have broad relevance to the taxonomy, biogeography, systematics and evolutionary concepts of most marine organisms, not just corals.

Traditional concepts of species. Species have long been considered to be the 'building blocks' of Nature, blocks that have a time and place of origin, which can evolve, and which maintain themselves as discrete entities by being reproductively isolated. This is a logical concept, but one which breaks down, for most species, when confronted with the realities of geographic variation. The alternative, where most species have none of these attributes, is initially less intuitive but ultimately explains very well what can actually be observed in the real world.

Concepts of what species are have long been debated, a debate driven as much by misinterpretation as information. The debate, however, has mostly been ignored by taxonomists (who often believe they 'know' what their species are for one reason or another) as well as most other biologists (who often consider it 'armchair philosophy' and thus largely irrelevant to the needs of reality). Be that as it may, species have variously been considered to be (a) self defining natural units, (b) human defined units composed of assemblages united by common descent, or (c) genetically defined units resulting from Darwinian natural selection and/or genetic (reproductive) isolation. The last involves a group of concepts, sometimes called the 'neo-Darwinian synthesis' or the 'biological species concept' (depending on a wealth of detail, see 'Glossary') which embodies the notions of the building blocks referred to above.

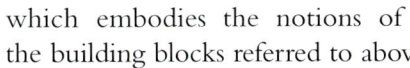
Nemenzophyllia turbida

'Species' in reality: summary. Many readers will use this book as a guide to a name. Others who wish to delve deeper may need to set aside their pre-conceptions about species in order to understand this chapter and the next. Summary points are as follows:

- Humans need 'species' as labelled units in order to communicate. As in this book, these species typically have a name, a distribution range, and all manner of morphological, ecological and other attributes. In Nature, however, living organisms (including corals) seldom form units of any kind (pp432-3). Many taxonomic decisions are, of necessity, arbitrary (pp430-2) because they must make units out of continua.

- 'Species' in this book are the nearest thing to recognisable biological units. Most are distinct (taxonomically definable) in one or more regions or countries but they become less distinct over wider geographic ranges (pp430-2). They are not units: given enough geographic space they intergrade morphologically and genetically with other species.

- There is no clear distinction between a species and a geographic subspecies, nor between a species and a natural 'hybrid'. Species can converge (forming geographic 'hybrids') or diverge (forming several geographically separated species).

   (a) Species within the same geographic location usually have continuous variation due to environmental gradients or genetic 'switches'

Previous page: Islands on an atoll. PALAU *Photograph: Roger Steene*

Opposite: *Porites porites* with polyps contracted (centre) and extended. CAYMAN ISLANDS *Photograph: Nancy Sefton*

(pp426-8). Single species are distinguished from two or more species by detailed taxonomic studies in many different environments.

(b) Species usually show wide variation over great geographic distance. This variation may be as great as variation due to environment. Single species are distinguished from two or more species by detailed taxonomic studies in many different geographic regions (p427).

(c) Species can be 'split' or 'lumped' according to the degree of continuity of variation. Such decisions may be arbitrary (p430).

(d) Reproductive isolation may be complete or partial and may vary geographically. This leads to the formation of geographic links among different species (pp430-2).

- Species are not necessarily reproductively (genetically) isolated, but syngameons (by definition, pp428-9) are. Many well defined species are syngameons, but most are components of syngameons. It is important to understand the difference between syngameons and species.

    (a) Syngameons are genetic units typically composed of many species, even genera. Species are morphological units.

    (b) Syngameons may have no distinct morphological characteristics. Species have, but usually only in limited geographic space.

    (c) Syngameons are not useful for taxonomic purposes. Species, to the maximum extent possible, are.

- Syngameons exist because their component species cannot maintain genetic cohesion over large geographic distances (p430). This cohesion is maintained, or broken, by ocean currents (pp438-41).

Variation in species

In considering what species are, in the broadest context, it is helpful to compare corals with plants (as done here with *Eucalyptus* p428). Like most plants, corals have morphological characteristics that vary according to the environment in which they grow, their geographic location, and genetic links among component populations.

Environmental variation

Most colonial corals show morphological variation within the one colony. Colonies of some species (such as the species illustrated, '2'-'5', opposite) have variable growth-forms which are at least partly genetically regulated independently of environment. However, most species exhibit variations that are clearly

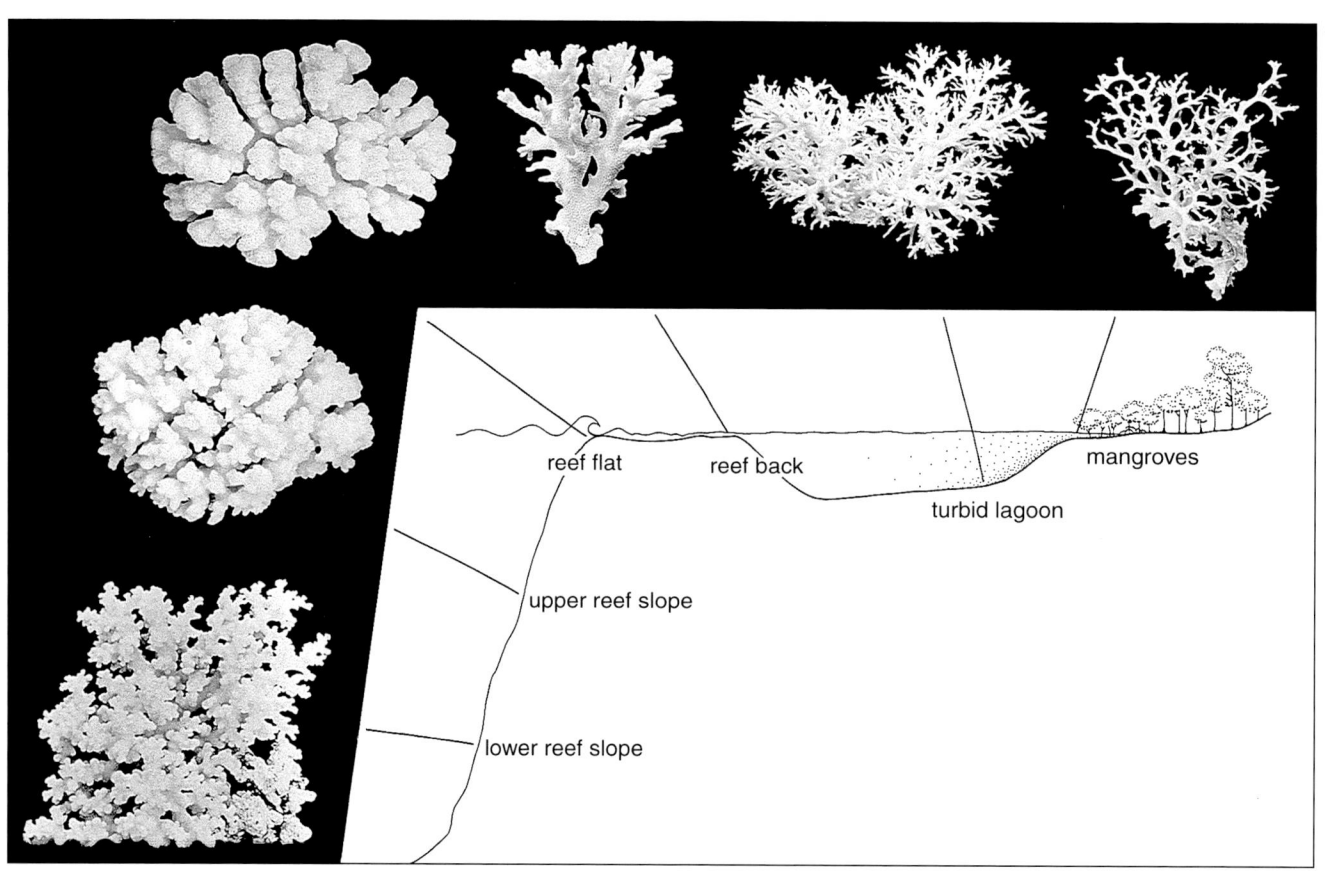

associated with different parts of the colony having different microenvironments. Differences among microenvironments include growing space, light availability, sedimentation and fish grazing. The morphological variations that result affect both growth-form (as with the *Pavona* illustrated below, bottom right) and corallite structure (as with the *Porites* illustrated p278).

Different colonies (or individuals) of all coral species vary along environmental gradients (see *Pocillopora* opposite). Divers can readily see this happening as they descend down a reef slope. This variation is also seen if a coral is transplanted from one habitat to another. It will usually change its appearance to reflect the conditions of the new habitat. Most plants do likewise; these responses are not under direct genetic control, although particular genotypes may be associated with particular morphologies.

The most important environmental factors controlling growth-form in corals are exposure to wave action, levels of illumination, sediment load and exposure to currents. The different morphologies that result must be accommodated by taxonomic descriptions and understood by persons identifying corals.

Geographic variation

There are two distinct categories of geographic variation in corals: those which are the outcome of environmental factors, and those which are genetic.

Geographic variation which is the outcome of environment is best seen in high latitude coral communities, where low temperature and non-reef habitats result in colonies which are distinct from their tropical reef dwelling counterparts. Likewise, there are whole geographic regions where the water is almost always clear (such as the Red Sea or Bahamas) and other regions where the water almost always contains sediment from the land (such as the south-east Asian coast). Corals from these regions commonly have distinct points of morphological detail which are probably due to environmental differences.

Geographic variation which is genetically based affects almost all species in some way or other. Again like most plants, the appearance of a single species changes (gradually or abruptly) from one region to the next. These changes may not concern the biologist who is interested in the species of a single country, but they will definitely concern anybody who tries to answer the question 'what are species?'. Of more practical importance, it will also concern anybody who tries to identify a species in one country from information about that species (including photographs) from another country. The taxonomic issues that arise are developed below.

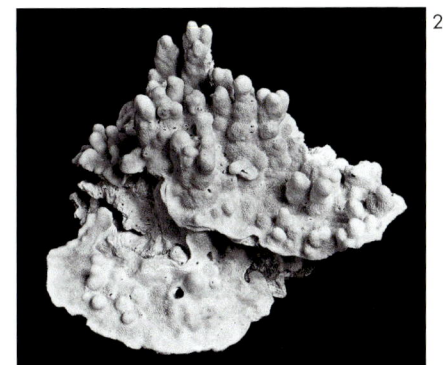

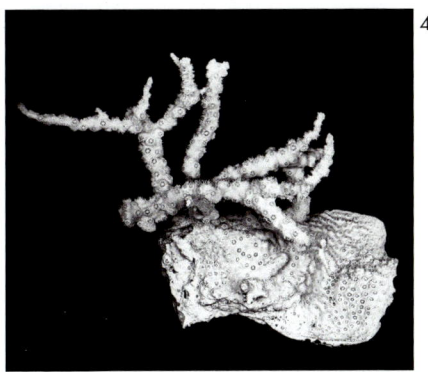

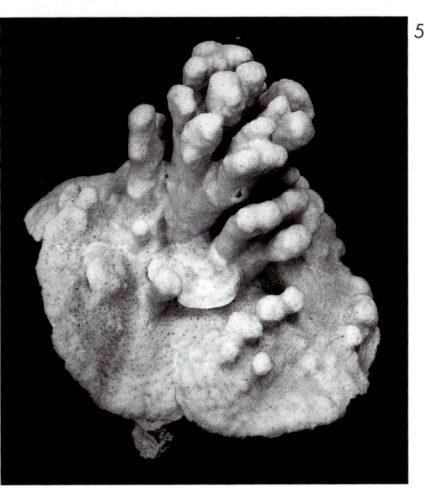

1 Variation in the skeletal structure of colonies of *Pocillopora damicornis* from a wide range of habitats at a single geographic location.

2-4 Just as in plants and other colonial animals, single colonies of corals may have two or more distinct growth-forms. These colonies of (2) *Porites annae* (3) *Merulina scabricula* and (4) *Echinopora horrida* have both flat plates and upright branches of various forms. Sometimes this variation appears to be controlled by genetic switches, turning different growth-forms on or off. Other colonies of these species may consist of plates only or branches only.

5 A small colony of *Pavona maldivensis* showing plate-like and columnar growth-forms combined. These two growth-forms are usually found in different colonies depending on the environment in which they occur. They were originally described as *Siderastrea maldivensis* Gardiner, 1905 and *Pavona (Pseudocolumnastrea) policata* Wells, 1954 and have been placed in different genera or subgenera by several taxonomists.

Genetic links among populations

It is common for a series of adjacent colonies of the same species (such as the *Lobophyllia* illustrated '4' opposite) growing in a uniform environment, to display a wide range of colours and to have a variety of morphological differences. In such cases the presence or absence of morphological continuities among colonies, or populations, can be used to distinguish groups of species (commonly called 'sibling species') from a single variable species (commonly called a 'polymorphic species'). The distinct populations of *Montipora* illustrated '5' opposite are an example. Although the differences seen are widespread, intermediate colonies are sometimes found. For this reason it is considered to be a single species.

Genetic bridges and barriers generate reticulate patterns in time and space, patterns which maintain the species' genetic heterogeneity (see 'Evolution of species', p437). The taxonomic issues that arise are of endless complexity. They inevitably lead to the conclusion that there are no fundamental differences between species and subspecies taxonomic levels.

Syngameons

The concept of the syngameon, first recognised by botanists and introduced to the marine world through corals, is important for the understanding of the evolutionary mechanisms of corals and the geographic patterns they form. A syngameon, by definition, is a reproductively isolated unit. In concept, so are neo-Darwinian species. In reality, syngameons are nothing like any of the concepts of species referred to above (p425) as they incorporate geographic variation and the spectrum of genetic links geographic variation creates among different species.

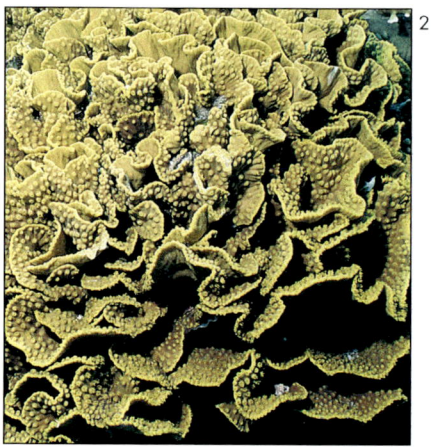

1 When exposed to strong light, *Turbinaria mesenterina* colonies such as this have highly convoluted growth-forms. In progressively deeper or more turbid water where light levels become reduced, colonies become less convoluted until they are flat sheets. PAPUA NEW GUINEA *Photograph: author*

2 *Turbinaria reniformis* is less convoluted than *T. mesenterina* (shown above), a distinction which is clear when the two species occur together. In south-east Asian coastal regions, a colony such as this would normally be restricted to shallow water, but regularly occurs to depths of 15 metres or more in the clear waters of the Red Sea. SINAI PENINSULA, EGYPT *Photograph: author*

3 There are many groups of plants that illustrate the nature of syngameons. Large genera, such as *Eucalyptus* (as it was known before the recent breakup of the genus) contain species that are able to cross-fertilise with other species. At any one place, any one species may be distinct from all other species. Over a very large area, such as the continent of Australia, relatively few species are always distinct: most intergrade, in one location or another, with other species. This creates a mosaic of morphological variations that is much larger than any individual component species. An individual species may be visible at a single location, but over an increasingly large area, that species will progressively lose its original (location specific) identity. Note that this variation cannot be accommodated in a taxonomic hierarchy. In the case of *Eucalyptus*, one major botanical revision resulted in 7 taxonomic levels: subgenera, sections, series, subseries, superspecies, species and subspecies. Only the highest of these levels (the subgenus) is reproductively isolated; various lower level taxa may be cross-fertilised. The whole genus of approximately 800 species consists of a relatively small (but unknown) number of truly reproductively isolated units. These are syngameons. TOWNSVILLE, AUSTRALIA *Photograph: author*

A syngameon then, may be a single species, or it may be a cluster of different species which have variable genetic links (genetic flow through cross-fertilisation) with other members of the syngameon. Where a syngameon contains several species, a single component species may be distinct at a single geographic location but, because it intergrades with other species at other locations, it will become submerged in a mosaic of variation at other locations. The geographic range and morphological variation of the single 'species' will depend on taxonomic decisions. These decisions will be arbitrary if they impose divisions in natural continua rather than reflect natural units. The syngameon as a whole is not morphologically visible unless its component species are determined genetically or experimentally (through cross-fertilisation trials) in every part of the species' possible distribution range.

Taxonomic issues

Before the establishment of taxonomic studies on reefs, many coral taxonomists thought that a specimen with a different growth-form implied a different species. Thus, thousands of supposedly 'new species' were described from specimens collected on expeditions and brought back to museums and universities for study. The advent of scuba diving was largely responsible for changing this. It allowed taxonomists to observe variation in corals more or less as botanists had long observed variation in trees. It allowed differences between species that grow together to be studied in great detail and it also allowed environmental variation within the same species, such as occurs down a reef slope, to be observed directly.

Geographic variation is much less simple to study as knowledge of it must be accumulated from separate studies in different countries: it cannot be directly observed. The main taxonomic issues are: (a) species become progressively less recognisable as single units with increasing geographic range, (b) taxonomists are forced to make arbitrary decisions, (c) synonymies vary geographically. In each case, the more detailed a taxonomic study is, the greater the problem becomes.

4 Colonies of *Lobophyllia hemprichii* growing together commonly exhibit a wide range of colour and corallite shapes and structures. This is regarded as a single species because complete continua can be found among many colonies. BALI, INDONESIA *Photograph: author*

5 These two colonies on a reef flat of the Great Barrier Reef are considered the same species (*Montipora digitata*) because colonies from other locations intergrade with them. Nevertheless, they represent populations with consistent differences. One colony may be descended from immigrants from Fiji 100,000 years ago, while the other may be descended from immigrants from the Philippines 5,000 years ago. These colonies may not interbreed in Australia, but relatives of both may interbreed in Papua New Guinea, where descendants may have immigrated from Micronesia 200,000 years ago. GREAT BARRIER REEF, AUSTRALIA *Photograph: Ben Stobart*

Taxonomists' view of geographic variation. Corals, along with other tropical marine life, may have a west-to-east distribution range that spans much of the tropical Indo-west Pacific, a distance of half the circumference of the earth. Over such a vast distance, species gradually change. They may do so to the point where variation within a species becomes indistinguishable from variation between species. These maps show how the same species, such as depicted in this book, might appear to different taxonomists working primarily in the Red Sea (top map), the central Indo-Pacific (middle map) and the south Pacific (bottom map). The taxonomist makes confident decisions in regions marked '1', is uncertain in '2', is very doubtful in '3' and does not recognise the species in '4'. In general, the more widespread and detailed the taxonomic study, the less certain the taxonomist will be.

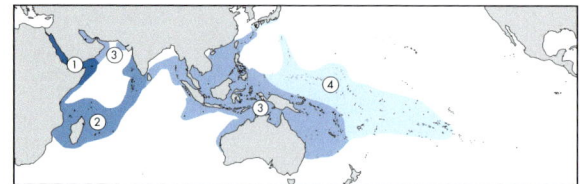

(a) the species is easily recognised in the Red Sea, becomes less clear in the central Indo-Pacific and appears to be a different species in the south Pacific.

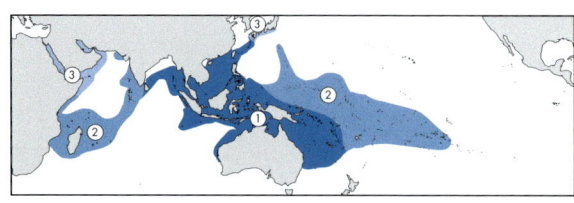

(b) the species is easily recognised in the central Indo-Pacific, becomes less clear in the Indian Ocean and south Pacific, and appears to be a different species in the Red Sea.

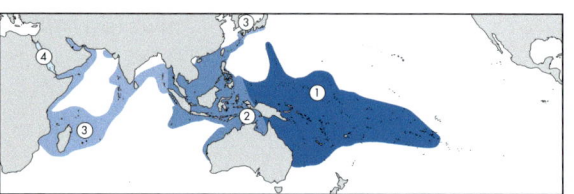

(c) the species is easily recognised in the south Pacific, becomes less clear in central Indo-Pacific and appears to be a different species in the Red Sea.

Taxonomic certainty and geographic range

With corals, as with most plants, most species *do* exist as more-or-less definable units in single geographic regions, such as the Red Sea, the Indonesian/Philippines archipelago, the Great Barrier Reef or the Caribbean. However, widespread species commonly show sufficient geographic variation that they could reasonably be divided into several separate 'sibling species' were it not for the fact that these smaller units form continua (see maps, above). Thus, for example, the majority of species of the Red Sea also occur in the central Indo-Pacific, but many are sufficiently different in the Red Sea that they would be considered to be distinct species if they were transplanted to the central Indo-Pacific.

Arbitrary decisions

In theory, taxonomy should accommodate biogeographic patterns. In practice, doing this creates issues like those created when a flat map is projected onto a sphere. The bigger the area of the map, the greater the distortion that results. It is possible to modify the flat map using different types of projections, but this only changes the nature of the distortion.

In traditional taxonomy, geographic variation within a species is accommodated by creating divisions within the species, such as varieties, races or geographic subspecies. In reality, geographic variation repeatedly

Arbitrary taxonomic decisions.

(a) Most corals of the Red Sea are distinct from their central Indo-Pacific counterparts. These may be given individual species status (left), or not (right), depending on the amount of regional distinctiveness.

(b) Many corals form a continuum, but not a cohesive unit, over a very wide range (left). If a part of this continuum is given individual species status, the remainder may form a more cohesive unit (right).

(c) Most corals that occur on rocky foreshores in subtropical regions are distinct from their tropical counterparts. The subtropical part of the continuum may be given individual species status (left), or not (right), depending on regional distinctiveness.

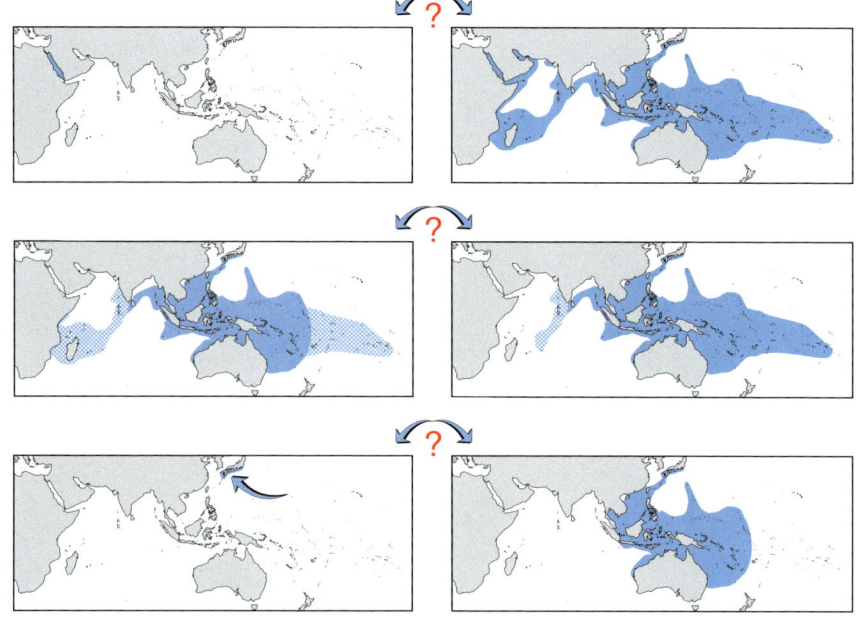

overrides the morphological boundaries of individual species. In other words, natural continua go beyond the taxonomic or morphological boundaries of individual species. This cannot be accommodated by creating divisions within species. The problem remains if the species unit is 'split' into smaller units or 'lumped' into larger units and it is not solvable by further or more detailed study. Ultimately, the only unit in Nature that is real is the syngameon, referred to above.

The reasons why syngameons are not used in operational taxonomy are (a) they can only be determined with any degree of certainty through exhaustive cross fertilisation studies in all geographic regions where their component species occur, (b) they are not likely to have distinguishing morphological characteristics and (c) they would include so many morphological species that they would need to be redivided into subunits of some kind in order to be useful.

This issue will always force taxonomists to make an arbitrary decision as to what a particular species is. Commonly encountered types of such decisions are illustrated opposite (below).

Some groups of species may be distinctive in some geographic regions and not in others. The outcome of detailed studies of these species may either be a single 'species complex' or a group of similar species. In either case, species descriptions and distribution maps artificially simplify the reality of the complex.

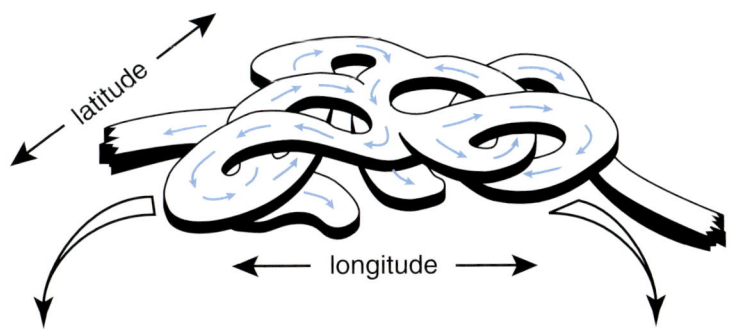

1, 2 Hypothetical representation of genetic connectivity in the distribution range of a species. The diagram represents genetic pathways created by ocean currents (blue arrows). In reality, a coral from one part of this range (eg. the Red Sea, left) may appear to be a different species to that from another part of the range (eg. Papua New Guinea, right). *Photographs: Roger Steene*

Geographic patterns and synonymies

One of the principle outcomes of any taxonomic revision is the formulation of 'synonymies'. A synonymy is usually a list of names that have been applied to a particular species together with the identity of the authors who used the names. With corals, synonymies typically include names used by different authors working in different countries at different times. They primarily focus on subspecies levels. These vary geographically just as species do, only more so. Therefore, the more comprehensive a synonymy is intended to be, the more likely it will be that it will include 'fuzzy' species boundaries and thus the more likely it will be that it includes arbitrary decisions.

General principles about species over large geographic ranges are: (1) their geographic boundaries interact with other species, (2) their morphological boundaries interact with other species, (3) their synonymy interacts with other species, (4) there are no definable distinctions between species and subspecies taxonomic levels.

Nature's organisation

The term 'species' can legitimately have a wide range of meanings. In this book, species are the most recognisable morphological units in Nature. Within a single region, species are morphologically distinguishable from other species and are genetically semi-isolated from other species. Over their full geographic range, species vary morphologically and genetically, and are not necessarily morphologically or genetically isolated from other species. Spatial patterns of species interact with temporal patterns to produce networks of genetic links. These links are not observable in single geographic regions and for this reason have not been recognised in any previous taxonomic study of corals.

In general, corals vary according to the spectrum of habitats they occupy, the size of their geographic range, and the extent of gene flow between local populations. In general, variation that is correlated with habitat (such as the variations of *Pocillopora damicornis*, illustrated p426) can be satisfactorily described and illustrated as a single species unit. However, genetic variation defies the use of units, if there are intergradations with other units. When this happens, the units can only have arbitrary boundaries. It happens with most marine life, including corals. The issues that arise are the common theme of the chapters of this volume.

(a) The distribution range of *Acropora cytherea*.

(b) The distribution range of *Acropora plana*.

(c) The distribution range of *Acropora bifurcata*.

Geographic links among species. These are three species which are distinct in the central Indo-Pacific, but these distinctions are blurred geographically. The result appears to be a 'complex' where component 'species' appear distinctive in some regions and not in others.

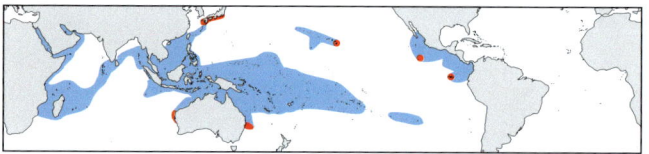

Units within species. *Porites lobata* is considered a single species although in reality it is a complex of smaller geographically separated units. If some of these geographic units (such as those indicated in red) were to be studied together without reference to specimens from other regions, each would appear to be a semi-distinct 'species'. However, when the species is studied throughout its range (indicated in blue), these differences become submerged in a broader spectrum of variation. These sorts of patterns are found in most well studied, widely distributed species.

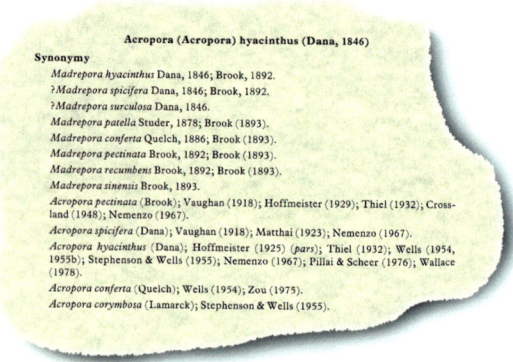

Synonymies. This is a typical 'synonymy' of a common coral *Acropora hyacinthus*, taken from the taxonomic monograph *Scleractinia of eastern Australia*. It gives details of who applied what name to this particular species in different parts of the Indo-Pacific. Synonymies are intended to accommodate geographic variation within species, but in so doing they must ultimately fail because variation in one part of the species' geographic range merges with variation among different species in other parts of the range. Some synonyms may therefore legitimately belong to more than one species.

There is a commonly held view that the species is the fundamental unit of nature and that all other units are artificial. It is also commonly believed that it is the species that evolves (each forming one fine branch of a Family Tree) and that they do so because each species is genetically (reproductively) isolated from other species. If this were true, this book, and others like it, would be relatively simple to write. But it is not true. Reproductive barriers must exist in one form or another in order to maintain distinctions between organisms that come into reproductive contact, but barriers between similar species (and this includes most plant and animal groups) are seldom impenetrable over large geographic areas or in evolutionary intervals of time. As a result, it is uncommon for species to exist as genetically isolated units. Most species which really are genetically isolated units (and are therefore syngameons) are probably both genetically uniform and morphologically distinctive (*Diploastrea heliopora*, for example). Such species probably have two particular evolutionary attributes: (a) they are incapable of hybridising with another species under any conditions and (b) they have sufficient dispersal capability to remain genetically cohesive (genetically mixed) under conditions which break up the distribution ranges of other species.

Nature, therefore, is mostly composed of continua in space and time, continua which will always defy human attempts to make taxonomic units. If we do so (as in the species units of this book) it must be remembered that these units are artificial: they lack discrete morphological boundaries and they also lack discrete geographic boundaries. Also, as our knowledge of coral biology increases, it is becoming clear that the ecological, physiological and reproductive characters of individual species

1 These two colonies of *Pocillopora damicornis* have morphological as well as colour distinctions. They may have had substantially different phylogenies and the species as a whole is a conglomeration of smaller, indistinct, units. LORD HOWE ISLAND, SOUTH-EAST AUSTRALIA
Photograph: Neville Coleman

also vary geographically. It is a human dilemma that we need taxonomic units at all to communicate information about Nature's organisation, when Nature is seldom divided into units of any kind.

Evolution of species

Reconstructing evolutionary change – geographic and genetic changes – ocean currents and genetic patterns – genetic repackaging – reticulate evolution – reticulate and neo-Darwinian evolution

In the previous chapter, it was noted that species have long been regarded as the fundamental units, or 'building blocks', of Nature – units that can be named, described and studied. It was also noted that when this concept is applied to corals over large geographic ranges, it breaks down. Thus, a species may only be distinct in limited geographic space. When that space is progressively increased (as, for example, when studying a particular species in one country, then in progressively more distant countries), it becomes progressively *not* that species.

The fundamental reason for this, as explained in the previous chapter, is that coral species exist in geographic space as interlinked patterns that change continuously so that variation *within* a single species becomes indistinguishable from variation *between* similar species. The majority of species do not exist as geographically or taxonomically definable units. It was noted that this creates a dilemma, for humans cannot easily communicate in terms of continua: they need discrete units of some form or other. In short, the units we have created in order to communicate are, at best, only an approximation of what actually occurs in Nature.

This chapter looks at the same problem, not in geographic space, but in evolutionary time. The issues that arise are not simply academic. They make it necessary to consider evolutionary change in a way that is different from that which has become generally accepted in both popular and scientific literature. This is 'reticulate evolution', a fundamentally distinct concept. The concept involves a different way of looking at what species are, the geographic patterns they make, and their change over evolutionary time. The concept itself is simple, once grasped.

Cycloseris sp.

Reconstructing evolutionary change. With corals, as with most other forms of life, evolution cannot be studied directly, it can only be inferred from other studies – palaeontology, taxonomy, biogeography and genetics. The first of these subjects, palaeontology, builds the 'big picture', as described in 'Geological history' (v1, p33). It shows something about how life was in the distant past and how it has changed, but it reveals nothing about the mechanisms of change. The last of these subjects, genetics, has the potential to reveal a great deal about evolutionary mechanisms, but this requires vast amounts of study – studies that are only just beginning. To a large extent, biologists depend on taxonomy and biogeography to provide an insight into how species evolve.

Because evolution can seldom be observed directly it must be reconstructed, as is commonly done using family 'trees'. Family trees are also one way of envisaging reticulate evolution. Such a tree might have a main trunk for the family, several large branches for

Previous page: A vegetated cay with surrounding reef. CHUUK ISLANDS, MICRONESIA *Photograph: Jim Maragos*

Opposite: A large colony of *Montipora aequituberculata*. HOUTMAN ABROLHOS ISLANDS, SOUTH-WEST AUSTRALIA *Photograph: author*

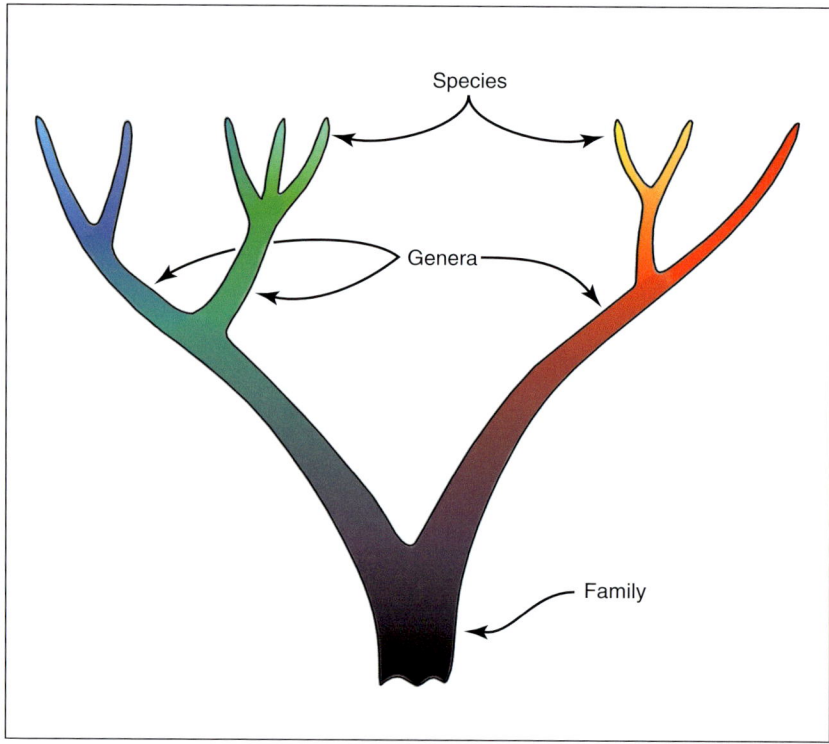

Traditional family tree. A 'typical' family tree (of a family with three genera and eight extant species) indicating some of the assumptions on which most taxonomy, biogeography and evolutionary theory is based. In contrast with the following two diagrams which illustrate reticulate evolution, this tree indicates that:

(a) species have a time (and therefore place) of origin. (The imagined circumstances of time and place of origin underpins most current biogeographic and evolutionary theories about the origins of species.) (b) species are genetically isolated, once formed. (c) taxonomic organisation is hierarchical. (d) evolution occurs by genetic changes within branches, not by gene flow between branches. (e) hybrids (not illustrated) are abnormal units. (f) extinction occurs only by termination of lineages. (g) species are units. If plotted on maps these units can undergo distribution changes, break up, and go extinct, but they will not form different units by lateral transfer (repackaging) of genes.

genera, and many fine branches for the species. Species that are currently alive will be the tips of the uppermost branches and these will be trimmed to uniform height, to represent present time. The branch tips, each representing a single species, can be converted into distribution maps, each map representing the present day distribution of one species (as in the species maps of this book). If a single branch (or species) is then sliced into a sequence of horizontal layers, and each layer is turned into a distribution map, each map will indicate the distribution of the species at progressively distant points in time. If the maps are viewed like the pages of a book, the pattern will change sequentially back in time. These changes will not be just distribution changes, they will also be genetic changes occurring in response to changes in ocean currents (see pp414-6). As a result, a species (or map) at one point in time is not the same as it is in another point in time: it has been genetically as well as geographically changed. These changes do not occur uniformly, they occur irregularly over the species' geographic range and evolutionary history. Geographic space and evolutionary time interact. The species may break apart and reform into a slightly different unit. This creates a 'reticulate' pattern in both geographic space and evolutionary time.

Ocean currents and reticulate patterns. When a species breaks apart as a result of weak ocean currents (see below) it may in time become many species. When it reforms as a result of strong ocean currents it may again become a single species or it may be more than one species. It may also contain genes from other species. This is reticulate 'repackaging', and it occurs constantly at all scales of space and time (see pp415-6). The repackaging is not confined to a single phylogeny or evolutionary clade, it involves many clades simultaneously.

Reticulate evolution. The illustration (opposite top) of reticulate evolution shows why this concept is sharply contrasted with the traditional view of evolutionary change. The diagram shows that species (a) have no time of origin, (b) have no place of origin, (c) are semi-arbitrary bits of continua rather than units, and (d) are continually repackaged in space and time.

Importantly, reticulate evolution is under physical environmental control, not biological control. The physical control changes patterns of genetic connections: it acts on genetic composition. This is again in sharp contrast with a major aspect of the traditional view where evolution is largely controlled by competition between species, resulting in morphological changes (such as longer legs or keener eyesight in a mammalian analogy). Reticulate evolution is therefore a mechanism of slow arbitrary change rather than a mechanism for progressive improvement.

As explained in the previous chapters of this volume, reticulate evolution is primarily driven by changes in surface circulation patterns causing changes to the dispersal patterns of larvae. If currents remained constant throughout evolutionary time, the oceans

Reticulate evolution. A hypothetical view of reticulate evolutionary change within a group of species (a syngameon). At the bottom (Time 0), the group forms three principal species each of which is distinct and widely dispersed by strong currents. At Time 1, the group forms many indistinct small species units that are geographically isolated because currents are weak. At Time 2, the group forms four species that are again widely dispersed by strong currents. Over the long geological interval to Time 3, the group has been repackaged several times.

Note: (a) no species has a time of origin, (b) rates of evolution and extinction are similar over the geological interval represented by the diagram, (c) there are no differences between mainstream species (represented by the thicker branches) and subspecies taxonomic levels (represented by thinner branches), (d) there are no differences between 'species' and 'hybrids', (e) the total amount of genetic information represented by this diagram has not greatly changed, but has been repackaged into different 'species' units, (f) extinction occurs through repackaging as well as terminations of lineages, (g) a single species at Time 0 and a single species at Time 3 may have few, if any, morphological distinctions. This will give an impression of evolutionary stability. Conversely, a single species at Time 0 and single species at Time 1 will give an impression of rapid evolutionary change. When combined, the overall impression may be one of 'punctuated equilibria', as is frequently observed in the fossil record.

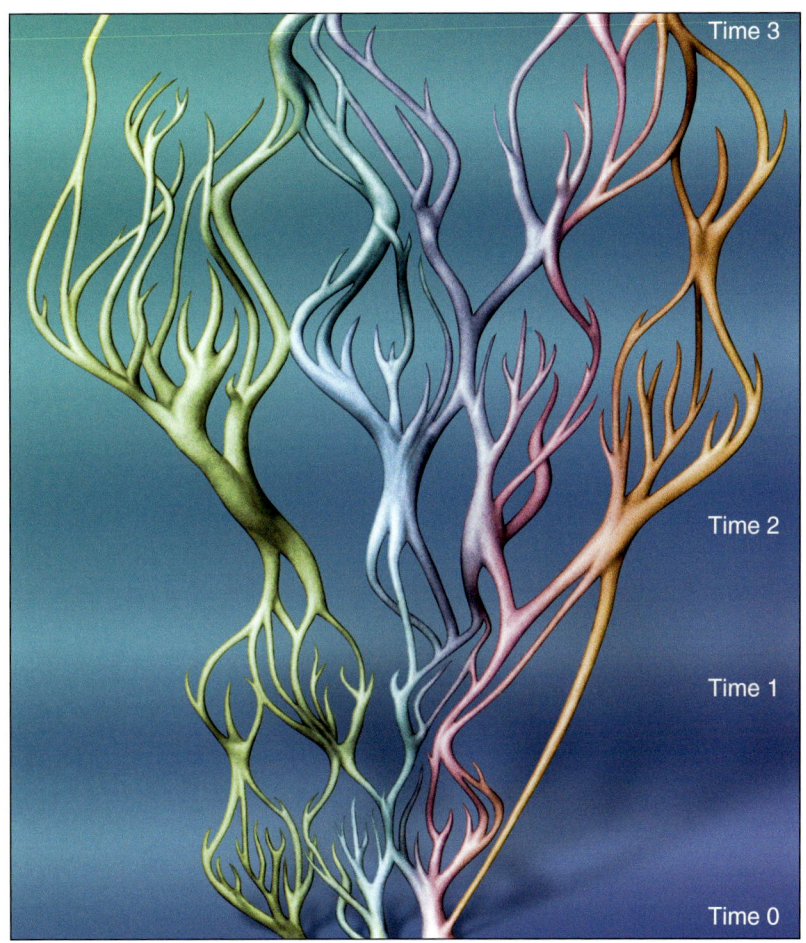

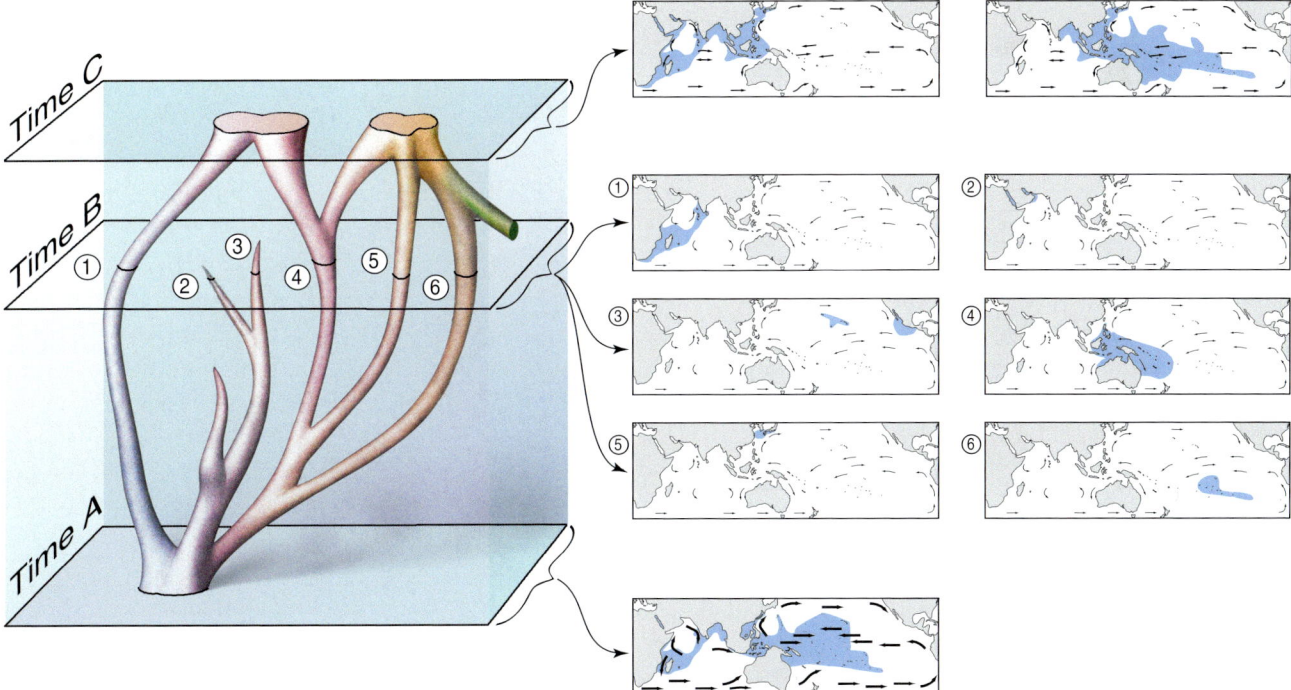

Reticulate evolution and distribution. A hypothetical view of evolutionary change in a small phylogeny (left) correlated with distribution patterns (maps, right). At the bottom (Time A), currents are strong (shown by thickness of arrows). The phylogeny exists as a single widespread species. At Time B, currents are weak and the phylogeny has broken up into six geographically isolated indistinct or 'sibling' species. (Two of these species (labled 2 and 3) are doomed to subsequent extinction.) At the top (Time C), currents are stronger again. The phylogeny has been repackaged into two widespread species which are reproductively isolated at this time. The species on the right is part of the phylogeny of another species (coloured green). Note: (a) there is no distinction between geographic (sympatric) and non-geographic (allopatric) origination, (b) evolution is driven primarily by environmental change, not biological competition.

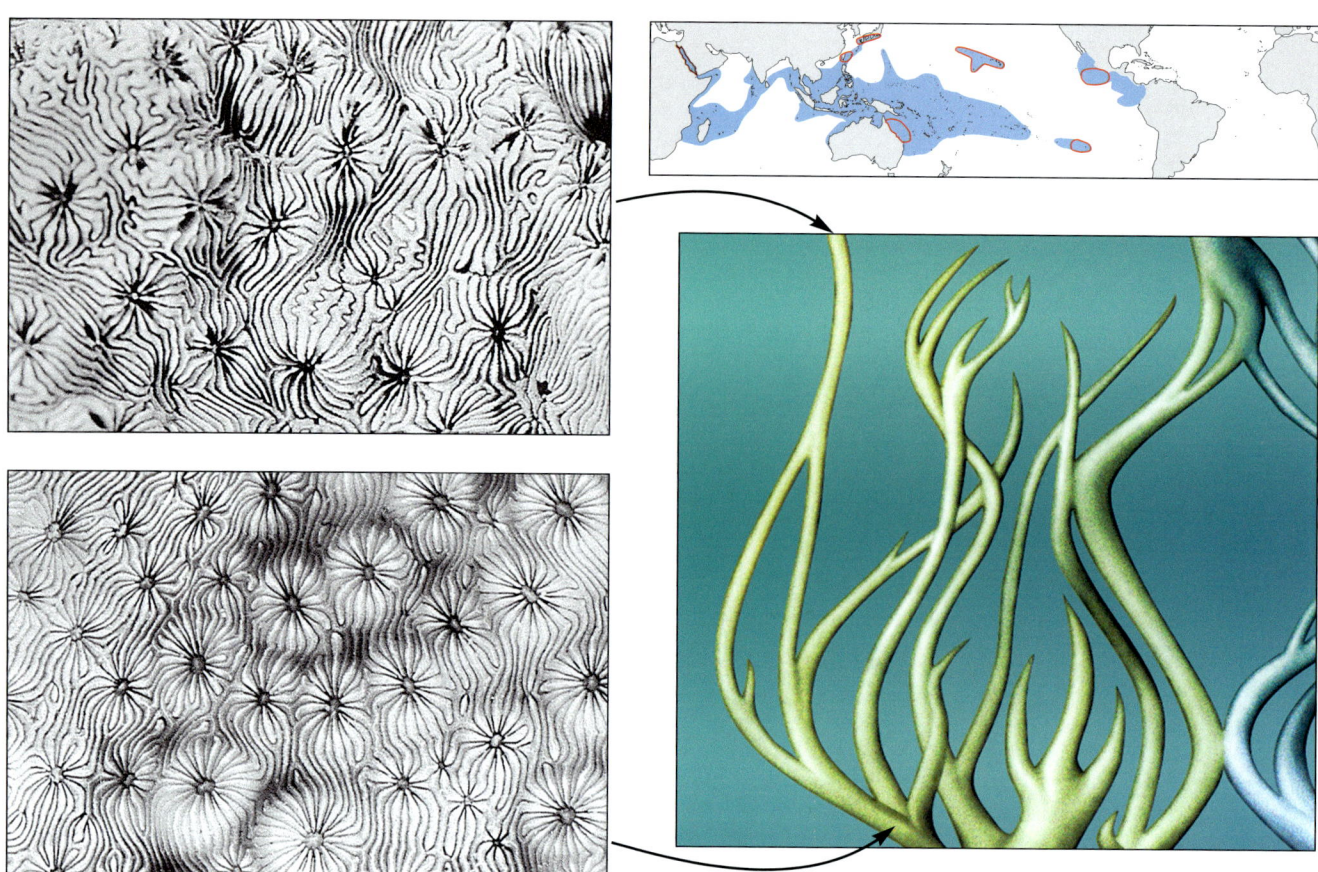

Origins of syngameons. Not all species are repackaged simultaneously or to a similar degree. Here, a single phylogeny (indicated by arrows) has sufficient dispersal capability combined with genetic isolation to be unaffected when other species over the same evolutionary interval have become repackaged or gone extinct. The photographs show one such species (*Pavona maldivensis*) at the present time (top) and as a two million year old fossil (bottom). At the present time the species spans the Indo-Pacific (see map) where it forms a series of semi-distinct geographically separated taxonomic units (those studied are indicated by red lines). This may represent a single syngameon, subject to the repackaging processes illustrated in the diagram p439 (bottom).

would be divisible into source areas (where larvae come from) and destination areas (where larvae go to) and there would be general uniformity in species and their distribution. However, with the exception of the most major currents, which are driven by the direction of rotation of the Earth, currents are not constant. Sea levels are known to fluctuate over 100 metres, oceanic passages are opened and closed by tectonic movements, and the Earth goes through cyclical climate changes due to variations in the tilt of its axis and variation in the shape of its orbital motion around the sun (Milankovitch cycles). These, and probably several other types of geo-climatic cycles, cause variations in ocean currents. These changes open and close genetic contacts: they generate reticulate patterns.

To understand how changing ocean currents affect genetic connectivity, it is helpful to imagine what would happen to dispersal if all the ocean currents stopped. Every reef, every island and every headland would be genetically isolated. In time, through the processes of Darwinian natural selection, augmented by genetic drift and mutations, the corals of each location would gradually become distinct from those of every other location. In time, every location would develop a unique fauna and every species would have a distribution range of just one location. There would be millions of species worldwide. If the very opposite is now imagined – where ocean currents are so strong and so variable that the corals of every location came into frequent genetic contact with those of every other location – every species would eventually become dispersed to every location. All species would be found everywhere they could grow and there would only be a small number of species worldwide. It is certain that these imaginary extremes never happened, but what has happened is that the Earth's climate has oscillated within these extremes, causing constant changes in dispersal, constant changes in genetic connectivity – causing reticulate patterns to arise.

The concept of reticulate evolution is highly explanatory of the observations about coral taxonomy and biogeography put forward in this book. It is also

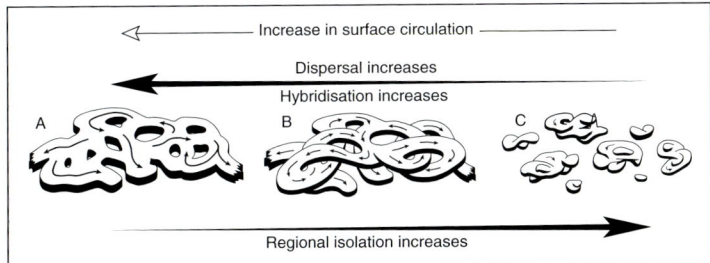

Effect of ocean currents on species fragmentation and cohesion. Hypothetical representations of the same gene pool under different current regimes. (a) currents are strong and the gene pool forms a single cohesive species, (b) currents are decreasing and the gene pool forms a single species but some parts of it are partly reproductively isolated (represented by overpasses), (c) currents are weak and the species is broken up into isolated pockets.

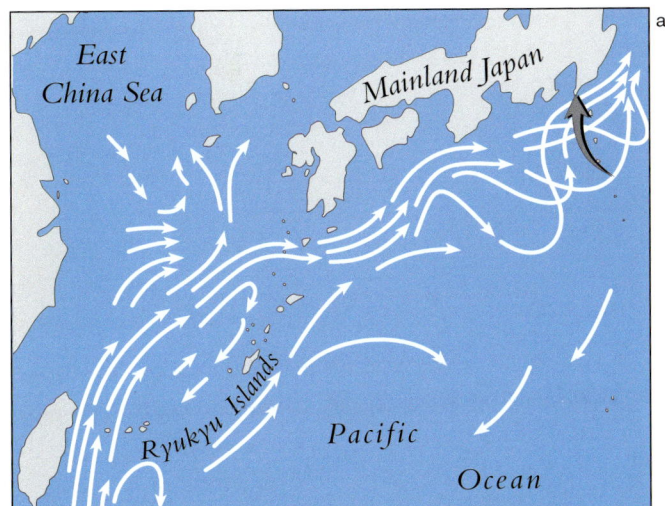
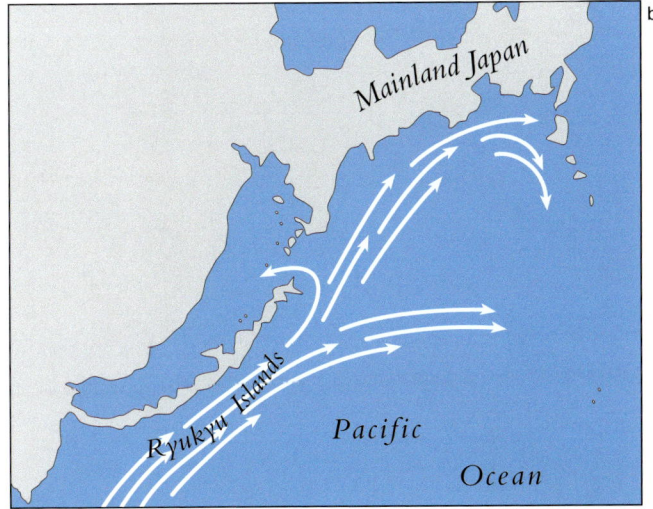

Changes in paths of dispersal over geological time. The Kuroshio, one of the world's strongest continental boundary currents, has undergone major changes in strength and direction. (a) Variations in currents south of mainland Japan occur in intervals of 10s to 100s of years and have been observed directly. Variations in intervals of 1,000s of years cannot be observed directly, but the extraordinary richness of a 5,000 year old fossil coral community at the world's northern-most limit of reef coral distribution (arrow) indicates that much stronger currents once came from the tropics than exist today. (b) Variations in currents over intervals of 10,000s to 100,000s of years can only be guessed from the drastic changes to the coastline that occurred with sea level changes combined with tectonic upheavals.

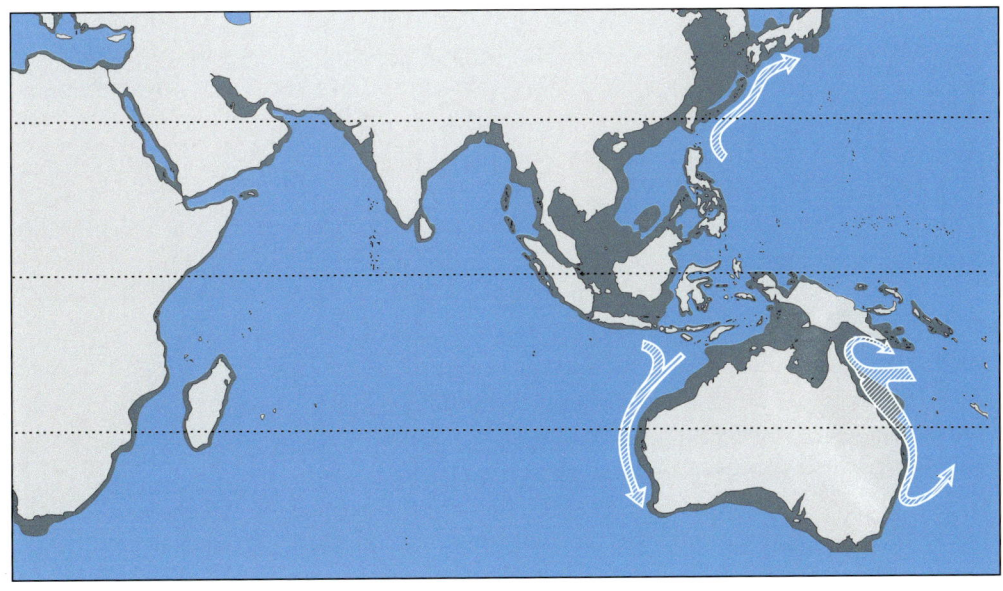

Sea level changes in the Indo-west Pacific. Over the past 18,000 years the coastline of the central Indo-Pacific has moved between the extremes shown here. Over longer intervals of time these sea level changes have been repeated and combined with substantial tectonic upheavals. These have caused major and repeated changes to the pathways of currents and hence pathways of larval dispersal.

strongly supported by what is known of coral reproduction and is starting to be supported by genetic studies.

Reticulate and neo-Darwinian evolution. How does reticulate evolution relate to the traditional view? The subject of evolution, now covered by hundreds of books since Darwin's (1859) epic *The Origin of Species*, is dominated by varying emphases on three broad notions: (a) that species exist in Nature as more-or-less reproductively isolated units of some kind, (b) that these units are comprised of individuals that compete in different ways or are modified by environment and (c) that the 'fittest' individuals survive. From these

1 Species compete for space. This community, near the world's northern-most reefs, exists only because local conditions are unfavourable for *Acropora*. It is a continuing source of propagules for more northerly (downstream) locations. RYUKYU ISLANDS, JAPAN *Photograph: author*

notions there has been an assumption that species have a time and place of origin (where phylogenies divide) and that they are units where geographic space and evolutionary time only interact in a non-reversible manner (where phylogenies divide irreversibly). The concepts of reticulate evolution and the traditional view of evolution are not compatible – they are two paradigms which become increasingly mutually exclusive with increasing amounts of space and time.

There is a point, however, where the paradigm of reticulate evolution interlinks with the traditional Darwinian view of competition and survival. Clearly, coral species compete, perhaps for the same resource (such as food, or living space), or in predator-prey relationships. These sorts of competition affect relative abundance, but only in specific environments, times and places. In other environments, times and places, the dynamics are likely to be different. Thus, one species of coral may continually outcompete another on the east coast of Africa, but this may have little bearing on how they compete on the Great Barrier Reef or how they compete on the east coast of Africa after an environmental upheaval or disease epidemic. For a start, the 'species' will not be the same in the two regions. Nevertheless, competition does affect survival.

The outcomes will initially be of local significance, but if repeated in many places, they can presumably cause evolutionary change within the species as a whole as well as total extinctions. They will not, however, create evolutionary change unless the competing species are, in fact, cohesive units of some kind.

The same issues apply to adaptations to environments: species can adapt to environmental pressures if they are cohesive units. If any given species is mobile enough to get its genes around the space it occupies in few enough generations to be able to maintain genetic cohesion, it will be able to adapt as a single unit. Clearly, this can happen with highly mobile organisms like most birds and large mammals. For other organisms, the larger the space they occupy and/or the less mobile they are, the less cohesive they will be. Marine organisms that rely on larval dispersal clearly have a minimal chance of cohesion.

It can generally be said that species that are alive today are only those small fragments that have not gone extinct. Ever changing reticulate systems are not prone to extinction and are resistant to major evolutionary movements. For this reason, evolutionary change large enough to be revealed in the fossil record is extremely

2 An *Acropora* releasing clouds of egg and sperm bundles into the water column. When eggs are fertilised they form planula larvae which can float on the sea surface for weeks, or even months, and still be competent to settle and commence growth as a coral. HOUTMAN ABROLHOS ISLANDS, SOUTH-WEST AUSTRALIA *Photograph: Bette Willis*

slow. At the present point in the Earth's history, sea levels have been approximately constant for many thousands of years and there has also been an extended interval of climatic stability. At present, the Earth is likely to be in a period of weak ocean currents that are providing the mechanism for repackaging species into small units. The stage in a repackaging sequence reached by each individual 'species' will be different as it will depend on genetic dissimilarity with other 'species', dispersal ability and abundance. These factors all vary geographically. The decision as to what an individual species is will, in concept, change with evolutionary time, but it will always be arbitrary.

Further reading: This account is based on the author's (1995) book *Corals in Space and Time*. This book indicates a multitude of sources of original information and gives references for further reading.

Corals of the World

Keys to genera and species

Keys to genera and species

The following keys are an artificial guide only, designed to be used with the text, maps and illustrations. Users of the keys should be aware that they are largely based on skeletal structures and generally do not accommodate environmental variation, geographic variation or general underwater appearance. Similar species suggested by the key and 'similar species' indicated in the text are likely to be different as the latter does reflect underwater appearance. Coral ID (Veron, Fenner and Stafford-Smith, in preparation), an electronic key offering multiple decision paths using characters of living corals, can be used instead of these keys.

Pectinia maxima

Families and genera within families are in alphabetical order. The position of the taxonomic group in the main text is given by the volume number in bold and the page number in normal text (e.g. **1**:374).

Family Acroporidae

no axial corallite
  corallites <2 mm diameter, columella absent
    branches without basal structures **Genus *Anacropora*** (**1**:168, see below)
    branches with basal structures **Genus *Montipora*** (**1**:62, see below)
  corallites obvious, columella present **Genus *Astreopora*** (**1**:434, see below)
axial corallites on branch ends **Genus *Acropora*** (**1**:176, see below)

Genus *Acropora*

colony without axial corallites Group 1 (**1**:184)
  colony submassive
    colony predominantly vertical
      corallites conspicuously exsert *A. elizabethensis* (**1**:188)
      corallites not conspicuously exsert *A. palifera* (**1**:186)
    colony predominantly horizontal *A. cuneata* (**1**:184)
  colony encrusting *A. crateriformis* (**1**:190)
colony with axial corallites
  colony with branches dominant
    radial corallites immersed Group 2 (**1**:192)
      coenosteum with tuberculae *A. togianensis* (**1**:192)
      coenosteum without tuberculae *A. cylindrica* (**1**:193)
    radial corallites exsert
      branches large
        branches irregular Group 3 (**1**:194)
          branches elongate, straight
            radial corallites very variable *A. variolosa* (**1**:197)
            radial corallites not very variable *A. hemprichii* (**1**:194)
          branches short, twisted
            axial corallites conspicuous *A. brueggemanni* (**1**:198)
            axial corallites not conspicuous
              branches highly fused *A. scherzeriana* (**1**:200)
              branches not highly fused *A. schmitti* (**1**:196)
      branches buffalohorn-like Group 4, *A. rudis* (**1**:201)
      branches elkhorn-like Group 5, *A. palmata* (**1**:202)
      branches staghorn-like
        branches mostly upright Group 6 (**1**:204)
          axial corallites distinctive
            axial corallites very exsert *A. grandis* (**1**:208)
            axial corallites moderately exsert
              radial corallites with rounded lips *A. formosa* (**1**:204)
              radial corallites with sharp lips *A. cervicornis* (**1**:206)
          axial corallites not distinctive *A. teres* (**1**:209)
        branches becoming prostrate
          radial corallites rasp-like Group 7 (**1**:210)
            colony mostly encrusting
              corallites minute *A. minuta* (**1**:210)
              corallites not minute *A. palmerae* (**1**:211)
            colony mostly branching
              centre of colony submassive
                peripheral branches short *A. pinguis* (**1**:212)
                peripheral branches well developed
                  branches distally flattened *A. irregularis* (**1**:214)
                  branches tapered *A. robusta* (**1**:216)
              centre of colony branching
                branches contorted *A. roseni* (**1**:218)
                branches tapered
                  branches highly fused *A. abrotanoides* (**1**:220)
                  branches not highly fused *A. nobilis* (**1**:222)
          radial corallites not obviously rasp-like Group 8 (**1**:224)
            branch ends conspicuously upturned
              branches compact *A. indonesia* (**1**:224)
              branches not compact

Previous page and opposite: Healthy coral communities dominated by *Acropora*. GREAT BARRIER REEF, AUSTRALIA *Photographs: Roger Steene*

             branches uneven *A. hoeksemai* (**1**:225)
           branches even
             branches large proximally *A. valenciennesi* (**1**:226)
             branches not large proximally *A. acuminata* (**1**:230)
         branch ends not conspicuously upturned
           branches irregular *A. kosurini* (**1**:231)
           branches straight *A. donei* (**1**:228)
       branches horizontal, interlocking
         radial corallites sharp-edged Group 9 (**1**:232)
           radial corallites immersed to nariform
             colony primarily prostrate *A. stoddarti* (**1**:232)
             colony not primarily prostrate *A. solitaryensis* (**1**:234)
           radial corallites tubular, appressed
             colony primarily prostrate *A. natalensis* (**1**:236)
             colony not primarily prostrate *A. divaricata* (**1**:238)
         radial corallites rounded Group 10 (**1**:240)
           basal branches mostly not distinguishable *A. branchi* (**1**:242)
           basal branches mostly distinguishable
             radial corallites in rows *A. glauca* (**1**:240)
             radial corallites not in rows *A. orbicularis* (**1**:244)
     branches middle-sized
       branches with conspicuous secondary branches Group 11 (**1**:245)
         branches and sub-branches distinct
           sub-branch ends abundant
             branches straight *A. florida* (**1**:248)
             branches curved *A. austera* (**1**:250)
           sub-branch ends not abundant
             corallites tubular on upper branches *A. wallaceae* (**1**:245)
             corallites appressed on upper branches *A. lovelli* (**1**:246)
         branches and sub-branches intergrade
           branches not highly fused
             branches twisted *A. forskali* (**1**:252)
             branches straight *A. lutkeni* (**1**:252)
           branches highly fused *A. seriata* (**1**:254)
       branches staghorn-like Group 12 (**1**:256)
         axial corallite large *A. abrolhosensis* (**1**:256)
         axial corallite not large
           radial corallites of uniform size *A. microphthalma* (**1**:258)
           radial corallites of different sizes *A. copiosa* (**1**:260)
       branches interlocking vertically
         radial corallites sharp-edged Group 13 (**1**:261)
           branches irregular *A. prolifera* (**1**:261)
           branches straight
             radial corallites with sharp lower lips
               radial corallites of one size *A. haimei* (**1**:263)
               radial corallites of variable size *A. yongei* (**1**:262)
             radial corallites cylindrical *A. pectinatus* (**1**:264)
         radial corallites irregular Group 14 (**1**:265)
           coenosteum coarse
             branches elongate *A. tortuosa* (**1**:265)
             branches compact *A. horrida* (**1**:266)
           coenosteum smooth
             radial corallites conical *A. rufus* (**1**:269)
             radial corallites not conical *A. vaughani* (**1**:268)
       branches interlock horizontally Group 15 (**1**:270)
         branches irregularly twisted *A. pruinosa* (**1**:270)
         branches straight
           axial corallites distinctive
             radial corallites aligned along branches
               corallites uniform along branches *A. tumida* (**1**:271)
               peripheral corallites exsert *A. parahemprichii* (**1**:274)
             radial corallites irregular *A. striata* (**1**:272)
           axial corallites not distinctive *A. sekiseiensis* (**1**:276)
         branches curved *A. akajimensis* (**1**:273)
     branches fine
       branches tubular Group 16 (**1**:277)
         branches irregularly twisted *A. meridiana* (**1**:280)
         branches straight
           radial corallites tubular
             branching pattern open *A. proximalis* (**1**:278)
             branching pattern compact *A. inermis* (**1**:281)
           radial corallites nariform *A. tizardi* (**1**:277)
       branches flat Group 17 (**1**:282)
         branches widely spaced *A. walindii* (**1**:287)
         branches closely spaced
           radial and axial corallites intergrade *A. elegans* (**1**:282)
           radial corallites distinct
             radial corallites small *A. simplex* (**1**:284)
             radial corallites conspicuous
               radial corallites primarily lateral *A. tenella* (**1**:285)
               radial corallites scattered *A. pichoni* (**1**:286)
 colony plate-like
   branches robust Group 18 (**1**:288)
     branches and sub-branches distinct
       branches widely spaced *A. plumosa* (**1**:288)
       branches closely spaced *A. pharaonis* (**1**:296)
     branches and sub-branches not distinct
       branches fused proximally
         branches laterally flattened *A. tutuilensis* (**1**:290)
         branches not laterally flattened
           branch ends upturned *A. downingi* (**1**:294)
           branch ends not upturned *A. clathrata* (**1**:292)
       branches fully fused *A. efflorescens* (**1**:298)
   branches fine Group 19 (**1**:299)
     corallites and branchlets intergrade *A. rambleri* (**1**:299)
     corallites and branchlets distinct
       axial corallites distinct
         branchlets upright
           axial corallites dome-shaped
             branches mostly fully fused *A. spicifera* (**1**:308)
             branches mostly distinct *A. hyacinthus* (**1**:306)
           axial corallites tubular *A. cytherea* (**1**:300)
         branchlets strongly inclined *A. bifurcata* (**1**:304)
         branchlets irregular *A. plana* (**1**:302)
       axial corallites indistinct *A. tanegashimensis* (**1**:310)
 colony digitate
   colony forms clumps
     branches cylindrical Group 20 (**1**:311)
       axial corallite small *A. ocellata* (**1**:312)
       axial corallite dome-shaped
         radial corallites not appressed *A. bushyensis* (**1**:311)
         radial corallites appressed
           branches radiate from a basal point *A. arabensis* (**1**:315)
           branches do not radiate *A. chesterfieldensis* (**1**:314)
     branches finger-like Group 21 (**1**:316)
       branches elongate, with sub-branches
         radial corallites of uniform size *A. torresiana* (**1**:316)
         radial corallites of two sizes *A. samoensis* (**1**:323)
       branches short
         axial corallites conspicuous *A. humilis* (**1**:318)
         axial corallites not conspicuous
           radial corallites increase in size *A. gemmifera* (**1**:324)
           radial corallites of uniform size
             radial corallites in rows *A. monticulosa* (**1**:320)
             radial corallites not in rows *A. retusa* (**1**:322)
           radial corallites irregular size *A. globiceps* (**1**:317)
   colony forms plates Group 22 (**1**:326)
     axial conspicuous
       radial corallites small *A. sarmentosa* (**1**:326)
       radial corallites not small
         branches taper slightly *A. digitifera* (**1**:328)
         branches taper strongly *A. japonica* (**1**:330)
     axial corallites inconspicuous *A. dendrum* (**1**:327)
     axial corallites large Group 23 (**1**:331)
       radial and axial corallites distinct *A. fastigata* (**1**:331)
       radial and axial corallites intergrade
         radial corallites bud obtusely *A. multiacuta* (**1**:332)
         radial corallites bud acutely *A. suharsonoi* (**1**:333)

                    radial corallites spiny Group 24 (**1**:334)
                        axial corallites exsert
                            radial corallites in rows *A. polystoma* (**1**:335)
                            radial corallites not in rows *A. massawensis* (**1**:336)
                        axial corallites not exsert *A. listeri* (**1**:334)
        colony corymbose to branching, radial corallites scale-like
            colony corymbose Group 25 (**1**:337)
                branchlets irregularly spaced
                    radial corallites compact *A. prostrata* (**1**:338)
                    radial corallites not compact *A. convexa* (**1**:337)
                branchlets uniformly spaced *A. millepora* (**1**:340)
            colony branching Group 26 (**1**:342)
                radial corallites of one size
                    branches sturdy *A. papillare* (**1**:345)
                    branches fine *A. loisetteae* (**1**:346)
                radial corallites of mixed sizes
                    radial corallites large, conspicuous *A. aspera* (**1**:342)
                    radial corallites small *A. pulchra* (**1**:344)
        colony forms clumps, branchlets well developed
            radial corallites appressed Group 27 (**1**:347)
                colony cushion-like plates
                    branches and sub-branches intergrade *A. mirabilis* (**1**:347)
                    branches and sub-branches distinct
                        radial corallites in a rosette *A. latistella* (**1**:348)
                        radial corallites not in a rosette *A. subulata* (**1**:350)
                colony not cushion-like plates
                    branchlets come from branches *A. kimbeensis* (**1**:352)
                    branchlets come from base
                        radial corallites nariform *A. azurea* (**1**:355)
                        radial corallites not nariform *A. nana* (**1**:354)
            radial corallites small Group 28 (**1**:356)
                radial corallites strongly appressed *A. aculeus* (**1**:356)
                radial corallites not strongly appressed *A. elegantula* (**1**:358)
            radial corallites with flaring lips Group 29 (**1**:359)
                radial corallites in a rosette
                    axial corallites long *A. tenuis* (**1**:360)
                    axial corallites short *A. vermiculata* (**1**:359)
                radial corallites not in a rosette
                    radial corallites widely spaced *A. insignis* (**1**:364)
                    radial corallites not widely spaced *A. selago* (**1**:362)
        colony forms plates Group 30 (**1**:366)
            branchlets with multiple axial corallites
                axial corallites larger than radial corallites *A. anthocercis* (**1**:368)
                axial and radial corallites similar
                    peripheral branchlets outwardly inclined
                        radial corallites with rounded lips *A. desalwii* (**1**:370)
                        radial corallites with sharp lips *A. parapharaonis* (**1**:367)
                    peripheral branchlets upright *A. willisae* (**1**:366)
            branchlets with single axial corallites
                axial and incipient axial corallites intergrade *A. batunai* (**1**:372)
                axial corallites distinct
                    radial corallites appressed
                        corallite openings nariform *A. microclados* (**1**:374)
                        corallite openings not nariform *A. macrostoma* (**1**:375)
                    radial corallites with flaring lips *A. lamarcki* (**1**:376)
        colony forms plate-like bushes
            axial corallites dominate colony shape Group 31 (**1**:378)
                axial corallites very large *A. lokani* (**1**:379)
                axial corallites middle-sized
                    corallites tubular *A. granulosa* (**1**:382)
                    corallites tapered *A. caroliniana* (**1**:380)
                axial corallites fine
                    corallites tubular *A. paniculata* (**1**:378)
                    corallites tapered *A. jacquelineae* (**1**:384)
            axial corallites not dominating Group 32 (**1**:386)
                radial corallites conspicuous
                    axial corallites dome-shaped
                        axial and radial corallites distinct *A. verweyi* (**1**:386)
                        axial and radial corallites similar

                            radial corallites widely spaced *A. squarrosa* (**1**:390)
                            radial corallites crowded *A. plantaginea* (**1**:391)
                    axial corallites tubular
                        branches distinct from axial corallites *A. rosaria* (**1**:394)
                        branches not distinct from axial corallites *A. loripes* (**1**:388)
                radial corallites not conspicuous *A. maryae* (**1**:392)
            corallites not dominating the colony structure
                radial corallites smooth-edged Group 33 (**1**:396)
                    axial corallites conspicuous
                        radial corallites not nariform *A. cophodactyla* (**1**:396)
                        radial corallites nariform *A. appressa* (**1**:397)
                    axial corallites not conspicuous *A. secale* (**1**:398)
                radial corallites sharp-edged, nariform Group 34 (**1**:400)
                    colony caespitose *A. cerealis* (**1**:402)
                    colony corymbose *A. nasuta* (**1**:400)
                radial corallites appressed Group 35 (**1**:404)
                    colony corymbose *A. valida* (**1**:404)
                    colony irregularly branched
                        corallites with sharp edges *A. variabilis* (**1**:406)
                        corallites with rounded edges *A. lianae* (**1**:407)
        colony thicket-like Group 36 (**1**:408)
            radial corallites of uniform size *A. parilis* (**1**:410)
            radial corallites of mixed sizes
                radial corallites with flaring lips *A. exquisita* (**1**:412)
                radial corallites appressed *A. kirstyae* (**1**:409)
            radial corallites vary down branches *A. gomezi* (**1**:408)
        colony forms tangles Group 37 (**1**:414)
            axial corallites distinctive
                branches very elongate
                    sub-branches frequent *A. derawanensis* (**1**:414)
                    sub-branches infrequent *A. filiformis* (**1**:418)
                branches short and twisted
                    branches prostrate *A. cardenae* (**1**:419)
                    branches upright *A. torihalimeda* (**1**:421)
            axial and incipient axial corallites intergrade
                radial corallites grouped *A. fenneri* (**1**:416)
                radial corallites widely spaced *A. russelli* (**1**:420)
        colony bottlebrush-like Group 38 (**1**:422)
            colony with distinct branches
                axial and incipient axial corallites intergrade
                    axial corallites very elongate
                        axial corallites have thick walls *A. turaki* (**1**:429)
                        axial corallites have thin walls *A. echinata* (**1**:426)
                    axial corallites not very elongate *A. awi* (**1**:422)
                axial corallites mostly distinct
                    corallites with truncated openings *A. navini* (**1**:431)
                    corallites with rounded openings *A. longicyathus* (**1**:430)
            colony bushy
                axial corallites and branches intergrade *A. speciosa* (**1**:424)
                branches distinct
                    sub-branches irregular
                        radial corallites short, crowded *A. carduus* (**1**:432)
                        radial corallites elongate, not crowded *A. subglabra* (**1**:428)
                    sub-branches increase in size proximally *A. elseyi* (**1**:433)

Genus *Anacropora*

branches robust (>10 mm thick)
    tips blunt
        coenosteum coarse, not patterned *A. spumosa* (**1**:171)
        coenosteum coarse, reticulate *A. reticulata* (**1**:172)
    tips pointed *A. puertogalerae* (**1**:170)
branches fine (<10 mm thick)
    lower lip to corallites conspicuous *A. spinosa* (**1**:173)
    lower lip to corallites inconspicuous *A. forbesi* (**1**:168)
branches very fine (<5 mm thick)
    branches twisted and tapered *A. pillai* (**1**:175)
    branches not twisted or tapered *A. matthai* (**1**:174)

Genus *Astreopora*

colony plate-like Group 1, *A. expansa* (**1**:434)
colony encrusting Group 2 (**1**:436)
    coenosteum coarse *A. incrustans* (**1**:437)
    coenosteum smooth *A. moretonensis* (**1**:436)
colony massive
    corallites small (<1.5 mm diameter) Group 3 (**1**:438)
        corallites immersed
            corallites with feathery openings *A. listeri* (**1**:439)
            corallites without feathery openings *A. randalli* (**1**:438)
        corallites on mounds *A. suggesta* (**1**:440)
    corallites large (>1.5 mm diameter) Group 4 (**1**:441)
        corallites conical, without dominating openings
            corallites evenly distributed
                coenosteum papillae form ridges *A. gracilis* (**1**:444)
                coenosteum papillae perpendicular *A. myriophthalma* (**1**:442)
                coenosteum papillae inclined *A. scabra* (**1**:441)
            corallites irregularly distributed *A. cucullata* (**1**:445)
        corallites not conical, with dominating openings
            corallites widely spaced *A. macrostoma* (**1**:447)
            corallites not widely spaced *A. ocellata* (**1**:446)

Genus *Montipora*

colony dominated by laminar growth-form
    radiating coenosteum ridges conspicuous Group 1 (**1**:66)
        corallites clearly visible
            corallites in rows between ridges *M. foliosa* (**1**:66)
            corallites irregular *M. cebuensis* (**1**:68)
        corallites barely visible
            laminae irregularly contorted *M. hodgsoni* (**1**:72)
            laminae flat *M. delicatula* (**1**:70)
    radiating coenosteum ridges not conspicuous Group 2 (**1**:73)
        tuberculae or papillae present
            tuberculae and papillae inconspicuous *M. florida* (**1**:74)
            tuberculae and papillae conspicuous
                corallites <1 mm *M. aequituberculata* (**1**:76)
                corallites >1 mm *M. crassituberculata* (**1**:78)
        tuberculae and papillae absent
            laminae thin, delicate *M. friabilis* (**1**:73)
            laminae thick, not delicate *M. capricornis* (**1**:80)
colony dominated by encrusting or massive growth-form
    corallites not very small (>1 mm diameter)
        coenosteum tuberculae prominent Group 3 (**1**:82)
            branches clearly developed
                coenosteum ridges clearly developed
                    ridges flame-shaped *M. confusa* (**1**:82)
                    ridges irregular *M. vietnamensis* (**1**:84)
                coenosteum ridges poorly developed
                    tuberculae aligned on ridges
                        peripheral tuberculae distinctive *M. undata* (**1**:86)
                        peripheral tuberculae not distinctive *M. circumvallata* (**1**:93)
                    tuberculae not aligned on ridges *M. saudii* (**1**:92)
            branches not clearly developed
                coenosteum has tuberculae *M. monasteriata* (**1**:88)
                coenosteum has papillae *M. tuberculosa* (**1**:90)
        coenosteum tuberculae not prominent
            thecal papillae clearly visible Group 4 (**1**:94)
                upgrowths present
                    papillae well developed
                        papillae intergrade with tuberculae
                            corallites distinct
                              coenosteum ridges present *M. lobulata* (**1**:95)
                              coenosteum ridges absent *M. efflorescens* (**1**:104)
                        corallites indistinct *M. flabellata* (**1**:99)
                    papillae distinct from tuberculae
                        papillae irregular in size *M. stilosa* (**1**:102)
                        papillae uniform in size *M. peltiformis* (**1**:100)
                    papillae not well developed
                        papillae compact *M. turtlensis* (**1**:96)
                        papillae widely spaced *M. dilatata* (**1**:98)
                upgrowths absent
                    corallites exsert
                        corallites large and conspicuous *M. nodosa* (**1**:110)
                        corallites not large
                            colonies massive *M. grisea* (**1**:94)
                            colonies explanate
                                colonies encrusting *M. verrilli* (**1**:107)
                                colonies not encrusting
                                  thecal papillae short *M. patula* (**1**:106)
                                  thecal papillae not short *M. effusa* (**1**:108)
                    corallites immersed
                        papillae of uniform size *M. informis* (**1**:112)
                        papillae of mixed size *M. corbettensis* (**1**:109)
            thecal papillae not clearly visible Group 5 (**1**:114)
                colonies distinctly columnar
                    coenosteum fine
                        corallites immersed *M. spongodes* (**1**:122)
                        corallites irregularly exsert *M. incrassata* (**1**:119)
                    coenosteum coarse *M. spumosa* (**1**:120)
                colonies not distinctly columnar
                    coenosteum coarse
                        surface irregular
                            corallites very conspicuous *M. cocosensis* (**1**:114)
                            corallites not very conspicuous *M. calcarea* (**1**:116)
                        surface smooth *M. mollis* (**1**:117)
                    coenosteum fine
                        corallites conspicuous *M. turgescens* (**1**:118)
                        corallites inconspicuous *M. orientalis* (**1**:114)
    corallites very small (<1 mm diameter) Group 6 (**1**:123)
        corallites occur in tuberculae
            tuberculae become verrucae-like *M. cryptus* (**1**:126)
            tuberculae not verrucae-like
                corallites microscopic *M. floweri* (**1**:124)
                corallites not microscopic *M. hoffmeisteri* (**1**:123)
        corallites do not occur in tuberculae *M. millepora* (**1**:125)
colony dominated by distinctive corallite or coenosteum characters
    corallites funnel-shaped (foveolate) Group 7 (**1**:127)
        colonies form branches *M. angulata* (**1**:127)
        colonies do not form branches
            corallites foveolate and non-foveolate *M. caliculata* (**1**:128)
            corallites all foveolate
                funnels very conspicuous *M. foveolata* (**1**:131)
                funnels not very conspicuous *M. venosa* (**1**:130)
    corallites not funnel-shaped
        coenosteum forms verrucae Group 8 (**1**:132)
            colonies form branches
                branches well defined *M. capitata* (**1**:144)
                branches not well defined *M. setosa* (**1**:137)
            colonies form flat plates
                verrucae irregularly fused *M. palawanensis* (**1**:132)
                verrucae fused into ridges
                    whole plate has ridges *M. mactanensis* (**1**:134)
                    plate margins only have ridges *M. verruculosus* (**1**:136)
            colonies submassive
                verrucae forms interlocking ridges *M. taiwanensis* (**1**:132)
                verrucae fused into nodules
                    corallites very conspicuous *M. verrucosa* (**1**:138)
                    corallites not very conspicuous
                        verrucae irregular *M. meandrina* (**1**:142)
                        verrucae not irregular *M. danae* (**1**:140)
colony dominated by branching growth-form
    branches thick Group 9 (**1**:146)
        branches becoming submassive *M. australiensis* (**1**:152)
        branches not submassive
            branches tall
                colonies with explanate bases

corallites with tall thecal papillae *M. hispida* (**1**:148)
corallites without tall thecal papillae *M. cactus* (**1**:150)
colonies without explanate bases *M. gaimardi* (**1**:146)
branches short and fused *M. hemispherica* (**1**:147)
branches twig-like
branches predominantly straight
coenosteum smooth Group 10 (**1**:153)
branches tubular
corallites with projecting lower lip *M. altasepta* (**1**:153)
corallites without lips
corallites in pits *M. samarensis* (**1**:156)
corallites not in pits *M. digitata* (**1**:154)
branches irregular *M. niugini* (**1**:158)
coenosteum forms ridges Group 11 (**1**:159)
branches with pointed ends
branches compact and fine *M. hirsuta* (**1**:159)
branches open *M. stellata* (**1**:160)
branches with rounded ends
corallites deeply pitted *M. porites* (**1**:162)
corallites not deeply pitted *M. malampaya* (**1**:163)
branches predominantly contorted Group 12 (**1**:164)
papillae well developed
papillae uniform *M. aspergillus* (**1**:167)
papillae irregular
colony surface spiny *M. echinata* (**1**:166)
colony surface not spiny *M. pachytuberculata* (**1**:166)
papillae not developed
corallites with projecting lower lips *M. spongiosa* (**1**:165)
corallites without projecting lower lips *M. kellyi* (**1**:164)

Family Agariciidae

colony not massive
corallite centres discernible
polyps aligned between collines
corallites without individual walls **Genus** *Agaricia* (**2**:170)
colony with horizontal plates and upright fronds Group 1 (**2**:171)
horizontal plates not well formed *A. tenuifolia* (**2**:171)
horizontal plates well formed *A. agaricites* (**2**:172)
colony consists of horizontal plates Group 2 (**2**:174)
valleys short, irregular
collines closely (<5 mm) spaced *A. undata* (**2**:175)
collines widely (>5 mm) spaced
>5 corallite centres per cm *A. fragilis* (**2**:174)
<5 corallite centres per cm *A. grahamae* (**2**:175)
valleys well formed *A. lamarcki* (**2**:176)
colony submassive Group 3, *A. humilis* (**2**:177)
corallites with individual walls **Genus** *Leptoseris* (**2**:202, see below)
polyps not aligned between collines **Genus** *Pavona* (**2**:178, see below)
corallite centres not discernible **Genus** *Pachyseris* (**2**:224)
colony primarily plate-like
plates primarily upright
columellae inconspicuous *P. gemmae* (**2**:224)
columellae well developed *P. rugosa* (**2**:226)
plates primarily horizontal *P. speciosa* (**2**:228)
colony bifurcated fronds
valleys shallow and even *P. foliosa* (**2**:230)
valleys irregular *P. involuta* (**2**:231)
colony massive
corallites in concavities **Genus** *Gardineroseris*, *G. planulata* (**2**:222)
corallites not in concavities **Genus** *Coeloseris*, *C. mayeri* (**2**:221)

Genus *Leptoseris*

colonies delicate fronds
fronds highly divided
fronds irregularly twisted *L. cailleti* (**2**:205)
fronds not irregularly twisted
fronds one corallite wide *L. papyracea* (**2**:204)
fronds more than one corallite wide *L. gardineri* (**2**:202)
fronds not highly divided
fronds flat sheets *L. amitoriensis* (**2**:207)
fronds not flat sheets *L. tubulifera* (**2**:206)
colonies not delicate fronds
corallites aligned or in valleys
corallites exsert
septo-costae alternate
corallites with rounded edges *L. explanata* (**2**:208)
corallites with sharp edges *L. cucullata* (**2**:214)
septo-costae sub-equal *L. solida* (**2**:211)
corallites immersed
central corallite distinguishable
peripheral corallites strongly inclined *L. scabra* (**2**:210)
peripheral corallites not strongly inclined *L. striata* (**2**:212)
central corallite not distinguishable
corallite openings rounded *L. foliosa* (**2**:219)
corallite openings not rounded
laminae have radiating ridges *L. mycetoseroides* (**2**:213)
laminae do not have radiating ridges *L. incrustans* (**2**:218)
corallites not aligned or in valleys
corallite openings rounded *L. hawaiiensis* (**2**:216)
corallite openings not rounded *L. yabei* (**2**:220)

Genus *Pavona*

colonies leafy or plate-like Group 1 (**2**:179)
colonies upright plates
plates short, thick
corallites aligned vertically *P. danai* (**2**:179)
corallites aligned horizontally *P. frondifera* (**2**:182)
plates not short, thick *P. cactus* (**2**:180)
colonies prostrate or encrusting
corallites large and conspicuous *P. explanulata* (**2**:184)
corallites not large and conspicuous *P. varians* (**2**:186)
colonies not leafy Group 2 (**2**:188)
colonies not submassive or columnar
corallites plocoid *P. maldivensis* (**2**:192)
corallites immersed *P. decussata* (**2**:194)
colonies submassive to columnar
colonies primarily submassive
corallites in short deep valleys *P. venosa* (**2**:190)
corallites not in valleys
colonies become massive *P. gigantea* (**2**:189)
colonies become nodular *P. diffluens* (**2**:188)
colonies become columnar
columns mostly cylindrical *P. clavus* (**2**:198)
columns laterally compressed *P. duerdeni* (**2**:200)
colonies explanate
corallites with large columellae *P. minuta* (**2**:196)
corallites with small columellae *P. bipartita* (**2**:197)

Family Astrocoeniidae

colonies encrusting, corallites inconspicuous
coenosteum style present **Genus** *Stylocoeniella* (**2**:4)
coenosteum styles prominent *S. armata* (**2**:4)
coenosteum styles not prominent
corallites immersed *S. guentheri* (**2**:6)
corallites conical *S. cocosensis* (**2**:8)
no coenosteum style **Genus** *Stephanocoenia*, *S. michelinii* (**2**:9)
colonies becoming branching, corallites conspicuous
septa with free margins **Genus** *Palauastrea*, *P. ramosa* (**2**:10)
septa fused with columella **Genus** *Madracis* (**2**:12, see below)

Genus *Madracis*

colonies not branching
  corallites have 6 primary septa *M. senaria* (**2**:14)
  corallites have 8 primary septa *M. formosa* (**2**:14)
  corallites have 10 primary septa
    second septal cycle well developed *M. pharensis* (**2**:12)
    second septal cycle not well developed
      colonies columnar *M. kirbyi* (**2**:16)
      colonies nodular *M. decactis* (**2**:18)
colonies branching
  branches irregularly contorted
    septa plunge steeply *M. asanoi* (**2**: 17)
    septa do not plunge steeply *M. asperula* (**2**:18)
  branches straight *M. mirabilis* (**2**:20)

Family Caryophylliidae

Genus *Heterocyathus*, *H. aequicostatus* (**2**:412)

Family Dendrophylliidae

colony attached to substrate
  colony with fronds or encrusting **Genus *Turbinaria*** (**2**:388, see below)
  colony composed of branches
    branches subdivide **Genus *Duncanopsammia*,** *D. axifuga* (**2**:405)
    branches do not subdivide **Genus *Balanophyllia*,** *B. europaea* (**2**:406)
colony not attached to substrate **Genus *Heteropsammia*,** *H. cochlea* (**2**:407)

Genus *Turbinaria*

colony not encrusting
  fronds not composed of fused corallites
    corallites >4 mm diameter
      corallites short, average 6 mm diameter *T. peltata* (**2**:390)
      corallites tubular, average 5 mm diameter *T. patula* (**2**:389)
    corallites <4 mm diameter
      fronds unifacial
        fronds not contorted
          colony surface smooth *T. reniformis* (**2**:396)
          colony surface not smooth *T. frondens* (**2**:392)
        fronds contorted *T. mesenterina* (**2**:394)
      fronds bifacial
        corallites conical *T. bifrons* (**2**:402)
        corallites not conical *T. conspicua* (**2**:403)
  fronds composed of fused corallites *T. heronensis* (**2**:404)
colony encrusting to submassive
  colony surface smooth *T. radicalis* (**2**:397)
  colony surface not smooth
    corallites conspicuously exsert *T. irregularis* (**2**:398)
    corallites not conspicuously exsert *T. stellulata* (**2**:400)

Family Euphyllidae

Note: characters of soft parts are used in this key.

colonies do not have vesicles extended during daytime
  colonies have V-shaped valleys **Genus *Catalaphyllia*,** *C. jardinei* (**2**:82)
  colonies do not have V-shaped valleys
    colonies have tentacles **Genus *Euphyllia*** (**2**:68, see below)
    colonies have mantles **Genus *Nemenzophyllia*,** *N. turbida* (**2**:84)
colonies have vesicles or mantles extended during daytime
  colonies not massive **Genus *Plerogyra*** (**2**:86)
    mantles extended during daytime *P. discus* (**2**:86)
    vesicles extended during the day
      colonies flabello-meandroid *P. sinuosa* (**2**:88)
      colonies phaceloid *P. simplex* (**2**:90)
  colonies massive **Genus *Physogyra*,** *P. lichtensteini* (**2**:92)

Genus *Euphyllia*

colonies phaceloid Group 1 (**2**:69)
  tentacles simple tubes
    primary septa strongly exsert
      tentacles elongate *E. glabrescens* (**2**:70)
      tentacles short *E. paraglabrescens* (**2**:72)
    primary septa not strongly exsert *E. cristata* (**2**:69)
  tentacles not simple tubes
    tentacles branch *E. paradivisa* (**2**:73)
    tentacles have anchor-shaped ends *E. paraancora* (**2**:74)
colonies flabello-meandroid Group 2 (**2**:76)
  tentacles are fleshy lobes *E. yaeyamaensis* (**2**:76)
  tentacles branch *E. divisa* (**2**:78)
  tentacles have anchor-shaped ends *E. ancora* (**2**:80)

Family Faviidae

colonies phaceloid
  corallites small (<5 mm diameter) **Genus *Cladocora*** (**3**:88)
    branches irregular *C. arbuscula* (**3**:90)
    branches form clumps *C. caespitosa* (**3**:88)
  corallites not small (>5 mm diameter) **Genus *Caulastrea*** (**3**:91, see below)
colonies flabello-meandroid **Genus *Erythrastrea*,** *E. flabellata* (**3**:98)
colonies massive or derived from massive
  budding intratentacular or meandroid
    colonies plocoid
      corallites not exsert **Genus *Favia*** (**3**:100, see below)
      corallites exsert **Genus *Barabattoia*** (**3**:132)
        costae alternate *B. laddi* (**3**:132)
        costae do not alternate *B. amicorum* (**3**:133)
    colonies cerioid to secondarily meandroid
      paliform lobes present
        paliform lobes not prominent **Genus *Favites*** (**3**:134, see below)
        paliform lobes prominent
          valleys <10 mm across **Genus *Goniastrea*** (**3**:156, see below)
          valleys >10 mm across
            ambulacral groove present
              colonies large **Genus *Colpophyllia*,** *C. natans* (**3**:210)
              colonies small **Genus *Manicina*,** *M. areolata* (**3**:99)
            ambulacral groove absent **Genus *Oulophyllia*** (**3**:195)
              valleys with less than 3 centres *O. bennettae* (**3**:200)
              valleys with more than 3 centres
                valleys deep, subsinuous *O. crispa* (**3**:196)
                valleys shallow, sinuous *O. levis* (**3**:198)
      paliform lobes absent or weakly developed
        paliform lobes spongy
          ambulacral groove absent **Genus *Platygyra*** (**3**:176, see below)
          ambulacral groove present **Genus *Diploria*** (**3**:206)
            ambulacral groove very distinct *D. labyrinthiformis* (**3**:206)
            ambulacral groove not distinct
              valleys irregular *D. clivosa* (**3**:209)
              valleys not irregular *D. strigosa* (**3**:208)
        paliform lobes wall-like **Genus *Leptoria*** (**3**:202)
          paliform walls solid *L. phrygia* (**3**:204)
          paliform walls not solid *L. irregularis* (**3**:202)
    colonies branching **Genus *Australogyra*,** *A. zelli* (**3**:194)
  budding extratentacular
    corallites small (<4 mm diameter)
      corallites crowded **Genus *Cyphastrea*** (**3**:240, see below)
      corallites not crowded **Genus *Plesiastrea*** (**3**:226)
        paliform lobes conspicuous *P. versipora* (**3**:226)
        paliform lobes not conspicuous *P. devantieri* (**3**:228)
    corallites middle-sized
      corallites plocoid

                colonies submassive
                    septa strongly alternate **Genus *Oulastrea*,** *O. crispata* (**3:**229)
                    septa do not alternate **Genus *Montastrea*** (**3:**212, see below)
                colonies massive to columnar **Genus *Solenastrea*** (**3:**250)
                    colonies massive *S. bournoni* (**3:**250)
                    colonies columnar *S. hyades* (**3:**251)
            corallites cerioid **Genus *Leptastrea*** (**3:**232, see below)
            corallites subplocoid **Genus *Parasimplastrea*,** *P. sheppardi* (**3:**239)
        corallites large, conspicuous
            corallites cerioid **Genus *Moseleya*,** *M. latistellata* (**3:**269)
            corallites plocoid **Genus *Diploastrea*,** *D. heliopora* (**3:**230)
    colonies explanate to branching **Genus *Echinopora*** (**3:**252, see below)

Genus *Caulastrea*

corallites >10 mm diameter
    corallites short *C. tumida* (**3:**94)
    corallites long *C. connata* (**3:**91)
corallites <10 mm diameter
    corallites straight
        septa of even thickness *C. echinulata* (**3:**97)
        septa of variable thickness *C. furcata* (**3:**92)
    corallites curved at colony margins *C. curvata* (**3:**96)

Genus *Cyphastrea*

colony massive
    12 primary septa
        costae equal or subequal
            septa irregularly exsert
                corallites widely spaced *C. agassizi* (**3:**248)
                corallites crowded
                      coenosteum spinules common *C. ocellina* (**3:**244)
                      coenosteum mostly smooth *C. japonica* (**3:**240)
            septa not irregularly exsert *C. serailia* (**3:**242)
        costae alternate *C. chalcidicum* (**3:**241)
    10 primary septa *C. microphthalma* (**3:**246)
    6 primary septa *C. hexasepta* (**3:**245)
colony branching *C. decadia* (**3:**249)

Genus *Echinopora*

colonies submassive
    septo-costae even *E. forskaliana* (**3:**264)
    septo-costae very uneven *E. robusta* (**3:**263)
colonies not submassive
    colonies do not form solid branches
        corallites >4 mm diameter
            corallites even *E. pacificus* (**3:**252)
            corallites uneven
                corallites up to 4.5 mm diameter *E. gemmacea* (**3:**258)
                corallites up to 7 mm diameter
                      corallites develop into branches *E. irregularis* (**3:**262)
                      corallites and branches discrete *E. hirsutissima* (**3:**260)
        corallites <4 mm diameter
            colony primarily explanate *E. lamellosa* (**3:**254)
            colony primarily tubular *E. ashmorensis* (**3:**256)
    colonies form solid branches
        branches are composed of single corallites
            coenosteum smooth *E. tiranensis* (**3:**265)
            coenosteum rough *E. fruticulosa* (**3:**257)
        branches not composed of single corallites
            coenosteum smooth *E. mammiformis* (**3:**266)
            coenosteum rough *E. horrida* (**3:**268)

Genus *Favia*

corallites small (<8 mm diameter) Group 1 (**3:**102)
    colonies columnar *F. stelligera* (**3:**102)
    colonies not columnar
        corallites irregular, immersed *F. fragum* (**3:**104)
        corallites regular, conical *F. laxa* (**3:**105)
corallites middle-sized (8-12 mm diameter) Group 2 (**3:**106)
    septa irregular in height
        paliform lobes well developed
            corallites inclined on the colony *F. truncatus* (**3:**113)
            corallites not inclined on the colony *F. albidus* (**3:**112)
        paliform lobes not well developed
            septa irregularly exsert *F. matthai* (**3:**106)
            septa not irregularly exsert *F. speciosa* (**3:**108)
    septa not irregular
        corallites conical *F. helianthoides* (**3:**110)
        corallites not conical
            corallites crowded, irregularly shaped *F. lacuna* (**3:**111)
            corallites not crowded or irregular *F. pallida* (**3:**114)
corallites large (>12 mm diameter) Group 3 (**3:**116)
    corallites irregularly shaped
        septa exsert *F. rotumana* (**3:**121)
        septa not exsert
            corallites exsert *F. vietnamensis* (**3:**127)
            corallites not exsert *F. rosaria* (**3:**119)
    corallites not irregularly shaped
        corallites exsert
            corallites conical
                costae strongly beaded *F. danae* (**3:**123)
                costae not strongly beaded *F. favus* (**3:**116)
            corallites not conical
                corallites compact *F. rotundata* (**3:**124)
                corallites not compact
                      paliform lobes conspicuous *F. maxima* (**3:**126)
                      paliform lobes not conspicuous *F. maritima* (**3:**130)
        corallites not exsert *F. veroni* (**3:**128)
            corallites compact or irregularly spaced *F. lizardensis* (**3:**120)
            corallites not compact, uniformly spaced
                primary septa exsert *F. leptophylla* (**3:**118)
                primary septa not exsert *F. marshae* (**3:**122)

Genus *Favites*

corallites very small (<6 mm diameter) Group 1 (**3:**136)
    corallites irregular in shape *F. stylifera* (**3:**136)
    corallites uniform in shape *F. micropentagona* (**3:**137)
corallites small (6-10 mm diameter) Group 2 (**3:**138)
    corallite angular
        paliform lobes well developed *F. pentagona* (**3:**138)
        paliform lobes weakly developed or absent
            septa very exsert *F. spinosa* (**3:**142)
            septa not exsert *F. acuticollis* (**3:**141)
    corallites rounded
        paliform lobes well developed *F. bestae* (**3:**140)
        paliform lobes absent *F. chinensis* (**3:**143)
corallites middle-sized (10-14 mm diameter) Group 3 (**3:**144)
    colony surface hillocky *F. halicora* (**3:**144)
    colony not hillocky
        septa irregularly exsert *F. russelli* (**3:**148)
        septa not irregular
            corallites angular *F. abdita* (**3:**146)
            corallites rounded *F. complanata* (**3:**150)
corallites large (>14 mm diameter) Group 4 (**3:**152)
    corallites rounded *F. vasta* (**3:**152)
    corallites angular
        septal teeth conspicuous *F. flexuosa* (**3:**154)
        septal teeth not conspicuous *F. paraflexuosa* (**3:**155)

Genus *Goniastrea*

colony predominantly monocentric
    corallites <5 mm diameter Group 1 (**3:**158)
        colonies massive

     corallites minute (<2 mm diameter) G. *minuta* (**3**:158)
     corallites not minute
      corallite walls rounded G. *edwardsi* (**3**:161)
      corallite walls acute G. *retiformis* (**3**:162)
    colonies branching G. *ramosa* (**3**:160)
   corallites >5 mm diameter Group 2 (**3**:164)
    corallites irregular G. *deformis* (**3**:167)
    corallites not irregular
     paliform lobes conspicuous
      corallites > 15 mm diameter G. *palauensis* (**3**:164)
      corallites < 15 mm diameter G. *aspera* (**3**:168)
     paliform lobes not conspicuous
      corallites aligned at colony periphery G. *peresi* (**3**:166)
      corallites not aligned on colony
       walls thin, angular G. *columella* (**3**: 165)
       walls thick, rounded G. *thecata* (**3**: 169)
colony predominantly meandroid Group 3 (**3**:170)
 colony lobed and/or has branchlets G. *pectinata* (**3**:174)
 colony massive or encrusting
  colony fully meandroid G. *australensis* (**3**:170)
  colony irregularly meandroid G. *favulus* (**3**:172)

Genus *Leptastrea*

corallites circular
 primary septa equal or subequal L. *aequalis* (**3**:235)
 primary septa not equal or subequal
  septa in three distinct cycles L. *inaequalis* (**3**:233)
  septa not in three distinct cycles L. *bottae* (**3**:234)
corallites angular
 primary septa very exsert L. *bewickensis* (**3**:232)
 primary septa not very exsert
  septa have plunging inner margins L. *transversa* (**3**:238)
  septa do not have plunging inner margins
   septa have granulated sides L. *pruinosa* (**3**:237)
   septa do not have granulated sides L. *purpurea* (**3**:236)

Genus *Montastrea*

corallites small (<5 mm diameter) Group 1 (**3**:213)
 corallites cylindrical M. *serageldini* (**3**:213)
 corallites conical
  paliform lobes well developed M. *salebrosa* (**3**:218)
  paliform lobes not well developed M. *annularis* (**3**:214)
 corallites not conical M. *curta* (**3**:216)
corallites middle-sized (5-8 mm diameter) Group 2 (**3**:219)
 septa irregular in size M. *annuligera* (**3**:220)
 septa uniform in size
  corallites compact M. *colemani* (**3**:219)
  corallites not compact M. *multipunctata* (**3**:221)
corallites large (>8 mm diameter) Group 3 (**3**:222)
 corallites conical, with narrow openings M. *cavernosa* (**3**:222)
 corallites not conical, with wide openings
  septo-costae irregular M. *valenciennesi* (**3**:224)
  septo-costae even M. *magnistellata* (**3**:225)

Genus *Platygyra*

colonies monocentric or have short valleys Group 1 (**3**:178)
 walls thick, rounded
  valleys irregularly contorted P. *yaeyamaensis* (**3**:184)
  valleys not contorted
   columellae well developed P. *crosslandi* (**3**:180)
   columellae weakly developed P. *pini* (**3**:178)
 walls not thick, rounded
  valleys mostly have several centres P. *ryukyuensis* (**3**:182)
  valleys mostly monocentric
   columellae well developed P. *carnosus* (**3**:184)
   columellae weakly developed P. *verweyi* (**3**:181)

colonies meandroid Group 2 (**3**:186)
 walls thick, rounded P. *lamellina* (**3**:192)
 walls not thick, rounded
  valleys contorted P. *contorta* (**3**:188)
  valleys not contorted
   septa irregularly exsert P. *daedalea* (**3**:191)
   septa not irregularly exsert
    top of wall acute P. *acuta* (**3**:190)
    top of wall not acute P. *sinensis* (**3**:186)

Family Fungiidae

not colonial
 free living
  central mouth dominant
   disc small, costae inconspicuous
    disc entire **Genus *Cycloseris*** (**2**:236, see below)
    disc partitioned into segments **Genus *Diaseris*** (**2**:248)
     septa thin and uniform D. *fragilis* (**2**:250)
     septa thick and wavy D. *distorta* (**2**:248)
   disc not small, costae conspicuous
    septal teeth not large lobes **Genus *Fungia*** (**2**:256, see below)
    septal teeth large lobes **Genus *Heliofungia***, H. *actiniformis* (**2**:254)
  axial furrow dominant **Genus *Ctenactis*** (**2**:286)
   axial mouth with single mouth
    tentacles with white tips C. *albitentaculata* (**2**:288)
    tentacles without white tips C. *echinata* (**2**:286)
   axial furrow with multiple mouths C. *crassa* (**2**:290)
 attached to substrate **Genus *Cantharellus*** (**2**:251)
  septa thin C. *noumeae* (**2**:252)
  septa thick
   disc button-like C. *doederleini* (**2**:251)
   disc irregular, encrusting C. *jebbi* (**2**:253)
colonial
 colony free living
  axial furrow distinct **Genus *Herpolitha*** (**2**:291)
   centres mostly restricted to axial furrow H. *weberi* (**2**:291)
   centres numerous outside axial furrow H. *limax* (**2**:292)
  axial furrow indistinct or absent
   septo-costae petaloid **Genus *Polyphyllia*** (**2**:294)
    septa grouped into parallel blocks P. *novaehiberniae* (**2**:294)
    septa radiate P. *talpina* (**2**:295)
   septa not petaloid
    corallites robust, crowded **Genus *Sandalolitha*** (**2**:296)
     colony strongly arched S. *robusta* (**2**:296)
     colony irregularly flat
      corallites evenly distributed S. *africana* (**2**:299)
      corallites restricted to central area S. *dentata* (**2**:298)
    corallites not robust or crowded
     colonies delicate domes **Genus *Zoopilus***, Z. *echinatus* (**2**:304)
     colonies not delicate domes **Genus *Halomitra*** (**2**:300)
      septal teeth smooth H. *pileus* (**2**:302)
      septal teeth ornamented H. *meierae* (**2**:300)
      septal teeth club-shaped H. *clavator* (**2**:301)
 colony attached to substrate
  colony mostly encrusting **Genus *Lithophyllon*** (**2**:306)
   central corallite distinguishable
    septo-costae thick, colony small L. *mokai* (**2**:306)
    septo-costae not thick, colony not small L. *lobata* (**2**:307)
   central corallite not distinguishable L. *undulatum* (**2**:308)
  colony mostly explanate **Genus *Podabacia*** (**2**:310)
   colony irregularly contorted P. *lankaensis* (**2**:315)
   colony not contorted
    colony with lobed margins P. *sinai* (**2**:314)
    colony with entire margins
     peripheral corallites strongly inclined P. *crustacea* (**2**:310)
     peripheral corallites not inclined P. *motuporensis* (**2**:312)

Genus *Cycloseris*

disc approximately circular
  disc strongly arched
    central arch not distinctive
      septa straight, symmetrical *C. cyclolites* (**2**:236)
      septa curved, not symmetrical *C. curvata* (**2**:240)
    central arch distinctive
      septa even within orders *C. costulata* (**2**:245)
      septa uneven *C. erosa* (**2**:241)
  disc generally flat
    septa not exsert around the mouth
      primary septa form radiating spokes *C. hexagonalis* (**2**:239)
      septa do not form radiating spokes *C. sinensis* (**2**:238)
    septa exsert around the mouth
      costae thin and even
        septo-costae alternate at the disc margin *C. vaughani* (**2**:244)
        septo-costae do not alternate *C. patelliformis* (**2**:246)
      costae thick and irregular *C. tenuis* (**2**:244)
  disc very thin and flat *C. colini* (**2**:247)
disc elliptical *C. somervillei* (**2**:242)

Genus *Fungia*

disc approximately circular
  septal teeth large and pointed
    disc mostly flat *F. scruposa* (**2**:259)
    disc not flat
      whole of upper surface arched
        septal teeth irregular *F. corona* (**2**:260)
        septal teeth uniform *F. klunzingeri* (**2**:266)
      central arch distinctive
        disc thick, heavy *F. horrida* (**2**:264)
        disc not thick *F. danai* (**2**:262)
  septal teeth saw-like, with a central rib *F. fungites* (**2**:268)
  septal teeth fine, rounded
    septa thick and wavy *F. granulosa* (**2**:276)
    septa not thick and wavy
      septal cycles even
        surface with axial mouth only
          septa teeth distinct *F. repanda* (**2**:272)
          septal teeth fine serrations *F. scabra* (**2**:274)
        surface with peripheral mouths *F. puishani* (**2**:277)
      septa cycles uneven
        septa in two distinct orders *F. fralinae* (**2**:271)
        septa not in two distinct orders
          central arch present *F. spinifer* (**2**:275)
          central arch absent *F. concinna* (**2**:270)
disc not distinctively circular
  disc irregularly shaped
    surface with peripheral mouths *F. taiwanensis* (**2**:278)
    surface without peripheral mouths *F. moluccensis* (**2**:284)
  disc with a regular shape
    tentacular lobes conspicuous *F. scutaria* (**2**:280)
    tentacular lobes inconspicuous or absent
      septa thick *F. paumotensis* (**2**:282)
      septa fine *F. seychellensis* (**2**:279)

Family Meandrinidae

colony not phaceloid
  colony meandroid
    valleys convoluted
      colony hemispherical **Genus *Ctenella***, *C. chagius* (**2**:123)
      colony columnar **Genus *Dendrogyra***, *D. cylindrus* (**2**:126)
    valleys not convoluted
      columellae present **Genus *Meandrina*** (**2**:120)
        colony submassive *M. meandrites* (**2**:120)
        colony not submassive *M. braziliensis* (**2**:122)
      columellae absent **Genus *Gyrosmilia***, *G. interrupta* (**2**:128)
  colony not meandroid
    colony plocoid **Genus *Dichocoenia***, *D. stokesi* (**2**:124)
    colony not plocoid **Genus *Montigyra***, *M. kenti* (**2**:129)
colony phaceloid **Genus *Eusmilia***, *E. fastigiata* (**2**:130)

Family Merulinidae

monticules developed **Genus *Hydnophora*** (**2**:364, see below)
monticules not developed
  colony consists of branches and/or laminae
    no basal laminae **Genus *Paraclavarina***, *P. triangularis* (**2**:374)
    with basal laminae **Genus *Merulina*** (**2**:376)
      colony consists of plates without branches *M. scheeri* (**2**:380)
      colony consists of plates with branches
        skeletal structures thick, blunt *M. ampliata* (**2**:378)
        skeletal structures thin, sharp *M. scabricula* (**2**:376)
  colony massive **Genus *Boninastrea*,** *B. boninensis* (**2**:382)
  colony columnar **Genus *Scapophyllia***, *S. cylindrica* (**2**:383)

Genus *Hydnophora*

colony entirely branching
  branches mostly <10 mm thick *H. rigida* (**2**:366)
  branches mostly >10 mm thick *H. grandis* (**2**:368)
colony not entirely branching
  colony with irregular upgrowths
    upgrowths column-like *H. pilosa* (**2**:364)
    upgrowths branch-like
      skeletal structures fine *H. bonsai* (**2**:369)
      skeletal structures coarse *H. exesa* (**2**:370)
  colony massive *H. microconos* (**2**:372)

Family Mussidae

Note: Measurements assume colonies are fully developed.

colonial
  corallites <12 mm diameter
    colony phaceloid **Genus *Blastomussa*** (**3**:4)
      corallites <7 mm diameter *B. merleti* (**3**:4)
      corallites >7 mm diameter *B. wellsi* (**3**:6)
    colony cerioid **Genus *Micromussa*** (**3**:8)
      corallites <5 mm diameter *M. diminuta* (**3**:9)
      corallites >5 mm diameter
        septa have 1-3 large teeth *M. amakusensis* (**3**:10)
        septa have a uniform series of teeth *M. minuta* (**3**:8)
  corallites >12 mm diameter
    colony cerioid to subplocoid (except *Mussismilia harttii*)
      septal teeth pointed **Genus *Acanthastrea*** (**3**:12, see below)
      septal teeth beaded **Genus *Mussismilia*** (**3**:32)
        corallites <10 mm diameter *M. braziliensis* (**3**:34)
        corallites >10 mm diameter
          colony phaceloid *M. harttii* (**3**:35)
          colony not phaceloid *M. hispida* (**3**:32)
    colony subplocoid to submeandroid **Genus *Isophyllia*** (**3**:36)
      colony subplocoid *I. rigida* (**3**:37)
      colony submeandroid *I. sinuosa* (**3**:36)
    colony phaceloid to flabello-meandroid
      corallites numerous **Genus *Lobophyllia*** (**3**:38, see below)
      corallites not numerous **Genus *Mussa***, *M. angulosa* (**3**:64)
    colony meandroid
      septal teeth very prominent **Genus *Symphyllia*** (**3**:52, see below)
      septal teeth not very prominent

            valleys mostly radiate **Genus *Mycetophyllia*** (**3**:72)
                colonies thin laminae
                    valleys distinct *M. ferox* (**3**:74)
                    valleys not distinct
                        ridges not formed *M. reesi* (**3**:75)
                        irregular radiating ridges formed *M. aliciae* (**3**:78)
                colonies not thin laminae
                    valleys well developed *M. danaana* (**3**:76)
                    valleys not well developed *M. lamarckiana* (**3**:73)
            valleys concentric **Genus *Australomussa***, *A. rowleyensis* (**3**:80)
    non-colonial
        septal teeth pointed **Genus *Scolymia*** (**3**:66)
            polyps <60 mm diameter *S. australis* (**3**:70)
            polyps >60 mm diameter
                outer wall of polyp slopes inward *S. cubensis* (**3**:66)
                outer wall of polyp slopes outward *S. vitiensis* (**3**:68)
        septal teeth lobed
            septal teeth very large **Genus *Cynarina***, *C. lacrymalis* (**3**:82)
            septal teeth not very large **Genus *Indophyllia***, *I. macassarensis* (**3**:81)

Genus *Acanthastrea*

corallites <15 mm diameter Group 1 (**3**:13)
    peripheral corallites not distinctive
        septa without very prominent teeth
            corallites not distinctly plocoid
                corallites subplocoid *A. subechinata* (**3**:13)
                corallites cerioid *A. hemprichii* (**3**:22)
            corallites distinctly plocoid
                primary septa equal *A. faviaformis* (**3**:24)
                primary septa not equal *A. regularis* (**3**:16)
        septa with very prominent teeth
            fleshy mantle covers septal spines
                corallites subplocoid *A. lordhowensis* (**3**:14)
                corallites plocoid *A. echinata* (**3**:18)
            septal spines protrude through mantle *A. brevis* (**3**:17)
    peripheral corallites distinctive
        central corallite distinctive *A. bowerbanki* (**3**:26)
        central corallite not distinctive *A. rotundoflora* (**3**:20)
corallites >15 mm diameter Group 2 (**3**:27)
    septa with <4 very prominent teeth *A. ishigakiensis* (**3**:30)
    septa with numerous equal teeth
        corallites subplocoid *A. maxima* (**3**:27)
        corallites cerioid *A. hillae* (**3**:28)

Genus *Lobophyllia*

corallites mostly phaceloid
    corallites <40 mm diameter
        septal teeth pointed
            corallites <20 mm diameter *L. diminuta* (**3**:39)
            corallites >20 mm diameter
                corallites closely compact *L. dentatus* (**3**:46)
                corallites not closely compact *L. corymbosa* (**3**:42)
        septal teeth thick lobes *L. pachysepta* (**3**:40)
    corallites >40 mm diameter *L. robusta* (**3**:50)
corallites becoming flabello-meandroid
    central and peripheral parts of colony similar
        colony flabello-meandroid *L. flabelliformis* (**3**:48)
        colony not fully flabello-meandroid
            corallites >45 mm diameter *L. serratus* (**3**:41)
            corallites <45 mm diameter *L. hemprichii* (**3**:44)
    central and peripheral parts of colony not similar *L. hataii* (**3**:47)

Genus *Symphyllia*

peripheral and central valleys similar
    valleys short, submeandroid *S. erythraea* (**3**:54)
    valleys not short, fully meandroid
        valleys <12 mm diameter
            septal teeth sharp *S. recta* (**3**:56)
            septal teeth blunt *S. wilsoni* (**3**:53)
        valleys >12 mm diameter
            walls with prominent ambulacral grooves *S. hassi* (**3**:52)
            walls without prominent ambulacral grooves
                valleys with single row of centres *S. radians* (**3**:58)
                valleys with two rows of centres *S. agaricia* (**3**:60)
peripheral and central valleys not similar *S. valenciennesi* (**3**:62)

Family Oculinidae

septa <2 mm exsert
    colony branching
        budding intratentacular **Genus *Oculina*** (**2**:98)
            branches short
                branches thick *O. patagonica* (**2**:99)
                branches thin *O. diffusa* (**2**:102)
            branches elongate
                corallites <5 mm diameter
                    septa in two cycles *O. valenciennesi* (**2**:100)
                    septa in three cycles *O. varicosa* (**2**:98)
                corallites >5 mm diameter *O. robusta* (**2**:101)
        budding extratentacular **Genus *Schizoculina*** (**2**:104)
            branches primarily upright *S. fissipara* (**2**:105)
            branches primarily prostrate *S. africana* (**2**:104)
    colony not branching **Genus *Simplastrea***, *S. vesicularis* (**2**:103)
septa >2 mm exsert **Genus *Galaxea*** (**2**:106, see below)

Genus *Galaxea*

colonies encrusting to massive
    corallites >5 mm diameter *G. fascicularis* (**2**:108)
    corallites <5 mm diameter
        septa in two cycles
            corallites >3 mm diameter *G. astreata* (**2**:110)
            corallites <3 mm diameter *G. paucisepta* (**2**:112)
        septa in three cycles *G. longisepta* (**2**:116)
colonies branching
    branches lobed or truncated
        septa not very exsert *G. cryptoramosa* (**2**:114)
        septa very exsert *G. acrhelia* (**2**:115)
    branches well defined *G. horrescens* (**2**:107)

Family Pectiniidae

corallites more conspicuous than coenostial structures
    non-colonial **Genus *Echinomorpha***, *E. nishihirai* (**2**:333)
    colonial
        coenostial pits present **Genus *Oxypora*** (**2**:334, see below)
        coenostial pits absent
            corallites not inclined **Genus *Echinophyllia*** (**2**:322, see below)
            corallites inclined **Genus *Mycedium*** (**2**:342, see below)
coenostial structures very conspicuous **Genus *Pectinia*** (**2**:348, see below)

Genus *Echinophyllia*

laminae thick becoming submassive
    corallites immersed *E. taylorae* (**2**:327)
    corallites exsert *E. orpheensis* (**2**:328)
laminae thin
    corallites immersed or nearly so
        septo-costae in three orders *E. costata* (**2**:330)
        septo-costae not in three orders
            columellae well developed *E. patula* (**2**:326)
            columellae not well developed *E. pectinata* (**2**:331)

corallites not immersed
  corallites <6 mm diameter *E. echinoporoides* (**2**:322)
  corallites >6 mm diameter
    central corallite inconspicuous or absent *E. aspera* (**2**:324)
    central corallite conspicuous *E. echinata* (**2**:332)

Genus *Mycedium*

colonies mostly flat laminae
  costal ridges from corallites conspicuous *M. umbra* (**2**:342)
  costal ridges from corallites not conspicuous *M. elephantotus* (**2**:344)
colonies mostly convoluted laminae
  corallites >6 mm diameter *M. mancaoi* (**2**:343)
  corallites <6 mm diameter
    corallites well developed *M. robokaki* (**2**:346)
    corallites superficial *M. steeni* (**2**:347)

Genus *Oxypora*

laminae with conspicuous costal ridges *O. crassispinosa* (**2**:334)
laminae without conspicuous costal ridges
  corallites indistinct *O. convoluta* (**2**:340)
  corallites distinct
    costal spines abundant *O. lacera* (**2**:336)
    costal spines uncommon or absent
      corallites not very exsert *O. glabra* (**2**:338)
      corallites exsert *O. egyptensis* (**2**:341)

Genus *Pectinia*

colonies not branching
  colony with tall bifacial laminae
    costae have well developed teeth *P. alcicornis* (**2**:356)
    costae do not have well developed teeth *P. maxima* (**2**:349)
  colony without tall bifacial laminae
    colony with radiating valleys
      valleys mostly straight *P. lactuca* (**2**:350)
      valleys mostly not straight *P. africanus* (**2**:353)
    colony without radiating valleys
      laminae with radiating walls *P. ayleni* (**2**:352)
      laminae without radiating walls *P. paeonia* (**2**:354)
colonies branching
  colonies with basal laminae *P. elongata* (**2**:360)
  colonies without basal laminae
    branches >3 mm diameter *P. teres* (**2**:358)
    branches <3 mm diameter *P. pygmaeus* (**2**:361)

Family Pocilloporidae

colonies have verrucae **Genus *Pocillopora*** (**2**:24, see below)
colonies do not have verrucae
  branches fine **Genus *Seriatopora*** (**2**:46, see below)
  branches robust **Genus *Stylophora*** (**2**:56, see below)

Genus *Pocillopora*

Note: Because of the wide variation within species due to environment as well as geographic location, groupings of species in this key may not indicate similarity observed underwater.

verrucae intergrade with branches *P. damicornis* (**2**:26)
verrucae do not intergrade with branches
  colony consists of upright branches
    colony with short compact branches
      verrucae distributed uniformly on branches
        verrucae longer than wide *P. verrucosa* (**2**:28)
        verrucae not longer than wide
          verrucae rounded, medium sized *P. meandrina* (**2**:30)
          verrucae angular, small *P. elegans* (**2**:34)
      verrucae distributed irregularly on branches *P. ligulata* (**2**:38)
    colony with elongate branches
      verrucae longer than wide
        branches irregular *P. capitata* (**2**:35)
        branches uniform *P. indiania* (**2**:37)
      verrucae not longer than wide
        branches become laterally fused *P. zelli* (**2**:36)
        branches not laterally fused *P. eydouxi* (**2**:44)
  colony consists of prostrate branches
    branches large and elongate
      branches laterally compressed
        branches evenly spaced *P. kelleheri* (**2**:32)
        branches irregular *P. woodjonesi* (**2**:43)
      branches not laterally compressed *P. molokensis* (**2**:42)
    branches not large and elongate
      branches uniform over whole colony *P. danae* (**2**:25)
      branches irregular over whole colony *P. effusus* (**2**:39)
  colony encrusting with irregular branches
    verrucae well developed
      branches compact *P. ankeli* (**2**:33)
      branches irregular *P. fungiformis* (**2**:40)
    verrucae not well developed *P. inflata* (**2**:41)

Genus *Seriatopora*

branches <2 mm diameter *S. dendritica* (**2**:46)
branches 2-5 mm diameter
  branches taper to a point *S. hystrix* (**2**:48)
  branches do not taper *S. guttatus* (**2**:50)
branches >5 mm diameter
  corallites aligned in rows
    branches long and not tapered *S. caliendrum* (**2**:54)
    branches short and tapered *S. stellata* (**2**:53)
  corallites not aligned in rows *S. aculeata* (**2**:52)

Genus *Stylophora*

colony with elongate branches
  branches robust (>5 mm diameter)
    branches upright *S. pistillata* (**2**:58)
    branches prostrate *S. danae* (**2**:63)
  branches not robust (<5 mm diameter)
    branches closely compacted *S. madagascarensis* (**2**:57)
    branches not closely compacted
      spiny corallite hoods present *S. kuehlmanni* (**2**:62)
      spiny corallite hoods absent *S. subseriata* (**2**:60)
colony without elongate branches
  colony with knobby branches *S. wellsi* (**2**:64)
  colony encrusting with mounded surface *S. mamillata* (**2**:65)

Family Poritidae

corallites <2 mm diameter
  septa fused in non-cyclical pattern **Genus *Porites*** (**3**:276, see below)
  septa not fused
    columella present **Genus *Stylaraea*, *S. punctata*** (**3**:346)
    columella absent **Genus *Poritipora*, *P. paliformis*** (**3**:347)
corallites >2 mm diameter
  skeleton robust, not very porous **Genus *Goniopora*** (**3**:348, see below)
  skeleton delicate, very porous **Genus *Alveopora*** (**3**:380, see below)

Genus *Alveopora*

colonies branching or branching columns
  corallites without a palisade of exsert spines
    septa fine, irregular
      columella tangle present *A. allingi* (**3**:384)

columella tangle absent *A. gigas* (**3**:380)
septa in comb rows *A. catalai* (**3**:382)
corallites with a palisade of exsert spines *A. verrilliana* (**3**:387)
colonies not branching or branching columns
  corallites >2 mm diameter
    colonies lobed
      columellae distinct
        septa fine, irregular in shape
          septa do not clearly alternate *A. spongiosa* (**3**:388)
          septa clearly alternate *A. daedalea* (**3**:390)
        septa in comb rows *A. excelsa* (**3**:394)
      columellae not distinct *A. marionensis* (**3**:385)
    colonies hemispherical
      septa fine, irregular in shape *A. fenestrata* (**3**:386)
      septa in comb rows
        comb rows cyclical *A. tizardi* (**3**:392)
        comb rows irregular *A. japonica* (**3**:392)
  corallites <2 mm diameter
    corallites circular *A. ocellata* (**3**:397)
    corallites not circular
      corallite structures coarse *A. viridis* (**3**:395)
      corallite structures fine *A. minuta* (**3**:396)

Genus *Goniopora*

colonies not encrusting
  corallites >5 mm diameter
    colony massive Group 1 (**3**:350)
      corallites with uneven walls *G. stokesi* (**3**:352)
      corallites with even walls
        septa arranged cyclically *G. djiboutiensis* (**3**:351)
        septa irregular *G. pendulus* (**3**:350)
    colony branching or columnar Group 2 (**3**:354)
      columella large
        septa irregular *G. columna* (**3**:356)
        septa uniform *G. sultani* (**3**:355)
      columella small *G. lobata* (**3**:354)
  corallites 3-5 mm diameter
    colony massive Group 4 (**3**:362)
      septal development varies among corallites *G. cellulosa* (**3**:363)
      septal development uniform among corallites
        corallites excavated
          primary septa not distinctive *G. norfolkensis* (**3**:362)
          primary septa distinctive
            primary septa form a delta *G. minor* (**3**:366)
            primary septa do not form a delta *G. tenuidens* (**3**:364)
        corallites shallow *G. pearsoni* (**3**:365)
    colony branching or columnar Group 5 (**3**:368)
      colonies columnar
        corallites shallow *G. planulata* (**3**:368)
        corallites excavated *G. ciliatus* (**3**:372)
      colonies branching
        branches short, thick *G. palmensis* (**3**:374)
        branches elongate
          branches cylindrical *G. eclipsensis* (**3**:371)
          branches laterally compressed *G. pandoraensis* (**3**:370)
  corallites <3 mm diameter Group 6 (**3**:375)
    colony lobed
      corallites shallow *G. stutchburyi* (**3**:377)
      corallites excavated *G. burgosi* (**3**:375)
    colony columnar *G. savignyi* (**3**:376)
    colony branching *G. fruticosa* (**3**:378)
colonies encrusting Group 3 (**3**:358)
  corallites shallow
    septa compact
      columella broad *G. somaliensis* (**3**:358)
      columella not broad *G. tenella* (**3**:360)
    septa widely spaced *G. albiconus* (**3**:361)
  corallites excavated *G. polyformis* (**3**:360)

Genus *Porites*

Note: This key is primarily based on corallite characters that are not observable underwater. Groupings of species are artificial and may not indicate general similarity.

colonies massive Groups 1 (**3**:280) and 2 (**3**:292)
  corallites <1 mm diameter
    central fossa conspicuous *P. murrayensis* (**3**:292)
    central fossa not particularly conspicuous *P. stephensoni* (**3**:293)
  corallites >1 mm diameter
    columella indistinct or absent
      corallites >1.5 mm diameter
        triplet fused *P. panamensis* (**3**:283)
        triplet not fused *P. mayeri* (**3**:289)
      corallites <1.5 mm diameter *P. densa* (**3**:294)
    columella present
      triplet fused
        colony surface knobby *P. somaliensis* (**3**:291)
        colony surface not knobby *P. lutea* (**3**:287)
      triplet not fused
        pali conspicuous
          coenosteum conspicuously granulated *P. echinulata* (**3**:290)
          coenosteum not conspicuously granulated
            corallites deeply excavated *P. myrmidonensis* (**3**:288)
            corallites not deeply excavated *P. australiensis* (**3**:286)
        pali present but not conspicuous
          lateral pairs of septa mostly fused *P. lobata* (**3**:284)
          lateral pairs of septa mostly not fused
            denticle on septa conspicuous *P. brighami* (**3**:295)
            denticle on septa not conspicuous *P. astreoides* (**3**:280)
        pali absent *P. solida* (**3**:282)
colonies thick columns or thick plates Group 3 (**3**:296)
  colonies thick columns
    corallites deeply excavated
      denticles well developed *P. pukoensis* (**3**:299)
      denticles not well developed *P. nodifera* (**3**:296)
    corallites not deeply excavated
      walls thick ridges *P. evermanni* (**3**:298)
      walls not thick ridges *P. columnaris* (**3**:300)
  colonies thick plates
    pali well developed *P. arnaudi* (**3**:301)
    pali irregularly developed *P. colonensis* (**3**:302)
colonies columns, laminae and branches Group 4 (**3**:303)
  septa irregular
    colony nodular *P. desilveri* (**3**:308)
    colony not nodular *P. heronensis* (**3**:306)
  septa not irregular
    triplet fused
      septa thick, wedge-shaped *P. okinawensis* (**3**:307)
      septa not thick, wedge-shaped
        walls acute *P. cocosensis* (**3**:311)
        walls not acute *P. aranetai* (**3**:303)
    triplet not fused
      8 pali present
        denticles in 2 concentric circles *P. vaughani* (**3**:308)
        denticles not in 2 concentric circles *P. annae* (**3**:310)
      <8 pali present *P. lichen* (**3**:304)
colonies composites of laminae and branches Group 5 (**3**:312)
  corallites <1 mm diameter
    columella present
      walls thin *P. cumulatus* (**3**:324)
      walls not thin *P. monticulosa* (**3**:314)
    columella absent
      colony surface patterned by ridges *P. rus* (**3**:312)
      colony surface not patterned by ridges *P. deformis* (**3**:323)
  corallites >1 mm diameter
    lateral septa fused
      walls contorted *P. latistella* (**3**:320)

    walls not contorted
     septa short *P. sillimaniana* (**3**:319)
     septa not short
      corallites shallow *P. horizontalata* (**3**:316)
      corallites excavated *P. eridani* (**3**:322)
   lateral septa not fused
    septa short *P. branneri* (**3**:325)
    septa not short *P. napopora* (**3**:318)
colony branching Group 6 (**3**:326)
 corallites <1 mm diameter
  columella present *P. ornata* (**3**:340)
  columella absent *P. flavus* (**3**:341)
 corallites 1–1.8 mm diameter
  corallites not deeply excavated
   walls irregular
    branches very contorted *P. rugosa* (**3**:342)
    branches not very contorted *P. nigrescens* (**3**:334)
   walls not irregular
    columella inconspicuous *P. divaricata* (**3**:329)
    columella conspicuous
     walls angular *P. harrisoni* (**3**:343)
     walls rounded
      denticles absent *P. attenuata* (**3**:330)
      denticles present
       walls thick *P. cylindrica* (**3**:332)
       walls not thick *P. compressa* (**3**:344)
  corallites deeply excavated
   septa well developed *P. negrosensis* (**3**:336)
   septa not well developed *P. profundus* (**3**:338)
 corallites >1.8 mm diameter
  walls irregular *P. tuberculosa* (**3**:331)
  walls not irregular
   branches not closely compact *P. porites* (**3**:326)
   branches closely compact *P. furcata* (**3**:328)

Family Rhizangiidae

Genus *Astrangia*, *A. poculata* (**2**:319)

Family Siderastreidae

corallite walls well defined
 colony plocoid **Genus *Horastrea***, *H. indica* (**2**:136)
 colony not plocoid
  colony cerioid
   septal teeth saw-like **Genus *Pseudosiderastrea***, *P. tayami* (**2**:134)
   septal teeth not saw-like **Genus *Siderastrea*** (**2**:138, see below)
  colony not cerioid **Genus *Anomastraea***, *A. irregularis* (**2**:137)
corallite walls not well defined
 corallites <3 mm diameter **Genus *Psammocora*** (**2**:144, see below)
 corallites >3 mm diameter **Genus *Coscinaraea*** (**2**:158, see below)

Genus *Coscinaraea*

colonies plates or laminae
 colonies plate-like
  corallites aligned in valleys
   valleys irregular *C. hahazimaensis* (**2**:159)
   valleys not irregular
    septo-costae alternate *C. mcneilli* (**2**:165)
    septo-costae do not alternate *C. marshae* (**2**:164)
  corallites not aligned in valleys
   septo-costae alternate *C. crassa* (**2**:166)
   septo-costae do not alternate *C. columna* (**2**:160)
 colonies laminate *C. wellsi* (**2**:167)
colonies not plates or laminae
 colonies hemispherical *C. monile* (**2**:158)
 colonies columnar *C. exesa* (**2**:162)

Genus *Psammocora*

colony branching
 colony surface smooth *P. contigua* (**2**:146)
 colony surface not smooth
  columellae prominent
   colony finely branched *P. obtusangula* (**2**:145)
   colony not finely branched *P. decussata* (**2**:144)
  columellae not prominent *P. stellata* (**2**:148)
colony not branching
 colony composed of plates and columns
  colony plate-like *P. explanulata* (**2**:156)
  colony becoming columnar *P. digitata* (**2**:154)
 colony encrusting to submassive
  colony surface smooth *P. superficialis* (**2**:150)
  colony surface not smooth
   corallites aligned in valleys
    valleys short, superficial *P. profundacella* (**2**:149)
    valleys not short
     valleys do not meander *P. haimeana* (**2**:152)
     valleys meander *P. nierstraszi* (**2**:153)
   corallites not aligned in valleys
    primary septa petaloid *P. verrilli* (**2**:151)
    primary septa not petaloid *P. vaughani* (**2**:157)

Genus *Siderastrea*

colonies small (<100 mm diameter) *S. glynni* (**2**:138)
colonies large (>100 mm diameter)
 corallites <4 mm diameter
  septa in three cycles *S. radians* (**2**:142)
  septa not in three cycles
   corallite walls acute *S. savignyana* (**2**:139)
   corallite walls not acute *S. stellata* (**2**:143)
 corallites >4 mm diameter *S. siderea* (**3**:140)

Family Trachyphylliidae

Genus *Trachyphyllia*, *T. geoffroyi* (**3**:272)

Common names

Common names are mostly well established for Caribbean corals (by Humann, 1993). In other regions they are unreliable, either because the same name is frequently applied to several distinct species (even genera), or because they do not discriminate between closely related species. Many common names of Indo-Pacific corals have been proposed by Ditlev (1980) and others, but these have not gained general acceptance. Common names are seldom as informative as Latin names and may never get widespread acceptance except as cultivars.

Montipora monasteriata

Anchor coral: *Euphyllia ancora*
Antler coral: *Pocillopora eydouxi*
Antler lettuce coral: *Pectinia alcicornis*
Artichoke coral: *Scolymia cubensis*
Birdsnest coral: *Seriatopora*
Black turret coral: *Tubastrea micrantha*
Bladder coral: *Plerogyra sinuosa*
Blue coral: *Heliopora coerulea*
Blue crust coral: *Porites branneri*
Bluetip coral: *Acropora loripes*
Blushing star coral: *Stephanocoenia michelinii*
Boulder coral: *Porites*
Boulder brain coral: *Colpophyllia natans*
Boulder star coral: *Montastrea annularis*
Bowl coral: *Turbinaria peltata*
Brain coral: *Symphyllia, Platygyra, Leptoria*
Branching anchor coral: *Euphyllia paraancora*
Briar coral: *Anacropora*
Brush coral: *Seriatopora*
Bubble coral: *Physogyra, Plerogyra*
Bush coral: *Stylophora*
Button coral: *Scolymia australis*
Cactus coral: *Pavona decussata*
Cabbage coral: *Merulina*
Candycane coral: *Caulastrea furcata*
Carnation coral: *Pectinia*
Castle coral: *Pachyseris rugosa*
Cats eye coral: *Cynarina*
Cauliflower coral: *Pocillopora damicornis*
Chalice coral: *Oxypora*
Chinaman's hat: *Zoopilus*
Christmas coral: *Acropora elseyi*
Club finger coral: *Stylophora*
Cluster coral: *Stylophora pistillata*
Corduroy coral: *Pachyseris*
Crater coral: *Trachyphyllia geoffroyi*
Crystal coral: *Galaxea*
Cup coral: *Turbinaria*
Daisy coral: *Alveopora, Goniopora*
Disk coral: *Heliofungia*
Dome coral: *Halomitra pileus*
Doughnut coral: *Scolymia vitiensis*
Eight-ray finger coral: *Madracis formosa*

Elegant coral: *Cataphyllia jardinei*
Elephant skin coral: *Pachyseris*
Elephant nose coral: *Mycedium*
Elkhorn coral: *Acropora palmata*
Elliptical star coral: *Dichocoenia stokesi*
Finger coral: *Porites cylindrica, P. porites*
Fire coral: *Millepora*
Flowerpot coral: *Alveopora, Goniopora*
Folded coral: *Trachyphyllia geoffroyi*
Fragile saucer coral: *Agaricia fragilis*
Frilly lettuce coral: *Pectinia lactuca*
Frogspawn coral: *Euphyllia divisa*
Fused staghorn coral: *Acropora prolifera*
Galaxy coral: *Galaxea*
Golfball coral: *Favia fragum*
Grahams sheet coral: *Agaricia grahamae*
Great star coral: *Montastrea cavernosa*
Grooved brain coral: *Diploria labyrinthiformis*
Hammer coral: *Euphyllia ancora*
Helmet coral: *Halomitra pileus*
Hibiscus coral: *Pectinia*
Honeycomb plate coral: *Porites colonensis*
Horn coral: *Hydnophora*
Ivory bush coral: *Oculina*
Jasmine coral: *Nemenzophyllia turbida*
Knob coral: *Cyphastrea*
Knobby brain coral: *Diploria clivosa*
Knobby cactus coral: *Mycetophyllia aliciae*
Knobby star coral: *Solenastrea hyades*
Lace coral: *Stylaster*
Lamarck's sheet coral: *Agaricia lamarcki*
Large ivory coral: *Oculina varicosa*
Leaf coral: *Pavona cactus*
Lesser starlet coral: *Siderastrea radians*
Lettuce coral: *Pectinia, Agaricia*
Lowrelief lettuce coral: *Agaricia humilis*
Lowridge cactus coral: *Mycetophyllia danaana*
Massive starlet coral: *Siderastrea siderea*
Maze coral: *Platygyra, Leptoria*
Mole coral: *Herpolitha, Polyphyllia*
Moon coral: *Favia, Astreopora*
Mushroom coral: *Fungia, Heliofungia*
Mustard hill coral: *Porites astreoides*
Needle coral: *Seriatopora hystrix*
Neptune's cap coral: *Halomitra pileus*
Orange turret coral: *Tubastrea faulkneri*
Organ pipe coral: *Tubipora musica*
Pacific rose coral: *Cynarina lacrymalis*
Pagoda coral: *Turbinaria mesenterina*
Palm lettuce coral: *Pectinia paeonia*
Pearl bubble coral: *Physogyra lichtensteini*

Phonograph coral: *Pachyseris speciosa*
Pillar coral: *Dendrogyra cylindrus*
Pineapple coral: *Favites*
Pore coral: *Porites*
Potato chip coral: leafy *Pavona*
Puffed coral: *Trachyphyllia geoffroyi*
Ridged cactus coral: *Mycetophyllia lamarckiana*
Ridgeless cactus coral: *Mycetophyllia reesi*
Robust ivory tree coral: *Oculina robusta*
Rose coral: *Manicina areolata*
Rough cactus coral: *Mycetophyllia ferox*
Rough star coral: *Isophyllia rigida*
Ruffled coral: *Merulina ampliata*
Scalpel coral: *Galaxea horrescens*
Scroll coral: *Turbinaria, Agaricia undata*
Sea mole coral: *Polyphyllia talpina*
Sinuous cactus coral: *Isophyllia sinuous*
Slipper coral: *Polyphyllia talpina*
Small bubble coral: *Physogyra lichtensteini*
Smooth flower coral: *Eusmilia fastigiata*
Smooth star coral: *Solenastrea bournoni*
Solitary disk coral: *Scolymia cubensis*
Spiny flower coral: *Mussa angulosa*
Staghorn coral: *Acropora*
Star coral: *Madracis pharensis*
Starburst coral: *Galaxea*
Starlet coral: *Siderastrea*
Star column coral: *Pavona clavus*
Sun coral: *Tubastrea*
Sunflower coral: *Heliofungia actiniformis*
Sunray lettuce coral: *Leptoseris cucullata*
Symmetrical brain coral: *Diploria strigosa*
Table coral: *Acropora*
Ten-ray star coral: *Madracis decactis*
Thin leaf lettuce coral: *Agaricia tenuifolia*
Thorn coral: *Stylocoeniella*
Tongue coral: *Herpolitha limax*
Tooth coral: *Galaxea, Lobophyllia*
Torch coral: *Caulastrea*
Trumpet coral: *Caulastrea*
Tube coral: *Cladocora arbuscula*
Turban coral: *Turbinaria*
Turret coral: *Tubastrea*
Vase coral: *Turbinaria*
Velvet coral: *Montipora*
Velvet branch coral: *Montipora*
Whisker coral: *Duncanopsammia*
White grape coral: *Euphyllia cristata*
Yellow pencil coral: *Madracis mirabilis*
Yellow scroll coral: *Turbinaria reniformis*
Zebra coral: *Oulastrea crispata*

Opposite: *Pavona explanata* forming tiers of plates. HOUTMAN ABROLHOS ISLANDS, SOUTH-WEST AUSTRALIA Photograph: author

Glossary

Volume numbers are in bold.
Terms are explained here as they are used in this book, not necessarily other publications.

acanthocauli: juvenile corals (mostly *Fungia*) attached to the substrate either directly or on stalks.

acolonial corals: solitary corals that do not form colonies.

allopatric speciation: the splitting of a widespread population into two or more isolates by a geological or ecological isolating barrier and subsequent differentiation into new species, or the dispersal of a few propagules across a pre-existing barrier and subsequent differentiation into new species.

ambulacral grooves: grooves along the top of common walls between adjacent corallites, see **1**:52-3.

ampullae: the swollen part of a canal in skeletal Hydrozoa that hold medusae produced by internal fertilisation.

anastomose: descriptive term for branches which re-fuse after having initially divided.

appressed corallites: corallites which are fused (partly or completely) with the coenosteum on one side so that their axis is approximately parallel with the coenosteum, see **1**:179.

aragonite skeletons: skeletons primarily composed of the aragonite form of calcium carbonate. All Scleractinia have aragonitic skeletons (cf. calcite skeletons).

arborescent colonies: colonies with a tree-like growth form, see **1**:56,179.

archaeocyaths: sponge-like metazoans that had skeletons. They were mostly restricted to the Cambrian Era, see **1**:33.

attachment scars: a scar-like patch on the central undersurface of free-living fungiids from where they were attached to the substrate as juveniles.

atolls: reefs and islands that are the remnants of submerged land masses, see **1**:23-6.

autotomy: a means of asexual reproduction by the break up of a parent polyp. Seen commonly in *Diaseris*, see **2**:248.

axial corallite: a corallite which forms the tip of a branch. Most *Acropora* have axial corallites whereas they only occur sporadically in other corals, see **1**:179.

axial furrow: a groove along the axis of the upper surface of some fungiids.

azooxanthellate corals: corals that do not have zooxanthellae. These are commonly found on reefs, but most are restricted to deep water, below the level of light penetration.

Eusmilia fastigiata

barrier reefs: reefs along continental shelf breaks or otherwise well separated from landmasses, see **1**:23-7.

basal plate: the first skeletal element deposited by a planula larva.

bifacial: describes plates which have corallites on both sides.

bifurcate: divide into two equal branches.

biodiversity: a term that has acquired many meanings, but can be considered synonymous with 'systematic diversity'. Biodiversity thus has the same relationship to taxonomic diversity as systematics has to taxonomy. Patterns of taxonomic diversity are indicative of patterns of biodiversity.

biogeography: the study of the distribution of life and the reason for that distribution. In practice, biogeography is divisible into observations of distributions and explanations for those observations.

bioherms: reefs or large reef-like structures built of calcium carbonate of biological origin, see **1**:27.

biological species concept: the concept that biological species, unlike other taxa, are units within which genes are (or can be) freely exchanged, but within which gene flow does not occur, at least under normal circumstances.

bleaching: expulsion of zooxanthellae by corals. Usually occurs as a result of environmental stress and frequently results in the death of the coral.

bottlebrush branching: a descriptive term for a branch with compact radial sub-branches, usually used for some *Acropora* species, see **1**:178.

branching colonies: any growth-form where branches are formed.

brooding: development of larvae within the coelenteron of an adult coral.

Opposite: Columns of *Coscinaraea exesa*. GREAT BARRIER REEF, AUSTRALIA *Photograph: Mary Stafford-Smith*

budding: a form of asexual reproduction where a 'parent' corallite forms one or more 'daughter' corallites, see **1**:54.

caespitose: a descriptive term for branches which interlock similarly in three dimensions, usually used for some *Acropora* species, see **1**:178.

calcite skeletons: skeletons composed of the calcite form of calcium carbonate. All Rugose corals and molluscs have calcitic skeletons (cf. aragonitic skeletons).

calice: the upper surface of a corallite bounded by the wall.

Cambrian: a geological Period of the Palaeozoic Era, see **1**:34.

Cenozoic: a geological Era, see **1**:35.

Central American Seaway: a former seaway between north and south America, now closed by the Isthmus of Panama.

central arch: a raised area surrounding the mouth of some solitary fungiids.

cerioid corals: massive corals that have corallites sharing common walls, see **1**:54-5.

chimeras: single larvae, polyps or colonies which have developed from more than one original embryo and which have more than one genotype.

chromosomes: thread-like structures in cell nuclei carrying genetic information in a linear sequence.

cilia: microscopic hair-like structures growing on the ectodermis of polyps or planulae and which aid mucous movement or locomotion (respectively).

clade: a phylogeny inferred to be monophyletic; groups of taxa sharing a closer common ancestry with one another than with members of any other clade.

coelenteron: the body cavity of a coelenterate, see **1**:47,52.

coenosteum: thin horizontal plates between corallites, see **1**:48-51.

coenosteum pit: the point of insertion, or commencement, of septa, mostly found in Pectiniidae and Fungiidae.

coenosteum style: prominent projections from the coenosteum usually associated with a single corallite, see *Stylocoeniella*, **2**:4.

collines: skeletal ridges composed of coenosteum which separate corallites.

colonial corals: corals composed of many individuals. There may be no clear distinction between single individuals with many mouths and colonies with individuals with single mouths, see **1**:54.

columellae: skeletal structures at the axis of corallites. May be 'spine-like', 'spire-like', a 'tangle' of rods, or 'spongy' (structured like a sponge although not soft), see **1**:48,50.

columnar colonies: colonies forming into one or more columns, see **1**:56,178.

commensal: a partner in a mutually beneficial relationship between two different types of organisms.

compact branching: where branches of a colony are close together.

continua: where there is no clear discontinuity in morphology, genotype or distribution.

coral: unless the context indicates otherwise (eg. rugose coral, soft coral) the word 'coral' is used in this book to mean 'hard' or 'stony' coral.

coralline algae: algae that form solid calcium carbonate accretions.

corallite: the skeleton of an individual polyp, see **1**:48-51.

corymbose: a descriptive term for colonies which have horizontal interlocking branches and have short upright branchlets, usually used for some *Acropora* species, see **1**:178.

cosmopolitan: with a worldwide distribution within habitat limits.

costae: radial skeletal elements outside the corallite wall, see **1**:48-50.

Cretaceous: a geological Period of the Mesozoic Era, see **1**:35.

Crown-of-thorns starfish (*Acanthaster planci*): large starfish which eat coral. Typically they occur in plague-like outbreaks which cause widespread destruction.

cycles of septa/septo-costae: where radial elements occur in a set sequence of size (6 primary, 6 secondary, 12 tertiary and so on), see **1**:49-50.

cyclosystem: a system of fine tubes that links the polyps of calcareous hydrozoans.

dactylopores: the external opening in the skeleton through which dactylozooids extend nematocyst spines.

dactylozooids: polyps of Hydrozoa specialised for food capture, which have no mouth but have elongate nematocyst spines.

deltas of septa: fusion of septa into a hexamerous pattern of spongy columella. Common in *Goniopora*.

dendrogram: a tree-like hierarchical classification with a single root and branching representing levels of dissimilarities of objects. In this book, the objects are localities and the dendrograms are measures of dissimilarities in coral species compositions.

depauperate: having a relatively small number of species.

Devonian: a geological Period of the Palaeozoic Era, see **1**:34.

digitate: a colony with short branches shaped like the upturned fingers of a hand.

dispersal: the process of movement of propagules resulting in dispersion.

dissepiments: blistery horizontal plates of calcium carbonate adjoining corallites, see **1**:51.

distal: remote from the centre, eg. the end of a branch.

diversity: the number of taxa in a group or place, see biodiversity.

ectodermis: the outer cell layer of a polyp, see **1**:48,52.

encrusting colonies: thin colonies which adhere closely, and are attached to, the substrate.

endemic: a species restricted to a specific region.

endemism: reflecting the proportion of species restricted to a specific region, see **3**:412-3.

endo-symbiotic: symbiosis where individuals of one organism (zooxanthellae in the case of extant corals) live within the cells of another.

entire: without substantial irregularities.

environmental variation: the variety of environmental parameters associated with a particular place.

Eocene: a geological Epoch of the Cenozoic Era, see **1**:35.

epitheca: a tissue-like layer of calcium carbonate that grows outside corallite walls. Originally derived from the basal plate, see **1**:48-9.

explanate corals: colonies which spread horizontally as branches fuse into a solid or near solid plate.

extant: now living.

extinct: no longer living.

extratentacular budding: where daughter corallites grow from the outside wall of parent corallites, see **1**:54.

family: the taxon level representing a group of related genera.

flabello-meandroid corals: corals which have valleys with walls that are separate from the walls of adjacent valleys, see **1**:54-5.

flaring corallites: with expanding (trumpet-like) curves to the outer corallite wall.

fossa: a cavity or hole in the skeleton.

Foraminifera: Protozoa of the Order Foraminiferida which are abundant in the plankton and benthos of all oceans.

foveolate corallites: corallites of some species of *Montipora* which are situated at the base of funnel-shaped depressions, see **1:65**.

free-living coral: corals that are not attached to the substrate.

fringing reefs: reefs which occur adjacent to a shoreline, see **1:24-5**.

fuzzy boundaries: geographic, morphological taxonomic or systematic boundaries that are not clearly defined.

gametes: sex (egg and sperm) cells.

gastrodermis: the inner cell layer which lines the coelenteron, see **1:48,52-3**.

gastropores: the external opening in the skeleton of Hydrozoans through which gastrozooids are extended during feeding.

gastrozooids: polyps specialised for feeding in Hydrozoa and other Cnidaria.

genotype: the set of genes possessed by an individual organism.

genus/genera: the taxon level representing a group of related species.

geographic variation: geographic variation in morphology which has a genetic basis.

glabrous: devoid of attached structures.

gonads: testes and ovaries. These are usually developed annually, see **1:48**.

granulated: covered with sand-like particles.

groove-and-tubercle structures: fine epithecal structures, the development of which is controlled by polychaete worms, see **1:51**.

hermaphrodite: individuals that are both male and female.

hermatypic: literally 'reef building' but commonly used as a descriptor for marine invertebrates that have photosynthetic plants living symbiotically within their tissues. Because the word is a misnomer, several terms including 'reef-building', 'symbiotic' and 'zooxanthellate', are used synonymously. Of these, the first two are ambiguous and the last is, at least theoretically, restricted to extant taxa.

hermatypic corals: zooxanthellate or reef-building corals: the corals included in this book.

holotype: the principal specimen on which a species name is based.

hybrid: an individual with parents of different species.

hybridisation: formation of a hybrid.

hydnophore: an alternative name for monticule, sometimes used with *Hydnophora*.

incipient axial corallites: corallites intermediate in development between radial and axial corallites of *Acropora*.

immersed corallites: corallites which are embedded in the surrounding coenosteum.

intratentacular budding: where daughter corallites grow from the inside wall of parent corallites, usually by division of the parent corallite, see **1:54**.

Jurassic: a geological Period of the Mesozoic Era, see **1:35**.

lateral pairs of septa: two pairs of septa on each side of *Porites* corallites, see **3:277-8**.

latitudinal attenuation: the progressive decrease in diversity along continental coastlines with increasing distance from the equator, see **3:414**.

macroalgae: algae that are of conspicuous size.

mass extinction: an extinction that is characterised by loss of many taxa in a geologically brief time period.

mass spawning: spawning events where many taxa spawn simultaneously, see **3:417-9**.

massive colonies: colonies which are solid and which are typically hemispherical or otherwise have approximately similar dimensions in all directions, see **1:56**.

meandroid colonies: massive colonies that have corallite mouths aligned in valleys such that there are no individual polyps, see **1:54-5**.

medusae: free-living sexual reproductive stage of Hydrozoa and jellyfish. Morphologically, these are bell-shaped: the upsidedown equivalent of polyps.

mesoglea: an initially non-cellular layer between the ectodermis and gastrodermis, see **1:48,52**.

Mesozoic: a geological Era, see **1:35**.

metamorphosis: the transformation of a planula larva into a polyp.

micro-atoll: A colony shaped like an atoll because low tide level permits only lateral growth, illustrated **3:287**.

microhabitat: a vague word indicating a particular type of habitat occupied by a coral colony.

migration: large-scale movement of a population. Synonymous with dispersal except implying an activity specific in time or space.

Milankovitch cycles: cycles of variation the earth's orbital motion including oscillation of the earth's axis and eccentricities of the earth's orbit around the sun.

Miocene: a geological Epoch of the Cenozoic Era, see **1:35**.

monospecific: describes a genus with one species only, or a coral community with one species only.

monticules: conical sections of common wall between corallites which have a secondary radial symmetry, see **1:51-2**.

mucous: gelatinous substance secreted by the ectodermis for protection, to aid the capture of food, or to remove sediment. Mucous is usually moved by cilia.

nariform: a radial corallite, usually of *Acropora*, shaped like an upsidedown 'roman' nose, see **1:178**.

nematocysts: microscopic stinging cells occurring individually in the ectodermis or grouped into wart-like clumps on tentacles, see **1:48,52-3**.

neo-Darwinian synthesis: a synthesis of Darwin's concept of species and Mendelian genetics. This encapsulates the notion that evolutionary change occurs within species as a result of natural selection acting on variation within populations, variations that ultimately arise from random mutations.

neoplasm: cancerous growths commonly found on corals, see **3:421**.

nomenclature, rules of: an international code for the naming of taxa, see **1:11-2**.

nominal species: species that exist in name only.

obligate association: an association between two very different types of organisms where one member of the partnership cannot live without the other.

Oligocene: a geological Epoch of the Cenozoic Era, see **1:35**.

Ordovician: a geological Period of the Palaeozoic Era, see **1:34**.

oral cone: a mound of soft tissue surrounding the mouth, see **1:48,52**.

oral disc: the soft tissue between the mouth and the surrounding tentacles, see **1:48,52**.

orders of septa/septo-costae: where radial elements occur in different sizes, but not as cycles, see **1:49-50**.

palaeobiogeographic patterns: biogeographic patterns as seen in the fossil record.

Palaeocene: a geological Epoch of the Tertiary Period, see **1**:35.

Palaeozoic: a geological Era, see **1**:34.

pali: upright skeletal rods or plates at the inner margin of septa formed by pourtàles plan fusions, see **1**:50.

paliform crown: a circle of paliform lobes surrounding the columella, see **1**:50.

paliform lobes: upright skeletal rods or plates at the inner margin of septa formed by upward growth of the septum, see **1**:48,50.

papillae: projections of coenosteum on the surface of many species of *Montipora* that are less than a corallite in width, see **1**:64.

paradigm: a well defined perspective on a major area of thought or knowledge.

Permian: a geological Period of the Palaeozoic Era, see **1**:34.

petaloid septa: primary septa which have a tapered or curved (tear-drop) shape because they are enclosed by other septa, illustrated **2**:132.

phaceloid corals: corals that have corallites of uniform height and adjoined towards their base, see **1**:54-5.

phylum: the taxon level representing a group of related families.

pinnule: small upright structures, usually columellae, which are cylindrical in shape.

planula larvae/planulae: larvae of coral.

plate tectonics: the drifting of continents over geological time creating major changes in the shape of land masses and oceans.

platform reefs: general term for reefs which are not clearly derived from sea level change or the proximity of land, see **1**:24,27.

Pleistocene: a geological Epoch of the Cenozoic Era, see **1**:35.

Pliocene: a geological Epoch of the Cenozoic Era, see **1**:35.

plocoid colonies: colonies which have conical corallites with their own walls, see **1**:54-5.

polymorphic species: species which have a wide range of morphological variation.

polyp: an individual coral including soft tissues and skeleton, see **1**:48-53.

polyploidy: possessing more than two entire chromosome complements.

pourtàles plan: a cyclical arrangement of septa created by a specific pattern of fusion, see **1**:49-50.

propagule: a sexually or asexually produced reproductive body capable of developing into an adult organism.

prostrate: a descriptive term for a colony which sprawls horizontally over the substrate.

Proterozoic: a geological Era before the Palaeozoic Era.

Protoatlantic Ocean: the precurser of the modern Atlantic Ocean, see **1**:37.

proximal: close to the centre, eg. the base of a branch.

radial corallite: corallites on the sides of branches as opposed to axial corallites on the tips of branches. The term is usually used with *Acropora* and *Anacropora*, see **1**:178.

radii: inconspicuous septal elements connecting septa with the columella. Used in the taxonomy of *Porites*, see **3**:278.

rafting: the transport of biota on floating objects. This is a means of dispersal of some corals.

rasp-like corallites: regularly arranged corallites with sharp edges reminiscent of a wood rasp.

reef-building corals: zooxanthellate or hermatypic corals: the corals included in this book.

reef flat: the flat intertidal parts of reefs that are exposed to wave action.

reef slope: the sloping parts of reefs below the reef flat.

reefs: limestone platforms of shallow tropical seas built by corals, coralline algae and other photosynthetic organisms or symbionts.

reticulate evolution: evolution dominated by sequential division and fusion of clades, see **3**:438-43.

reticulate repackaging: the sequential division and fusion of phylogenies so that the genetic complement of species varies over evolutionary time, see **3**:438-41.

reticulation: interbreeding that creates reticulate patterns within and among species over large geographic areas or in evolutionary time..

rudists: a large group of Mesozoic bivalves that dominated reefs throughout much of the Cretaceous and which became extinct at the close of the Cretaceous.

rugose corals: a major group of non-scleractinian corals that became extinct at the close of the Palaeozoic Era.

satellite colonies: colonies that develop within the tissue of parent colonies and which have their own unattached skeletons. Best seen in *Goniopora stokesi*, see **3**:352.

scale-like corallites: corallites forming a pattern reminiscent of the pattern of fish scales.

scleractinian corals: 'hard' corals which have limestone skeletons and which belong to the order Scleractinia.

sea level change: change in sea level relative to the land, due to global change in ocean height primarily due to the extent of polar glaciation and/or upward or downward movement of land masses.

septa: radial skeletal elements projecting inwards from the corallite wall, see **1**:48-50.

septo-costae: radial skeletal elements crossing the corallite wall, composed of both septa and costae, see **1**:48-9.

septal teeth: sharp tooth-like or lobed structures along the margins of septa.

sibling species: similar species that are assumed to be the product of relatively recent speciation.

Silurian: a geological Period of the Palaeozoic Era, see **1**:34.

similarity, measures of: quantitative measure of the similarity between different faunal regions, see **3**:413.

solitary corals: corals composed of single individuals. There may be no clear distinction between single individuals with many mouths and colonies with individuals which have single mouths, see **1**:54.

spat: pinhead-sized single corallites that form immediately after metamorphosis of planula larvae.

spawning: the release of gametes into the water column.

species: a general term with a wide range of meanings. In this book (and most others), species are morphological units recognised by taxonomists. Within a single region they are morphologically distinguishable from other species and genetically semi-isolated from other species. Over their full geographic range, most vary morphologically and genetically to the extent that they intergrade with other species, see Index.

spinule: a spine of near microscopic size.

staghorn: common name for arborescent *Acropora*.

sterome: skeletal infilling derived from the thickening of septa to provide most of the content of corallite walls in some coral families, see **1**:48-9,51.

striae: a string-like arrangement of skeletal elements or soft tissue.

stolons: horizontal polyp outgrowths from which daughter polyps are budded. Common in *Astrangia*.

stromatolites: mounds of limestone formed by the growth of blue-green algae. Common in the Proterozoic Era and still extant.

stromatoporoids: sponge-like organisms that were major builders of Palaeozoic reefs.

sub-: a prefix meaning 'less than' or 'not quite'.

symbiosis: the close association between two organisms where there is substantial mutual benefit.

sympatric speciation: the formation of a new species in the same geographic region as the parent species.

synapticulae: rods linking septa, either forming a network or, in some coral families, contributing to the content of corallite walls, see **1**:48-9.

syngameon: a complex of species that can interbreed. Such a complex may have no well defined morphological characteristics, see Index.

synonymy: a list of names considered by a taxonomist to apply to a given taxon other than the name by which the taxon should be known.

systematics: study of the genetic relationship between taxa.

tabulate corals: a major group of non-scleractinian corals that became extinct at the close of the Palaeozoic Era.

taxon: a taxonomic unit.

taxonomy: study of the morphological relationship between taxa and the naming of taxa.

tentacles: tubular extensions of the polyp. The interior of the tentacles is continuous with the coelenteron, see **1**:48,52.

tentacular lobe: a lobe at the beginning (point of insertion) of a septum. Commonly found in *Fungia* where each lobe supports a single tentacle.

tethyan: originating in the Tethys Sea.

Tethys Sea: the ancient tropical sea that once connected the Indian and Atlantic Oceans, see **1**:37,40-2.

thicket: a descriptive term for colonies composed of closely compacted upright branches.

Triassic: a geological Period of the Mesozoic Era, see **1**:35.

trident: pattern of fusion of the ventral septa of some *Porites* where the septa are linked by a cross-bar.

triplet, of septa: the three ventral septa of *Porites* corallites, see **3**:278.

tuberculae: projections of coenosteum on the surface of many species of *Montipora* that are more than a corallite in width, see **1**:64.

type locality: the place where a species was originally described from.

type species: the species that a genus is primarily based on.

type specimens: the specimens that a species was originally described from. A single or principle specimen is the holotype.

unifacial: describes plates which have corallites on one side only.

verrucae: mounds of coenosteum on the surface of many species of *Montipora* and *Pocillopora* that are wider than a corallite, see **1**:64 and **2**:24.

vesicles: large grape-like sacs that are expanded during the day in some Euphyllidae.

vicariance: the process that occurs when a formerly continuous population is divided by a barrier and evolves into two or more species. Also the reverse of this process.

zooxanthellae: photosynthetic algae (dinoflagellates) that can occur symbiotically in animal tissue.

zooxanthellate corals: corals that have zooxanthellae.

References

Allen GR and Steene R (1994). *Indo-Pacific coral reef guide book.* Tropical Reef Res, Singapore. 378pp.

Amaral FD (1992). *Sobre* Favia Leptophylla *Verrill, 1868 (Cnidaria, Scleractinia).* Iheringia Sér Zool Porto Alegre **73**:117-118.

Amaral FD (1994). *Morphological variation in the reef coral* Montastrea cavernosa *in Brazil.* Coral Reefs **13**:113-117.

Bernard H (1897). *The genus* Montipora. *The genus* Anacropora. Cat Madreporarian Corals British Mus **III**:192pp.

Best MB and Suharsono (1991). *New observations on scleractinian corals from Indonesia. 3 Species belonging to the Merulinidae with new records of* Merulina *and* Boninastrea. Zool Mededelingen **65**:333-342.

Birkeland C (1997) (ed). *Life and death of coral reefs.* Chapman and Hall, New York. 536pp.

Borel-Best M and Hoeksema BW (1987). *New observations on scleractinian corals from Indonesia. 1 Free living species belonging to the Faviina.* Zool Mededelingen **61**:387-403.

Borneman E (2000). *Aquarium corals: husbandry, selection and natural history.* Microcosm, Vermont. 303pp.

Brook G (1891). *Descriptions of new species of Madreporaria in the British Museum.* Ann Mag Nat Hist **8**:458-471.

Brook G (1892). *Preliminary descriptions of new species of Madreporaria in the collections of the British Museum.* Part II Ann Mag Nat Hist **10**:451-465.

Brook G (1893). *The genus* Madrepora. Cat Madreporarian Corals British Mus **I**:212pp.

Brüggemann F (1877). *Notes on stony corals in the British Museum. III Revision of recent solitary Mussaceae.* Ann Mag Nat Hist **40**:300-312.

Budd AF and Guzmán HM (1994). Siderastrea glynni, *a new species of scleractinian coral (Cnidaria: Anthozoa) from the eastern Pacific.* Proc Biol Soc Washington **107**:591-599.

Opposite: *Montipora cebuensis* forming tiers of irregular plates. PAPUA NEW GUINEA Photograph: author

Cairns SD (1982). *Stony corals (Cnidaria: Hydrozoa, Scleractinia) of Carrie Bow Cay, Belize.* Smithsonian Contrib Mar Sci **12**:271-302.

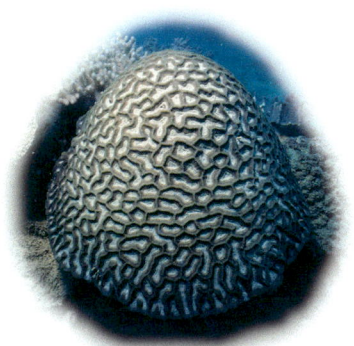

Oulophyllia bennettae

Carpenter KE, Harrison PL, Hodgson G, Alsaffar AH and Alhazeem H (1997). *The corals and coral reef fishes of Kuwait.* Kuwait Inst Sci Res and Environment Public Authority. 166pp.

Chevalier JP (1971). *Les scléractiniaires de la Mélanésie Française. I Expédition Française sur les récifs coralliens de la Nouvelle-Calédonie.* Singer-Polignac, Paris. 307pp.

Chevalier JP (1975). *Les scléractiniaires de la Mélanésie Française. II Expédition Française sur les récifs coralliens de la Nouvelle-Calédonie.* Singer-Polignac, Paris. 407pp.

Chevalier JP and Beauvais L (1987). *Classification en sius-orderes des Scléractiniaires.* In: Grassé PP (ed). *Traité de Zoologie.* Masson. **3**:679-764.

Claereboudt M (1990). Galaxea pauciseptata *nom. nov. (for* G. pauciradiata*), rediscovery and redescription of a poorly known scleractinian species (Oculinidae).* Galaxia **9**:1-18.

Claereboudt M and Hoeksema BW (1987). Fungia (Verrillofungia) spinifer *spec. nov., a new scleractinian coral (Fungiidae) from the Indo-Malayan region.* Zool Mededelingen **61**:303-309.

Coles S (1996). *Corals of Oman.* Keech, Samdani and Coles (publ), Thorns, UK. 106pp.

Colin PL (1978). *Caribbean reef invertebrates and plants.* TFH Publ. 512pp.

Colin PL and Arneson C (1995). *Tropical Pacific invertebrates.* Coral Reef Press, Beverley Hills, USA. 296pp.

Crossland C (1948). *Reef corals of the South African coast.* Ann Natal Mus **XI(2)**:169-205.

Crossland C (1952). *Madreporaria, Hydrocorallinae, Heliopora and Tubipora.* Scientific Rept Great Barrier Reef Exped 1928-29 **VI(3)**:85-257.

Dai CF (1989). *Scleractinia of Taiwan. I Families Astrocoeniidae and Pocilloporidae.* Acta Oceanog Taiwanica **22**:83-101.

Dai CF and Lin CH (1992). *Scleractinia of Taiwan. III Family Agariciidae.* Acta Oceanog Taiwanica **28**:80-101.

Dana JD (1846). *Zoophytes.* United States Exploring Exped 1838-1842 **7**:740pp.

Darwin C (1842). *Structure and distribution of coral reefs.* Smith Elder, London. 355pp.

Darwin C (1859). *On the origin of species by means of natural selection, or the preservation of favoured races in the struggle for life.* Murray, London. 703pp.

Delbeek JC and Sprung J (1994). *The reef aquarium.* Ricordea Publ, Florida. 544pp.

Dinesen ZD (1980). *A revision of the coral genus* Leptoseris *(Scleractinia: Fungina: Agariciidae).* Mem Queensland Mus **20(1)**:182-235.

Ditlev H (1980). *A field-guide to the reef-building corals of the Indo-Pacific.* W Backhuys, Rotterdam. 290pp.

Dubinsky Z (1990) (ed). *Coral reefs.* In: Goodall DW (ed) *Ecosystems of the World,* 25. Elsevier, Amsterdam. 550pp.

Eguchi M (1968). *The hydrocorals and scleractinian corals of Sagami Bay collected by HM the Emperor of Japan.* Maruzen, Tokyo: C1-C74.

Ehrenberg CG (1834). *Beiträge zur physiologischen kenntniss der corallenthiere im Allgemeinen und besonders des Rothen Meeres.* Abhand Akad Wiss der DDR: 250-380.

Fenner DP (1993). *Species distinctions among several Caribbean stony corals.* Bull Mar Sci **53(3)**:1099-1116.

Fenner DP (1999). *New observations on the stony coral (Scleractinia, Milleporidae, Stylasteridae) species of Belize (Central America) and Cozumel (Mexico).* Bull Mar Sci **64**:143-154.

Fosså SA and Nilsen AJ (1998). *The modern coral reef aquarium.* Birgit Schmettkamp Verlag, Bornheim **2**:479pp.

Glynn PW (1999). Pocillopora inflata, *a new species of scleractinian coral (Cnidaria: Anthozoa) from the Tropical Eastern Pacific.* Pacific Sci **53(2)**:168-180.

Haime J (1852). *Catalogue raissonné des fossiles nummulitiques du comté de Nice.* Bull Geol Soc France Ser 2, **7**:249.

Harrison PL and Wallace CC (1990). *Reproduction, dispersal and recruitment of scleractinian corals.* In: Dubinsky Z (ed). *Ecosystems of the World.* Elsevier, Amsterdam. **10**:133-207.

Head SM (1983). *An undescribed species of* Merulina *and a new genus and species of siderastreid coral from the Red Sea.* J Nat Hist **17**:419-435.

Hetzel B and Castro CB (1994). *Corals of southern Bahia.* Editora Nova Fronteira, Rio de Janeiro. 189pp.

Hodgson G (1985). *A new species of* Montastrea *(Cnidaria, Scleractinia) from the Philippines.* Pacific Sci **39(3)**:283-290.

Hodgson G and Carpenter K (1995). *Scleractinian corals of Kuwait.* Pacific Sci **49(3)**:227-246.

Hodgson G and Ross MA (1981). *Unreported scleractinian corals from the Philippines.* Proc Fourth Int Coral Reef Symp, Manila **2**:171-175.

Hoegh-Guldberg O (1999). *Coral bleaching, climate change and the world's coral reefs.* Australian J Mar Freshwater Res **50**:839-866.

Hoeksema BW (1989). *Systematics and ecology of mushroom corals (Scleractinia: Fungiidae).* Zool Verhandelingen **254**:471pp.

Hoeksema BW (1993a). *Mushroom corals (Scleractinia: Fungiidae) of Madang Lagoon, northern Papua New Guinea: an annotated check list with the description of* Cantharellus jebbi *spec. nov.* Zool Mededelingen **67**:1-19.

Hoeksema BW (1993b). *Historical biogeography of* Fungia *(Pleuractis) spp (Scleractinia: Fungiidae), including a new species from the Seychelles.* Zool Mededelingen **67**:639-654.

Hoeksema BW and Borel-Best MB (1984). Cantharellus noumeae *(gen. nov., spec. nov.), a new scleractinian coral (Fungiidae) from New Caledonia.* Zool Mededelingen **58**:323-328.

Hoeksema BW and Borel-Best MB (1991). *New observations on scleractinian corals from Indonesia. 2 Sipunculan-associated species belonging to the genera* Heterocyathus *and* Heteropsammia. Zool Mededelingen **65**:221-245.

Hoeksema BW and Dai CF (1991). *Scleractinia of Taiwan. II Family Fungiidae (including a new species).* Bull Inst Zool Academia Sinica (Taipei) **30**:203-228.

Hoffmeister JE (1925). *Some corals from American Samoa and the Fiji Islands.* Papers Mar Biol Carnegie Inst Washington **22**:90pp.

Holliday L (1989). *Coral reefs.* Salamander Books, London. 204pp.

Horst, van der CJ (1921). *The Madreporaria of the Siboga Expedition. II Madreporaria Fungida.* Siboga-Expeditie **XVIb**:53-98.

Humann P (1993). *Reef coral identification.* New World Publ, Jacksonville. 240pp.

Johnston C (1992) (ed). *Crown-of-thorns starfish on the Great Barrier Reef: reproduction, recruitment and hydrodynamics.* CSIRO Publ, Melbourne. 146pp.

Klunzinger CB (1879). *Die Korallenthiere des Rothen Meeres.* Gutmann, Berlin. 188pp.

Laborel J (1967). *A revised list of Brazilian scleractinian corals and description of a new species.* Postilla **107**:1-14.

Laborel J (1969). *Madréporaires et hydrocoralliaires récifaux des cotes Brésiliennes systématique, écologie, répartition verticale et géographique.* Annales de l'Institut Oceanog **47**:171-229.

Laborel J (1974). *West African reef corals, an hypothesis on their own origin.* Proc Second Int Coral Reef Symp **1**:425-443.

Lamarck JBP (1816). *Historie des animaux sans vertebres.* Verdière, Paris **2**:568pp.

Lamberts AE (1982). *The reef coral* Astreopora *(Anthozoa, Scleractinia, Astrocoeniidae): a revision of the taxonomy and description of a new species.* Pacific Sci **36(1)**:83-105.

Lamberts AE (1984). *The reef coral* Lithactinia *and* Polyphyllia *(Anthozoa, Scleractinia, Fungiidae): a study of morphological, geographical, and statistical differences*. Pacific Sci **38(1)**:12-27.

Maragos JE (1977). *Order Scleractinia*. In: Devaney DM and Eldredge LG (eds) *Reef and Shore Fauna of Hawaii*. Bishop Mus Press, Honolulu. pp158-241.

Marenzeller, von E (1901). *Ostafrikanische steinkorallen*. Naturwiss Mitteilungen, Hamburg. **18**:117-134

Marenzeller, von E (1906). *Riffkorallen*. In: *Expedition* SM "Pola" *in das Rote Meer*. Zool Ergebnisse XXVI Denkschriften Kaiserlichen Akad Wiss, Wien **80**:97pp.

Matthai G (1914). *A revision of the recent colonial Astreidae possessing distinct corallites*. Trans Linnean Soc Second Ser (Zool) **17**:140pp.

Matthai G (1928). *A monograph of the recent Meandroid Astraeidae*. Cat on Madreporarian Corals British Mus (Nat Hist) **7**:288pp.

McManus JW and Vergara SG (1998) (eds). *ReefBase*. ICLARM and WCMC, Manila. CD-ROM.

Milne Edwards H and Haime J (1849). *Recherches sur les polipiers. Monographie des Astréides. Suite 1 Astréens agglomérés*. Ann Sci Nat (Zool) (3). **11**:233-312.

Milne Edwards H and Haime J (1860). *Historie Naturelle Coralliaires* **3**:560pp.

Mojetta A and Ghisotti A (1994). *Flora e fauna del Mediterraneo*. Arnoldo Mondadori Editore. 318pp.

Moll H and Borel-Best M (1984). *New scleractinian corals (Anthozoa: Scleractinia) from the Spermonde Archipelago, South Sulawesi, Indonesia*. Zool Mededelingen **58**:47-58.

Moorsel, van GWNW (1983). *Reproductive strategies in two closely related stony corals (*Agaricia *Scleractinia)*. Mar Ecology Prog Ser **13**:273-283.

Nemenzo F (1955). *Systematic studies on Philippine shallow-water scleractinians. I Sub-order Fungiida*. Nat Appl Sci Bull Univ Philippines **15**:3-84.

Nemenzo F (1959). *Systematic studies on Philippine shallow-water scleractinians. II Suborder Faviida*. Nat Appl Sci Bull Univ Philippines **18**:1-21.

Nemenzo F (1967). *Systematic studies on Philippine shallow-water scleractinians. VI Suborder Astrocoeniina (*Montipora *and* Acropora*)*. Nat Appl Sci Bull Univ Philippines **20**:1-141.

Nemenzo F (1971). *Systematic studies on Philippine shallow-water scleractinians. VII Additional forms*. Nat Appl Sci Bull Univ Philippines **23**:142-185.

Nemenzo F (1976). *Some new Philippine scleractinian reef corals*. Nat Appl Sci Bull Univ Philippines **28**:229-276.

Nemenzo F (1979). *Astrocoeniid and faviid reef corals from Central Philippines*. Philippine J Biol **8**:37-50.

Nemenzo F (1980). *New species and new records of stony corals from west-central Philippines*. Philippine J Sci **108**:1-25.

Nemenzo F (1986). *Guide to Philippine flora and fauna. Corals*. Nat Res Management Centre and Univ Philippines. 273pp.

Nemenzo F and Montecillo E (1981). *Four new scleractinian species from Argangasa Islet (Surigao del sur Province, Philippines)*. Philippine Sci **18**:120-128.

Nishihira M (1988). *Field guide to hermatypic corals of Japan*. Tokai Univ Press (in Japanese). 241pp.

Nishihira M and Veron JEN (1995). *Hermatypic corals of Japan*. Kaiyusha Publ, Tokyo (in Japanese). 440pp.

Ortmann A (1889). *Beobachtungen an Steinkorallen von der Sudküste Ceylons*. Zool Jahrb Abteilung Allgemeine zool Physiol Tiere **4**:493-590.

Ortmann A (1892). *Die korallriffe von Dar-es-Salaam und Umgegend*. Zool Jahrb Abteilung Allgemeine zool Physiol Tiere **6**:631-670.

Peters EC, Cairns SD, Pilson MEQ, Wells JW, Jaap WC, Lang JC, Vasleski CE and Gollahon L St P (1988). *Nomenclature and biology of* Astrangia poculata (= A. danae, = A. astreiformis) *(Cnidaria: Anthozoa)*. Proc Biol Soc Washington **101**:234-250.

Pichon M (1971). *Comparative study of the main features of some coral reefs of Madagascar, La Réunion and Mauritius*. In: Stoddart DR and Yonge CM (eds) *Regional variation in Indian Ocean coral reefs*. Symp Zool Soc London **28**:185-216.

Pillai CSG (1973). *A review of the genus* Anacropora *Ridley, (Scleractinia, Acroporidae) with the description of a new species*. J Mar Biol Assoc India **15**:296-301.

Pillai CSG and Scheer G (1973). *Bemerkungen uber einige riffkorallen von Samoa und Hawaii*. Zool Jahrb Abteilung Systrmatik Oekologie Geographie Tiere **100**:466-476.

Pillai CSG and Scheer G (1976). *Report on the stony corals from the Maldive Archipelago*. Zoologica (Stuttgart) **43**:83pp.

Pourtalés, de LF (1871). *Deep-sea corals*. Illustrated Cat Mus Comparative Zoölogy 2, **4**:93pp.

Quelch JJ (1886). *Report on the reef corals collected by HMS* Challenger *during the years 1873-1876*. Rept Sci Results Voyage of HMS Challenger. Zoology **3**:203pp.

Randall RH and Myers RF (1983). *The corals. Guide to the coastal resources of Guam*. Univ Guam Press **2**:128pp.

Reed JK (1980). *Distribution and structure of deep-water* Oculina varicosa *coral reefs off central eastern Florida*. Bull Mar Sci **30**:667-677.

Richmond RH and Hunter CH (1990). *Reproduction and recruitment of corals; comparisons among the Caribbean, the tropical Pacific, and the Red Sea*. Mar Ecol Prog Ser **60**:185-203.

Riegl B (1995). *Description of four new species in the hard coral genus* Acropora *Oken, 1815 (Scleractinia: Astrocoeniina: Acroporidae) from South-East Africa*. Zool J Linnean Soc **113**:229-247.

Roos PJ (1971). *The shallow-water stony corals of the Netherlands Antilles*. Studies on the Fauna of Curaçio and other Caribbean Islands **130**:108pp.

Scheer G and Pillai CSG (1974). *Report on the Scleractinia from the Nicobar Islands*. Zoologica (Stuttgart) **122**:75pp.

Scheer G and Pillai CSG (1983). *Report on the stony corals from the Red Sea*. Zoologica (Stuttgart) **133**:98pp.

Scott PJB (1984). *The corals of Hong Kong*. Hong Kong Univ Press, Hong Kong. 112pp.

Sheppard CRC (1985). *Reefs and coral assemblages of Saudi Arabia*. Fauna of Saudi Arabia **7**:37-58.

Sheppard CRC and Salm SV (1988). *Reefs and coral communities of Oman, with a description of a new coral species (Order Scleractinia, Genus Acanthastrea)*. J Nat Hist **22**:263-279.

Sheppard CRC and Sheppard ALS (1991). *Corals and coral communities of Arabia*. Fauna of Saudi Arabia **12**:170pp.

Sheppard CRC, Dinesen ZD, and Drew EA (1983). *Taxonomy, ecology and physiology of the geographically restricted coral* Ctenella chagius *Matthai*. Bull Mar Sci **33**:905-918.

Shirai S (1980). *Ecological encyclopaedia of the marine animals of the Ryukyu Islands*. Okinawa Kyoiku Shupann (in Japanese). 636pp.

Sprung J (1999). *Corals. A quick reference guide*. Ricordea Publ, Florida. 240pp.

Squires DF (1958). *Stony corals from the vicinity of Bimini, Bahamas, British West Indies*. Bull American Mus Nat Hist **115**:215-262.

Studer T (1877). *Übersicht der Steinkorallen aus der Familie der Madreporaria aporosa, Eupsammia and Turbinaria, welche auf Reise SMS Gazelle um die Erde gesammelt wurden*. Monatsberichte Deutchen Akad Wiss Berlin **1877**:625-655.

Turnπek D (1997). *Mesozoic corals of Slovenia*. ZRC Sazu, Ljubljana. 513pp.

Umbgrove JHF (1939). *Madreporaria from the Bay of Batavia*. Zool Mededelingen **22**:64pp.

Vaughan TW (1907). *Recent Madreporaria of the Hawaiian Islands and Laysan*. US National Mus Bull **59**:427pp.

Vaughan TW (1918). *Some shoal-water corals from Murray Islands, Cocos-Keeling Islands and Fanning Islands*. Papers Dep Mar Biol Carnegie Inst Washington **9**: 51-234.

Veron JEN (1985). *New scleractinia from Australian coral reefs*. Rec Western Australian Mus **12**:147-183.

Veron JEN (1986). *Corals of Australia and the Indo-Pacific*. Angus and Robertson, Sydney. 644pp.

Veron JEN (1990a). *New Scleractinia from Japan and other Indo-West Pacific Countries*. Galaxea **9**:95-173.

Veron JEN (1990b). *Re-examination of the reef corals of Cocos (Keeling) Atoll*. Rec Western Australia Mus **14**:553-581.

Veron JEN (1992). *Hermatypic corals of Japan*. Australian Inst Mar Sci Monogr. Ser **IX**:86pp.

Veron JEN (1993). *A biogeographic database of hermatypic corals: species of the central Indo-Pacific, genera of the world*. Australian Inst Mar Sci Monogr Ser **X**:433pp.

Veron JEN (1995). *Corals in space and time; the biogeography and evolution of the Scleractinia*. UNSW Press, Sydney. 321pp.

Veron JEN (in preparation). *New species described in 'Corals of the World'*. Australian Inst Mar Sci Monogr Ser.

Veron JEN, Fenner D and Stafford-Smith MG (in preparation). *Coral ID*. Australian Inst Mar Sci, Townsville. CD-ROM.

Veron JEN and Hodgson G (1989). *Annotated checklist of the hermatypic corals of the Philippines*. Pacific Sci **43**:234-287.

Veron JEN and Marsh LM (1988). *Hermatypic corals of Western Australia. Records and annotated species list*. Rec Western Australian Mus, Suppl **29**:136pp.

Veron JEN and Pichon M (1976). *Scleractinia of Eastern Australia. Part 1, Families Thamnasteriidae, Astrocoeniidae, Pocilloporidae*. Australian Inst Mar Sci Monogr Ser **I**:86pp.

Veron JEN and Pichon M (1980). *Scleractinia of Eastern Australia. Part 3, Families Agaraciidae, Siderastreidae, Fungiidae, Oculinidae, Merulinidae, Mussidae, Pectiniidae, Caryophylliidae, Dendrophylliidae*. Australian Inst Mar Sci Monogr Ser **IV**:471pp.

Veron JEN and Pichon M (1982). *Scleractinia of Eastern Australia. Part 4, Family Poritidae*. Australian Inst Mar Sci Monogr Ser **V**:210pp.

Veron JEN and Wallace C (1984). *Scleractinia of Eastern Australia. Part 5 Family Acroporidae*. Australian Inst Mar Sci Monogr Ser **VI**:485pp.

Veron JEN, Pichon M and Wijsman-Best M (1977). *Scleractinia of Eastern Australia. Part 2, Families Faviidae, Trachyphylliidae*. Australian Inst Mar Sci Monogr Ser **III**:233pp.

Verrill AE (1864). *A list of the corals and polyps sent by the Museum of Comparative Zoology to other institutions in exchange, with annotations*. Bull Mus Comparative Zool, Harvard Univ **1**:29-60.

Verrill AE (1866). *Synopsis of the polyps and corals of the North Pacific Exploring Expedition, 1853-1856, with descriptions of some additional species from the west coast of North America*. Communications Essex Inst **5**:17-50.

Verrill AE (1872). *Names of the species of corals*. In: Dana JD *'Corals and coral islands'*. New Haven **1**:379-388.

Wallace CC (1978). *The coral genus* Acropora *(Scleractinia: Astrocoeniina: Acroporidae) in the central and southern Great Barrier Reef Province*. Mem Queensland Mus **18**:273-319.

Wallace CC (1994). *New species and a new species group of the coral genus* Acropora *from Indo-Pacific locations*. Invertebrate Taxonomy **8**:961-988.

Wallace CC (1997). *New species and new records of recently described species of the coral genus* Acropora *(Scleractinia: Astrocoeniina: Acroporidae) from Indonesia*. Zool J Linnean Soc **120**:27-50.

Wallace CC (1999). *Staghorn corals of the world.* CSIRO Publ, Melbourne. 422pp.

Wallace CC and Dai C-F (1997). *Scleractinia of Taiwan. IV Review of the coral genus* Acropora *from Taiwan.* Zool Studies **36**:288-324.

Wallace CC and Wolstenholme J (1998). *Revision of the coral genus* Acropora *(Scleractinia: Astrocoeniina: Acroporidae) from Indonesia.* Zool J Linnean Soc **123**:199-384.

Walton Smith FG (1971). *Atlantic reef corals* (2nd edn). Univ Miami Press. 164pp.

Wells JW (1950). *Reef corals from Cocos Keeling Atoll.* Bull Raffles Mus **22**:29-52.

Wells JW (1954). *Recent corals of the Marshall Islands.* US Geol Survey Professional Papers **260**:384-486.

Wells JW (1956). *Scleractinia.* In: Moore RC (ed) *Treatise on invertebrate palaeontology.* Geol Soc America and Univ Kansas Press. **F**:328-440.

Wells JW (1961). *Notes on Indo-Pacific scleractinian corals. Part 3: A new reef coral from New Caledonia.* Pacific Sci **15**:189-191.

Wells JW (1962). *Two new scleractinian corals from Australia.* Rec Australian Mus, Sydney **25**:239-241.

Wells JW (1966a). *Evolutionary development of the scleractinian family Fungiidae.* In: Rees WJ (ed) *The Cnidaria and their evolution.* Symp Zool Soc London **16**:223-246.

Wells JW (1966b). *Notes on Indo Pacific scleractinian corals. Part 4: A second species of* Stylocoeniella. Pacific Sci **20**:203-205.

Wells JW (1971a). *Notes of Indo-Pacific scleractinian corals. Part 7:* Catalaphyllia, *a new genus of reef corals.* Pacific Sci **25**:368-371.

Wells JW (1971b). *Note on the scleractinian corals* Scolymia lacera *and S. cubensis in Jamaica.* Bull Mar Sci **21**:960-963.

Wells JW (1973a). *Two new hermatypic scleractinian corals from the West Indies.* Bull Mar Sci **23**:925-932.

Wells JW (1973b). *New and old scleractinian corals from Jamaica.* Bull Mar Sci **23**:16-58.

Wells JW (1983). *Annotated list of the Scleractinian corals of the Galapagos.* In: Glynn PW and Wellington TM (eds) *Corals and coral reefs of the Galapagos Islands.* Univ California Press, Berkeley. 330pp.

Wells JW (1985). *Notes on Indo-Pacific Scleractinian corals. Part 11: A new species of* Acropora *from Australia.* Pacific Sci **39**:338-339.

Wells S (1988) (ed). *Corals reefs of the world.* UNEP and IUCN, **1**:373pp, **2**:389pp, **3**:329pp.

Wijsman-Best M (1972). *Systematics and ecology of New Caledonian Faviidae (Coelenterata Sceractinia).* Bijdragen Dierkunde **42**:1-90.

Wijsman-Best M (1973). *A new species of the Pacific coral genus* Blastomussa *from New Caledonia.* Pacific Sci **27**:154-155.

Wijsman-Best M (1974). *Biological results of the Snellius Expedition. XXV Faviidae collected by the Snellius Expedition. 1 The Genus* Favia. Zool Mededelingen **48**:249-261.

Wijsman-Best M (1976). *Biological results of the Snellius Expedition. XXVII Faviidae collected by the Snellius Expedition II The genera* Favites, Goniastrea, Platygyra, Oulophyllia, Leptoria, Hydnophora *and* Caulastrea. Zool Mededelingen **50**:45-63.

Wijsman-Best M (1977). *Indo-Pacific coral species belonging to the subfamily Montastreinae Vaughan and Wells, 1943 (Scleractinia, Coelenterata). Part 1: The genera* Montastrea *and* Plesiastrea. Zool Mededelingen **52**:81-97.

Wijsman-Best M (1980). *Indo-Pacific coral species belonging to the subfamily Montastreinae Vaughan and Wells, 1943 (Scleractinia-Coelenterata). Part II: The genera* Cyphastrea, Leptastrea, Echinopora *and* Diploastrea. Zool Mededelingen **55**:235-263.

Wood EM (1983). *Corals of the world.* TFH Publ, Neptune City. 256pp.

Wood R (1999). *Reef evolution.* Oxford Univ Press. 414pp.

Yabe H and Sugiyama T (1936). In: Yabe H, Sugiyama T and Eguchi M (1936). *Recent reef-building corals from Japan and the South Sea Islands under the Japanese mandate.* Sci Rept Tohoku Univ Second Ser (Geology) **1**:166pp.

Yabe H and Sugiyama T (1937). *Two new species of reef-building corals from Yoron-zima and Amami-O-sima.* Proc Imperial Acad, Tokyo. **13**:425-429.

Yabe H and Sugiyama T (1941). *Recent reef building corals from Japan and the south sea islands under the Japanese mandate.* Sci Rept Tohoku Univ Second Ser (Geology) **2**:67-91.

Yabe H, Sugiyama T and Eguchi M (1936). *Recent reef-building corals from Japan and the South Sea Islands under the Japanese mandate.* Sci Rept Tohoku Univ Second Ser (Geology) **1**:166pp.

Zibrowius H (1980). *Les scléractiniaires de la Méditerranée et de l'Atlantique nord-oriental.* Mémoires De L'Inst Oceanog **11**:1-227.

Zlatarski VN (1990). Porites colonensis, *a new species of stony coral (Anthozoa: Scleractinia) off the Caribbean Coast of Panama.* Proc Biol Soc Washington **103**:257-264.

Zlatarski VN and Estalella NM (1982). *Les Scléractiniaires de Cuba avec des données sur les organismes associés.* Editions Acad Bulgare Sci Sofia. 472pp.

Zou RL (1980). *Studies on the corals of the Xisha Islands. IV Two new hermatypic scleractinian corals.* Nanhai Studia Marina Sinica **1**:117-118.

Index

Volume numbers are in bold. Minor references are in small italics.

abdita, Favites **3:***134*,146-7
abrolhosensis, Acropora **1:***209*,256-7
abrotanoides, Acropora **1:**220-1,*253*
abundance *1:4,***3:**442
Acanthaster planci **1:**16
Acanthastrea: key to species **3:**456
Acanthastrea **3:**12
    amakusensis **3:***10-1*
    bowerbanki **3:**26
    brevis **3:**17
    echinata **3:**18-9
    erythraea **3:***54*
    faviaformis **3:**24-5
    hemprichii **3:**22-3
    hillae **3:***13*,28-9
    ishigakiensis **3:**30-1
    lordhowensis **3:***13*,14-5
    maxima **3:**27
    minuta **3:***8*
    regularis **3:**16
    rotundoflora **3:**20-1
    subechinata **3:**13
acanthocauli **2:**256,*271*,**3:***420*-1
Acanthophyllia deshayesiana **3:***82*
acolonial corals **1:**54
Acrhelia **2:***106,107,115*
    horrescens **2:***107*
acrhelia, Galaxea **2:**115
Acropora: corallite structure **1:**178-9
Acropora: diversity contours **1:**176
Acropora: growth-forms **1:**178-9
Acropora: key to species **3:**447-9
Acropora: type species **1:**176
Acropora **1:***40*,176-83
    abrolhosensis **1:***209*,256-7
    abrotanoides **1:**220-1,*253*

Previous page: A shallow water community of *Acropora* and *Montipora*. HOUTMAN ABROLHOS ISLANDS, SOUTH-WEST AUSTRALIA Photograph: author

Opposite: *Javania exserta* (Flabellidae), a delicate azooxanthellate beauty in deep water. PALAU Photograph: Pat Colin

aculeus **1:**356-7
acuminata **1:**230
akajimensis **1:**273
anthocercis **1:**368-9
appressa **1:**397
arabensis **1:**315
aspera **1:**342-3
austera **1:**250-1
awi **1:**422-3
azurea **1:**355
batunai **1:**372-3
bifurcata **1:**304-5,**3:***432*
branchi **1:**242-3
brueggemanni **1:**198-9
bushyensis **1:**311
cardenae **1:**419
carduus **1:**432
caroliniana **1:**380-1
cerealis **1:**402-3
cervicornis **1:**206-7
chesterfieldensis **1:**314-5
clathrata **1:**292-3
convexa **1:**337
cophodactyla **1:**396
copiosa **1:**260
corymbosa **1:***367,376*
crateriformis **1:**190-1
cuneata **1:**184-5
cylindrica **1:**193
cytherea **1:**300-1,*304*,**3:***432*
danai **1:***220*
dendrum **1:**327
derawanensis **1:**414-5
desalwii **1:**370-1
digitifera **1:**328-9
divaricata **1:**238-9
donei **1:**228-9
downingi **1:**294-5
echinata **1:**426-7
efflorescens **1:**298
elegans **1:**282-3
elegantula **1:**358
elizabethensis **1:**188-9
elseyi **1:**433
eurystoma **1:***360*
exquisita **1:**412-3
fastigata **1:**331

fenneri **1:**416-7
filiformis **1:**418
florida **1:**248-9
formosa **1:**204-5
forskali **1:**252
gemmifera **1:***320*,324-5
glauca **1:**240-1
globiceps **1:**317
gomezi **1:**408
grandis **1:**208,*262*
granulosa **1:**382-3
haimei **1:**263
halmaherae **1:***410*
hemprichii **1:**194-5
hoeksemai **1:**225,*305*
horrida **1:***262*,266-7
humilis **1:**318-9,*325*
hyacinthus **1:**306-7,**3:***432*
indiana **1:***345*
indonesia **1:**224
inermis **1:**281
insignis **1:***161,279*,364-5
intermedia **1:***222*
irregularis **1:***213*,214-5,*218*
jacquelineae **1:**384-5
japonica **1:***29*,330
kimbeensis **1:**352-3
kirstyae **1:**409
kosurini **1:**231
lamarcki **1:**376-7
latistella **1:**348-9
lianae **1:**407
listeri **1:**334
loisetteae **1:**346
lokani **1:**379
longicyathus **1:**430
loripes **1:**388-9
lovelli **1:***189*,246-7
lutkeni **1:**252-3
macrostoma **1:**375
maryae **1:**392-3
massawensis **1:**336
meridiana **1:**280
microclados **1:**374
microphthalma **1:**258-9
millepora **1:**340-1
minuta **1:**210

mirabilis **1:**347
monticulosa **1:**320-1,*325*
mossambica **1:***322*
multiacuta **1:**332
muricata **1:***176*
nana **1:**354
nasuta **1:**400-1
natalensis **1:**236-7
navini **1:**431
nobilis **1:**222-3
ocellata **1:**312-3
orbicularis **1:**244
palifera **1:***27,*186-7
palmata **1:**202-3
palmerae **1:**211
paniculata **1:**378
papillare **1:**345
parahemprichii **1:**274-5
parapharaonis **1:**367
parilis **1:**410-1
pectinatus **1:**264
pharaonis **1:**296-7
pichoni **1:**286
pinguis **1:**212-3
plana **1:**302-3,**3:***432*
plantaginea **1:**391
plumosa **1:**288-9
polystoma **1:**335
prolifera **1:**261
prostrata **1:**338-9
proximalis **1:**278-9
pruinosa **1:**270
pulchra **1:**344-5
rambleri **1:**299
retusa **1:**322
robusta **1:**216-7
rosaria **1:**394-5
roseni **1:**218-9
rudis **1:**201
rufus **1:**269
russelli **1:**420-1
samoensis **1:**323
sarmentosa **1:**326
scherzeriana **1:**200
schmitti **1:**196-7
secale **1:**398-9
sekiseiensis **1:**276
selago **1:**362-3
seriata **1:**254-5
simplex **1:**284
solitaryensis **1:**234-5
sordiensis **1:***397*
spathulata **1:***340*
speciosa **1:**424-5
spicifera **1:***210,*308-9,*344*
squarrosa **1:**390
stoddarti **1:**232-3
striata **1:**272,*275*
subglabra **1:**428
subulata **1:**350-1
suharsonoi **1:**333
sukarnoi **1:***214*
tanegashimensis **1:**310
tenella **1:**285
tenuis **1:**360-1
teres **1:**209
tizardi **1:**277

togianensis **1:**192
torihalimeda **1:**421
torresiana **1:**316
tortuosa **1:**265
tumida **1:**271
turaki **1:**429
tutuilensis **1:**290-1
valenciennesi **1:**226-7
valida **1:**404-5
variabilis **1:**406
variolosa **1:**197
vaughani **1:**268-9
vermiculata **1:**359
verweyi **1:**386-7
walindii **1:**287
wallaceae **1:**245
wardii **1:***317*
willisae **1:**366
yongei **1:**262-3
Acroporidae: diversity contours **1:**61
Acroporidae: key to species **3:**447
Acroporidae **1:***36-7,49,50,*61
actiniformis, Heliofungia **2:**254-5
active dispersal **3:**416
aculeata, Seriatopora **2:**52
aculeus, Acropora **1:**356-7
acuminata, Acropora **1:**230
acuta, Platygyra **3:**190
acuticollis, Favites **3:**141
aequalis, Leptastrea **3:**235
aequicostatus, Heterocyathus **2:**412-3
aequituberculata, Montipora **1:***69,*76-7,**3:***436*
affinity, species **3:**413,430-3
africana, Sandalolitha **2:**299
africana, Schizoculina **2:**104
africanus, Pectinia **2:**353
Agaricia **2:**170
    agaricites **2:**172-3,**3:***86*
    fragilis **2:**174
    grahamae **2:**175
    humilis **2:**177
    lamarcki **2:***viii,173,*176
    tenuifolia **2:**171,*172*
    undata **2:**175
Agariciidae: key to species **3:**451
Agariciidae **1:***42,48-9,***2:**169
agaricia, Symphyllia **3:**60-1
agaricites, Agaricia **2:**172-3,**3:***86*
agassizi, Cyphastrea **3:**248
age of genera **1:**42
akajimensis, Acropora **1:**273
albiconus, Goniopora **3:**361
albidus, Favia **3:**112
albitentaculata, Ctenactis **2:**288-9
alcicornis, Pectinia **2:**356-7
Alcyonacea **3:***399,406-7*
algal symbiosis **1:***21*
aliciae, Mycetophyllia **3:**78-9
allingi, Alveopora **3:**384
allopatric origination **3:**439,*463*
allopatric patterns **3:**439
altasepta, Montipora **1:***151,*153
Alveopora: key to species **3:**457-8
Alveopora **1:***53,***3:***349,*380
    allingi **3:**384
    catalai **3:**382-3
    daedalea **3:**390-1

excelsa **3:**394
fenestrata **3:**386,*387*
gigas **3:**380-1
japonica **3:**392
marionensis **3:**385
minuta **3:**396
ocellata **3:**397
spongiosa **3:**388-9
tizardi **3:**392-3
verrilliana **3:**387
viridis **3:**395
amakusensis, Acanthastrea **3:***10-1*
amakusensis, Micromussa **3:***10-1*
ambulacral grooves **1:**51-2
americana, Phyllangia **2:***410*
amicorum, Barabattoia **3:**133
amitoriensis, Leptoseris **2:**207
ampliata, Merulina **2:***377,*378-9
ampullae, reproductive **3:***400,402*
Anacropora: key to species **3:**449
Anacropora **1:**168
    forbesi **1:**168-9
    matthai **1:**174
    pillai **1:**175
    puertogalerae **1:**170-1
    reticulata **1:**172
    spinosa **1:**173
    spumosa **1:**171
ancora, Euphyllia **2:***68,*80-1,*226*
angulata, Montipora **1:**127
angulosa, Mussa **3:**64-5,*66*
ankeli, Pocillopora **2:**33
annae, Porites **3:**310-1,*427*
annularis, Montastrea **3:***87,*214-5
annuligera, Montastrea **3:**220
Anomastraea **2:**137
    irregularis **2:**137
anthocercis, Acropora **1:**368-9
Anthozoa **3:***399*
Antillia **3:***271*
anus **1:**47
appressa, Acropora **1:**397
appressed corallites **1:**179
aquaria **1:**14-6
arabensis, Acropora **1:**315
aragonitic skeletons **1:***34-5,40,***3:**404
aranetai, Porites **3:**303
arborescent colonies **1:**56,179
archaeocyaths **1:***33*
Archohelia **2:**96
    rediviva **2:**96-7
areolata, Manicina **3:**99
armata, Stylocoeniella **2:***2,*4-5
arnaudi, Porites **3:**301
arbuscula, Cladocora **3:**90
arnoldi, Porites **3:**301
asanoi, Madracis **2:**17
asexual reproduction **1:***14-5,***3:**420-1
ashmorensis, Echinopora **3:**256
aspera, Acropora **1:**342-3
aspera, Echinophyllia **2:***323,*324-5
aspera, Goniastrea **3:**168-9
aspergillus, Montipora **1:**167
asperula, Madracis **2:**18
Aspidosiphon corallicola **2:***407,412*
Astrangia **2:**318,319
    astreiformis **2:***319*
    danae **2:***319*

poculata **2**:319
woodsi **2**:*318*
Astrangiidae **2**:317
astreata, Galaxea **2**:*106*,110-1
astreiformis, Astrangia **2**:*319*
astreoides, Porites **3**:280-1
Astreopora: key to species **3**:450
Astreopora **1**:434
   cucullata **1**:445
   expansa **1**:434-5
   explanata **1**:*434*
   gracilis **1**:444
   incrustans **1**:437
   listeri **1**:439
   macrostoma **1**:447
   moretonensis **1**:436
   myriophthalma **1**:*437*,442-3
   ocellata **1**:446
   randalli **1**:438
   scabra **1**:441
   suggesta **1**:440
Astreosmilia **3**:*91*
   connata **3**:*91*
Astrocoeniidae: key to species **3**:451
Astrocoeniidae **1**:*36-7,50*,**2**:3
atolls **1**:23-6
attenuata, Porites **3**:330
attenuation of diversity **3**:*412*,414
attenuation, latitudinal **3**:414
austera, Acropora **1**:250-1
australensis, Goniastrea **3**:*14,156*,170-1
australiensis, Montipora **1**:152
australiensis, Porites **3**:*277,286*
australis, Scolymia **3**:70-1
Australogyra **3**:194
   zelli **3**:194
Australomussa **3**:80
   rowleyensis **3**:80
autotomy **2**:248,**3**:421
awi, Acropora **1**:422-3
axial corallites **1**:178
axifuga, Duncanopsammia **2**:405
axillaris, Cyathelia **2**:*96-7*
ayleni, Pectinia **2**:352
ayleni, Physophyllia **2**:*352*
azooxanthellate corals **1**:*27*,**2**:*3,67,95,317,385, 409*,**3**:*399*
azurea, Acropora **1**:355
bairdiana, Balanophyllia **2**:*387*
Balanophyllia **2**:406
   bairdiana **2**:*387*
   europaea **2**:406
Banks, Joseph **1**:*11*
Barabattoia **3**:*87*,132
   amicorum **3**:133
   laddi **3**:132
barrier reefs **1**:23-7
basal plate **1**:51
batunai, Acropora **1**:372-3
bennettae, Oulophyllia **3**:200-1
bestae, Favites **3**:140
bewickensis, Leptastrea **3**:232
bifrons, Turbinaria **2**:402
bifurcata, Acropora **1**:304-5,**3**:*432*
biogeography **3**:411-21,438-9
bioherms **1**:27
bipartita, Pavona **2**:197

Blastomussa **3**:4
   loyae **3**:*4*
   merleti **3**:4-5
   wellsi **3**:6-7
bleaching **1**:16-7
blue coral **3**:404-5
Boninastrea **2**:382
   boninensis **2**:382
boninensis, Boninastrea **2**:382
bonsai, Hydnophora **2**:369
bottae, Leptastrea **3**:234
bottlebrush colonies **1**:178
boundaries
   geographic **3**:429-33
   morphological **3**:429-30
   species **3**:429-33
bournoni, Solenastrea **3**:250
bowerbanki, Acanthastrea **3**:26
branchi, Acropora **1**:242-3
branching colonies **1**:56
branneri, Porites **3**:325
braziliensis, Meandrina **2**:122
braziliensis, Mussismilia **3**:34
brevis, Acanthastrea **3**:17
breviserialis, Colpophyllia **3**:*210*
brighami, Porites **3**:295
brooding **3**:417-9
brueggemanni, Acropora **1**:198-9
budding **1**:54,**3**:*135*
burgosi, Goniopora **3**:375
bushyensis, Acropora **1**:311
cactus, Montipora **1**:150-1
cactus, Pavona **2**:180-1,*183*
caespitosa, Cladocora **3**:88-9
caespitose colonies **1**:178
cailleti, Leptoseris **2**:205
calcarea, Montipora **1**:116
calcitic skeletons **1**:3
caliculata, Montipora **1**:128-9
caliendrum, Seriatopora **2**:*47,48*,54-5
cancers **3**:421
Cantharellus **2**:251
   doederleini **2**:251
   jebbi **2**:253
   noumeae **2**:252
capitata, Montipora **1**:144-5
capitata, Pocillopora **2**:35
capricornis, Montipora **1**:80-1
cardenae, Acropora **1**:419
carduus, Acropora **1**:432
Caribbean corals **3**: 412-3
carnosus, Platygyra **3**:184
caroliniana, Acropora **1**:380-1
Caryophyllia **2**:*411*
Caryophylliidae **1**:*49*,**2**:409-11
catalai, Alveopora **3**:382-3
Catalaphyllia **2**:82
   jardinei **2**:82-3
Caulastrea: key to species **3**:453
Caulastrea **3**:91
   connata **3**:91
   curvata **3**:96
   echinulata **3**:97
   furcata **3**:92-3
   tumida **3**:94-5
cavernosa, Montastrea **3**:*87*,222-3
cebuensis, Montipora **1**:68-9,**3**:*468*

cellulosa, Goniopora **3**:363
Cenozoic corals **1**:41-2
Central American Seaway **1**:42-3
centres of diversity **3**:412,415
cerealis, Acropora **1**:402-3
cerioid colonies **1**:54-5
cervicornis, Acropora **1**:206-7
chagius, Ctenella **2**:123
chalcidicum, Cyphastrea **3**:241
characters
   families **1**:4
   genera **1**:4
   species **1**:4
chesterfieldensis, Acropora **1**:314-5
chimeras **3**:421
chinensis, Favites **3**:143
ciliatus, Goniopora **3**:372-3
circulation patterns **3**:414-6,438-41
circumvallata, Montipora **1**:93
Cladocora **3**:88
   arbuscula **3**:90
   caespitosa **3**:88-9
clathrata, Acropora **1**:292-3
clavator, Halomitra **2**:301
clavus, Pavona **2**:198-9
climate change **1**:16
clivosa, Diploria **3**:209
cochlea, Heteropsammia **2**:407
cocosensis, Montipora **1**:114-5
cocosensis, Porites **3**:311
cocosensis, Stylocoeniella **2**:8
coelenteron **1**:47,52
Coeloseris **2**:221
   mayeri **2**:221
coenosteum **1**:48-51
   pits **3**:*464*
   styles **3**:*464*
coerulea, Heliopora **3**:404-5
Colangia **2**:*318*
   immersa **2**:*318*
colemani, Montastrea **3**:219
colini, Cycloseris **2**:247
collines **3**:*464*
colonensis, Porites **3**:302
colonial corals **1**:54
colony formation **1**:54-55
colour **1**:4
Colpophyllia **3**:210
   breviserialis **3**:*210*
   natans **3**:*73,207*,210-1
columella, Goniastrea **3**:165
columellae **1**:48,50,*54*
   of *Porites* **3**:278
columna, Coscinaraea **2**:160-1
columna, Goniopora **3**:*349,356-7*
columnar colonies **1**:56,178
columnaris, Porites **3**:300
common names **3**:461
competition **3**:441-2
complanata, Favites **3**:150-1
compressa, Porites **1**:*12*,**3**:344-5
concepts of species **1**:11-3,**3**:425-6,441-3
concinna, Fungia **2**:270
confusa, Montipora **1**:82-3
connata, Astreosmilia **3**:*91*
connata, Caulastrea **3**:91
connectivity, genetic **3**:415,428-9,438-42

Conotrochus **1**:*49*
consagensis, Phyllangia **2**:*410*
conservation **1**:14-7
conspicua, Turbinaria **2**:403
contigua, Psammocora **2**:146-7
continua in Nature **3**:425-33,438-40
contorta, Platygyra **3**:188-9
contours of diversity **3**:412
convexa, Acropora **1**:337
convoluta, Oxypora **2**:340
cophodactyla, Acropora **1**:396
copiosa, Acropora **1**:260
coral/corals (see also subject entries)
    acolonial **1**:54
    affinities of **3**:413,430-3
    azooxanthellate *1:27*,**2**:*3,67,95,317,385,409*, **3**:*399*
    Cenozoic **1**:41-2
    colonial **1**:54
    Cretaceous **1**:40
    diversity (see 'diversity')
    Eocene **1**:41
    families of **1**:36,*40-1*
    free-living **3**:*465*
    hermatypic **3**:*465*
    high latitude **1**:28-9,**3**:*414,441*
    Jurassic **1**:36-7,40
    Mesozoic **1**:36-40
    Miocene **1**:41-2
    Oligocene **1**:41
    Palaeogene **1**:41
    Palaeozoic **1**:33-7
    reefs **1**:21-31,33-41
    reproduction *1:14-5*,**3**:*416-20*
    rugose *1:35*,**3**:*399,407*
    sex of **3**:417
    solitary **1**:*49*,54
    structure **1**:47-57
    tabulate *1:34-5*,**3**:*399,407*
    Triassic **1**:36-9
    zooxanthellate **1**:53
coralline algae *1:25,27*
corallite structure **1**:48-51
    Acropora **1**:178-9
    Montipora **1**:64-5
    Porites **3**:277-8
corbettensis, Montipora **1**:109
corona, Fungia **2**:260-1
corymbosa, Acropora **1**:*367,376*
corymbosa, Lobophyllia **3**:42-3
corymbose colonies **1**:178
Coscinaraea: key to species **3**:459
Coscinaraea **2**:158
    columna **2**:160-1
    crassa **2**:166
    exesa **2**:162-3,**3**:*462*
    hahazimaensis **2**:159
    marshae **2**:164
    mcneilli **2**:165
    monile **2**:158-9
    vaughani **2**:*157*
    wellsi **2**:167
costae **1**:48-50
    of *Fungia* **2**:258
costata, Echinophyllia **2**:330
costulata, Cycloseris **2**:245
crassa, Coscinaraea **2**:166

crassa, Ctenactis **2**:290
crassispinosa, Oxypora **2**:334-5
crassituberculata, Montipora **1**:78-9
crateriformis, Acropora **1**:190-1
Cretaceous
    corals **1**:40
    environment **1**:40
    reefs **1**:37,40
crispa, Oulophyllia **3**:*195*,196-7
crispata, Oulastrea **3**:229
cristata, Euphyllia **2**:69
cross fertilisation **3**:419-20,438-9
crosslandi, Platygyra **3**:180
crown-of-thorns starfish **1**:16
crustacea, Podabacia **2**:310-1
cryptoramosa, Galaxea **2**:114
cryptus, Montipora **1**:126
Ctenactis **1**:*54*,**2**:286
    albitentaculata **2**:288-9
    crassa **2**:290
    echinata **2**:*235*,286-7
Ctenella **2**:123
    chagius **2**:123
cubensis, Scolymia **3**:66-7
cucullata, Astreopora **1**:445
cucullata, Leptoseris **2**:214-5
Culicia **2**:*316,318*
cumulatus, Porites **3**:324
cuneata, Acropora **1**:184-5
currents, ocean **3**:415-6,438-41
curta, Montastrea **3**:*212*,216-7
curvata, Caulastrea **3**:96
curvata, Cycloseris **2**:240
Cyathelia **2**:*96-7*
    axillaris **2**:*96-7*
cycles of radial elements **1**:49-50
cyclolites, Cycloseris **2**:236-7,*249*
Cycloseris: key to species **3**:455
Cycloseris **1**:*54*,**2**:236
    colini **2**:247
    costulata **2**:245
    curvata **2**:240
    cyclolites **2**:236-7,*249*
    erosa **2**:241
    hexagonalis **2**:239
    patelliformis **2**:246
    sinensis **2**:238
    somervillei **2**:242-3
    tenuis **2**:244
    vaughani **2**:244
cyclosystems **3**:*400,402*
cylindrica, Acropora **1**:193
cylindrica, Porites **3**:*274,314,330,332-3,335*
cylindrica, Scapophyllia **2**:383
cylindrus, Dendrogyra **2**:126-7
Cynarina **1**:*53*,**3**:*82*
    lacrymalis **3**:82-3
Cyphastrea: key to species **3**:453
Cyphastrea **3**:240
    agassizi **3**:248
    chalcidicum **3**:241
    decadia **3**:249
    hexasepta **3**:245
    japonica **3**:240
    microphthalma **3**:246-7
    ocellina **3**:244
    serailia **3**:*87*,242-3

cytherea, Acropora **1**:300-1,*304*,**3**:*432*
dactylopores **3**:*400,402*
dactylozooids **3**:*400,402*
daedalea, Alveopora **3**:390-1
daedalea, Platygyra **3**:*177*,191
damicornis, Pocillopora **2**:26-7,*29*,**3**:*427*
Dana, James *1:11*
danaana, Mycetophyllia **3**:*73*,76-7
danae, Astrangia **2**:*319*
danae, Favia **3**:*100*,123
danae, Montipora **1**:140-1
danae, Pocillopora **2**:25
danae, Stylophora **2**:63
danai, Acropora **1**:*220*
danai, Fungia **2**:*258*,262-3
danai, Pavona **2**:179
Darwin, Charles *1:11,23*,**3**:*414,441*
Darwinian evolution **3**:441-3
decactis, Madracis **2**:18-9
decadia, Cyphastrea **3**:249
decussata, Pavona **2**:*178*,194-5
decussata, Psammocora **2**:144
deformis, Goniastrea **3**:167
deformis, Porites **3**:323
delicatula, Montipora **1**:70-1
Deltocyathus **2**:*411*
dendritica, Seriatopora **2**:46-7,*367*
Dendrogyra **2**:126
    cylindrus **2**:126-7
Dendrophyllia **2**:*386*
Dendrophylliidae: key to species **3**:452
Dendrophylliidae **1**:*49-51*,**2**:*385*
dendrum, Acropora **1**:327
densa, Porites **3**:294
dentata, Sandalolitha **2**:298
dentatus, Lobophyllia **3**:46
denticles, *Porites* **3**:278
derawanensis, Acropora **1**:414-5
desalwii, Acropora **1**:370-1
deshayesiana, Acanthophyllia **3**:*82*
desilveri, Porites **3**:308
Desmophyllum **2**:*411*
devantieri, Plesiastrea **3**:228
Devonian reefs **1**:32-5
Diaseris **2**:248
    distorta **2**:*240*,248-9
    fragilis **2**:250
Dichocoenia **1**:*43*,**2**:124
    stellaris **2**:*124*
    stokesi **2**:124-5
diffluens, Pavona **2**:188
diffusa, Oculina **2**:102
digitata, Montipora **1**:*151*,154-5,**3**:*429*
digitata, Psammocora **2**:154-5
digitate colonies **1**:178
digitifera, Acropora **1**:328-9
dilatata, Montipora **1**:98
diminuta, Lobophyllia **3**:39
diminuta, Micromussa **3**:9
Diploastrea **3**:230
    heliopora **3**:*84*,230-1,*433*
Diploria **1**:*43*,**3**:206
    clivosa **3**:209
    labyrinthiformis **3**:*87*,206-7
    strigosa **3**:208
discus, Plerogyra **2**:86-7
dispersal **3**:416-20

larval **3**:417-9
long-distance **3**:416-20
timing of **3**:418
dissepiments **1**:51
Distichopora **3**:*398-9*,402-3
violacea *3:398*
distorta, Diaseris **2**:*240*,248-9
distribution
maps **1**:6,**3**:430,432
patterns **1**:6,**3**:411-5
ranges **1**:6,**3**:430,432
reefs **1**:22
temperature correlation *1:28-9*,**3**:414-5
divaricata, Acropora **1**:238-9
divaricata, Porites **3**:329
diversity **3**:411-2,415-6
attenuation of **3**:*412*,414
Cenozoic **1**:41-2
centres of **3**:412,415
family **1**:36,**3**:412
generic *1:42*,**3**:412
habitat **3**:414-5
Mesozoic **1**:37
patterns of **3**:411-5
species **3**:412,415-6
subspecies *3:413*
diversity contours **3**:412
Acropora **1**:176
Acroporidae **1**:61
families **3**:412
Faviidae **3**:85
genera **3**:412
Montipora **1**:62
Porites **3**:276
Poritidae **3**:275
species **3**:412
divisa, Euphyllia **2**:*68*,78-9
djiboutiensis, Goniopora **3**:*349*,351
doederleini, Cantharellus **2**:251
donei, Acropora **1**:228-9
doubtful species *1:16*,**3**:429-32
downingi, Acropora **1**:294-5
drop-out sequence **3**:*412*,414
duerdeni, Pavona **2**:200-1
duerdeni, Porites *3:344*
Duncanopsammia **2**:405
axifuga **2**:405
earliest Scleractinia **1**:37
East Australian Current **3**:414
echinata, Acanthastrea **3**:18-9
echinata, Acropora **1**:426-7
echinata, Ctenactis **2**:*235*,286-7
echinata, Echinophyllia **2**:332
echinata, Montipora **1**:166
echinatus, Zoopilus **2**:*234*,304-5
Echinomorpha **2**:333
nishihirai **2**:333
Echinophyllia: key to species **3**:456-7
Echinophyllia **2**:322
aspera **2**:*323*,324-5
costata **2**:330
echinata **2**:332
echinoporoides **2**:322-3
maxima *2:349*
nishihirai *2:333*
orpheensis **2**:328-9
patula **2**:326-7

pectinata **2**:331
taylorae **2**:327
Echinopora: key to species **3**:453
Echinopora **3**:252
ashmorensis **3**:256
ehrenbergi *2:349*
forskaliana **3**:264-5
fruticulosa **3**:257
gemmacea **3**:258-9
hirsutissima **3**:260-1
horrida **3**:268
irregularis **3**:262
lamellosa **3**:254-5,*259*
mammiformis **3**:266-7
pacificus **3**:252-3
robusta **3**:263
tiranensis **3**:265
echinoporoides, Echinophyllia **2**:322-3
echinulata, Caulastrea **3**:97
echinulata, Porites **3**:290,*296*
eclipsensis, Goniopora **3**:*349*,371
ectodermis **1**:48,52
edwardsi, Goniastrea **3**:161
efflorescens, Acropora **1**:298
efflorescens, Montipora **1**:104-5
effusa, Montipora **1**:108
effusus, Pocillopora **2**:39
egg and sperm bundles **3**:*173*,417-9,*443*
eggs **3**:417-9
Eguchipsammia *2:386*
egyptensis, Oxypora **2**:341
ehrenbergi, Echinopora *2:349*
elegans, Acropora **1**:282-3
elegans, Pocillopora **2**:34-5
elegantula, Acropora **1**:358
elephantotus, Mycedium **2**:*320*,*342*,344-5
elizabethensis, Acropora **1**:188-9
elongata, Pectinia **2**:360
elseyi, Acropora **1**:433
embryos **3**:418-9
encrusting colonies **1**:56,178
endemics **3**:412-3
environmental
control of evolution **3**:438-41
cues, spawning **3**:418-9
variation **3**:426-7
Eocene
corals **1**:41
reefs **1**:41
epitheca **1**:48-9,51
eridani, Porites **3**:322
erosa, Cycloseris **2**:241
erythraea, Acanthastrea *3:54*
erythraea, Symphyllia **3**:54-5
Erythrastrea **3**:98
flabellata **3**:98
Euphyllia: key to species **3**:452
Euphyllia *1:54*,**2**:68
ancora **2**:*68*,80-1,*226*
cristata **2**:69
divisa **2**:*68*,78-9
glabrescens **2**:*68*,70-1
paraancora **2**:74-5
paradivisa **2**:*66*,73
paraglabrescens **2**:72
yaeyamaensis **2**:76-7
Euphyllidae: key to species **3**:452

Euphyllidae *1:49,51*,**2**:67
Subfamily **2**:67
type genus **2**:67
eupsammides, Heteropsammia *2:407*
europaea, Balanophyllia **2**:406
eurysepta, Plerogyra *2:88*
eurystoma, Acropora *1:360*
Eusmilia *1:43*,**2**:130
fastigiata **2**:130-1
Eusmiliinae, Subfamily **2**:67
evermanni, Porites **3**:298
evolutionary repackaging **3**:438-43
evolution **3**:437-43
control of **3**:438-41
Darwinian **3**:441-3
rates of **3**:439-40
reticulate **3**:437-43
excelsa, Alveopora **3**:394
exserta, Javania *3:476*
exerta, Plerogyra *2:92*
exesa, Coscinaraea **2**:162-3,*3:462*
exesa, Hydnophora **2**:*362*,370-1
expansa, Astreopora **1**:434-5
experimental aquaria **1**:16
explanata, Astreopora *1:434*
explanata, Leptoseris **2**:208-9
explanate colonies **1**:56
explanulata, Pavona **2**:*168*,184-5,*196*
explanulata, Psammocora **2**:*132*,156
exquisita, Acropora **1**:412-3
external fertilisation **3**:417-9
extinction **3**:438-42
end Cretaceous **1**:41
end Triassic **1**:37
mass **1**:37,40-1
rates of **3**:439-42
extratentacular budding **1**:54
eydouxi, Pocillopora **2**:*31*,*43*,44-5
families of corals **1**:36,*40-1*
families, origins of **1**:36,*40-1*,**3**:411-2
family diversity **1**:36,**3**:412
family tree of Scleractinia **1**:36-7
family trees *3:437-8*
fascicularis, Galaxea **2**:*94*,*106*,108-9
fastigata, Acropora **1**:331
fastigiata, Eusmilia **2**:130-1
faulkneri, Tubastrea *2:386*
faveolata, Montastrea *3:214*
Favia: key to species **3**:453
Favia *1:54*,**3**:100-1
albidus **3**:112
danae **3**:*100*,123
favus **3**: *100*,116-7
fragum **3**:104
gravida *3:104*
helianthoides **3**:110-1
lacuna **3**:111
laxa **3**:105
leptophylla **3**:118
lizardensis **3**:120
maritima **3**:130-1
marshae **3**:122
matthaii **3**:106-7
maxima **3**:126
pallida **3**:114-5
rosaria **3**:119
rotumana **3**:121

481

rotundata **3**:124-5
speciosa **3**:108-9
stelligera **3**:102-3
truncatus **3**:113
veroni **3**:128-9
vietnamensis **3**:127
faviaformis, Acanthastrea **3**:24-5
Faviidae: diversity contours **3**:85
Faviidae: key to species **3**:452-3
Faviidae *1:42,49,54*,**3**:85-7
Faviina, origins of *1:40*
Favites: key to species **3**:453
Favites *1:54*,**3**:134-5
  abdita **3**:*134*,146-7
  acuticollis **3**:141
  bestae **3**:140
  chinensis **3**:143
  complanata **3**:150-1
  flexuosa **3**:*134*,154
  halicora **3**:144-5
  melicerum **3**:*140*
  micropentagona **3**:137
  paraflexuosa **3**:155
  pentagona **3**:138-9
  peresi **3**:*166*
  russelli **3**:148-9
  spinosa **3**:142
  stylifera **3**:136
  vasta **3**:*134*,152-3
favulus, Goniastrea **3**:172-3
favus, Favia **3**:*100*,116-7
fenestrata, Alveopora **3**:386,*387*
fenneri, Acropora **1**:416-7
ferox, Mycetophyllia **3**:74
fertilisation **3**:417-20,438-9
field guides *1:7*,**3**:469-73
filiformis, Acropora **1**:418
fire coral **3**:400-1
fissipara, Schizoculina **2**:105
flabellata, Erythrastrea **3**:98
flabellata, Montipora **1**:99
Flabellidae *1:49*
flabelliformis, Lobophyllia **3**:48-9
flabello-meandroid colonies **1**:54-5
flavus, Porites **3**:341
flexuosa, Favites **3**:*134*,154
florida, Acropora **1**:248-9
florida, Montipora **1**:74-5
floweri, Montipora **1**:124
foliosa, Leptoseris **2**:219
foliosa, Montipora **1**:66-7
foliosa, Pachyseris **2**:230
forbesi, Anacropora **1**:168-9
formosa, Acropora **1**:204-5
formosa, Madracis **2**:14-5
forskali, Acropora **1**:252
forskaliana, Echinopora **3**:264-5
foveolata, Montipora **1**:131
foveolate corallites **1**:65
fragilis, Agaricia **2**:174
fragilis, Diaseris **2**:250
fragum, Favia **3**:104
fralinae, Fungia **2**:271
franksi, Montastrea **3**:*214*
free-living corals **3**:*465*
friabilis, Montipora **1**:73
fringing reefs **1**:24-5

frondens, Turbinaria **2**:392-3
frondifera, Pavona **2**:182-3
fruticosa, Goniopora **3**:378-9
fruticulosa, Echinopora **3**:257
Fungia: key to species **3**:455
Fungia: skeletal structure **2**:258
Fungia *1:53,54*,**2**:256-8,**3**:*417*
  concinna **2**:270
  corona **2**:260-1
  danai **2**:*258*,262-3
  fralinae **2**:271
  fungites **2**:*258*,268-9
  granulosa **2**:276
  horrida **2**:264-5
  klunzingeri **2**:266-7
  moluccensis **2**:284-5
  paumotensis **2**:282-3
  puishani **2**:277
  repanda **2**:*258*,272-3
  scabra **2**:274
  scruposa **2**:259
  scutaria **2**:*258*,280-1
  seychellensis **2**:279
  spinifer **2**:275
  taiwanensis **2**:278-9
Fungiacyathidae **2**:*235*
Fungiacyathus **2**:*235*
  marenzelleri **2**:*235*
  stephanus **2**:*235*
fungiformis, Pocillopora **2**:40
Fungiidae: key to species **3**:454
Fungiidae *1:49,54*,**2**:233-5
Fungiina, origins of *1:40*
fungites, Fungia **2**:*258*,268-9
furcata, Caulastrea **3**:92-3
furcata, Porites **3**:328
fuzzy boundaries **3**:429-33
gaimardi, Montipora **1**:146
Galaxea: key to species **3**:456
Galaxea **2**:97,106,**3**:*417*
  acrhelia **2**:115
  astreata **2**:*106*,110-1
  cryptoramosa **2**:114
  fascicularis **2**:*94,106*,108-9
  horrescens **2**:107
  longisepta **2**:116-7
  paucisepta **2**:112-3
gardineri, Leptoseris **2**:202-3,*206*
Gardineroseris *1:42*,**2**:222
  planulata **2**:222-3
gastrodermis **1**:48,52-3
gastropores **3**:*400,402*
gastrozooids **3**:*400,402*
gemmacea, Echinopora **3**:258-9
gemmae, Pachyseris **2**:224-5
gemmifera, Acropora **1**:*320*,324-5
generic
  ages **1**:42
  diversity **1**:42,**3**:412
genetic
  connectivity **3**:415,428-9,438-42
  isolation **3**:412-3,433,438-41
  links **3**:428-9
  variation **3**:426-9
geoffroyi, Trachyphyllia **3**:*270*,272-3
geographic
  boundaries **3**:429-33

isolation **3**:415,430,438-40
  patterns *1:6*,**3**:411-5,430
  reticulation **3**:429-31,439
  subspecies **3**:425,430-2
  variation **3**:427-32,437
geological time charts **1**:34-5
gigantea, Pavona **2**:189
gigas, Alveopora **3**:380-1
glabra, Oxypora **2**:338-9
glabrescens, Euphyllia **2**:*68*,70-1
glauca, Acropora **1**:240-1
globiceps, Acropora **1**:317
glossary **3**:463-7
glynni, Siderastrea **2**:138
gomezi, Acropora **1**:408
gonads **1**:48,53
Goniastrea: key to species **3**:453-4
Goniastrea *1:54*,**3**:156-7
  aspera **3**:168-9
  australensis **3**:*14,156*,170-1
  columella **3**:165
  deformis **3**:167
  edwardsi **3**:161
  favulus **3**:172-3
  minuta **3**:158-9
  palauensis **3**:164-5
  pectinata **3**:*157*,174-5
  peresi **3**:166
  ramosa **3**:160
  retiformis **3**:162-3
  thecata **3**:169
Goniopora: key to species **3**:458
Goniopora *1:52,53*,**3**:348-9
  albiconus **3**:361
  burgosi **3**:375
  cellulosa **3**:363
  ciliatus **3**:372-3
  columna **3**:*349*,356-7
  djiboutiensis **3**:*349*,351
  eclipsensis **3**:*349*,371
  fruticosa **3**:378-9
  lobata **3**:*349*,354-5,*371*
  minor **3**:366-7
  norfolkensis **3**:362-3
  palmensis **3**:374
  pandoraensis **3**:*348-9,351*,370,*371*
  pearsoni **3**:365
  pendulus **3**:350
  planulata **3**:368-9
  polyformis **3**:360
  savignyi **3**:376
  somaliensis **3**:358-9
  stokesi **3**:352-3
  stutchburyi **3**:377
  sultani **3**:355
  tenella **3**:360
  tenuidens **3**:364
Goreaugyra memoralis **2**:*119*
gracilis, Astreopora **1**:444
grahamae, Agaricia **1**:175
grandis, Acropora **1**:208,*262*
grandis, Hydnophora **2**:368
granulosa, Acropora **1**:382-3
granulosa, Fungia **2**:276
gravida, Favia **3**:*104*
grisea, Montipora **1**:94
groove-and-tubercle structures **1**:51

growth-forms **1**:56-7
 Acropora **1**:178-9
guentheri, Stylocoeniella **2**:6-7
guttatus, Seriatopora **2**:50-1
Gyrosmilia **2**:128
 interrupta **2**:*118*,128-9
habitat diversity **3**:414-5
habitats **1**:4
hahazimaensis, Coscinaraea **2**:159
haimeana, Psammocora **2**:152
haimei, Acropora **1**:263
halicora, Favites **3**:144-5
halmaherae, Acropora *1:410*
Halomitra **2**:300
 clavator **2**:301
 meierae **2**:300
 pileus **2**:*232*,302-3
harrisoni, Porites **3**:343
harttii, Mussismilia **3**:35
hassi, Symphyllia **3**:52
hataii, Lobophyllia **3**:47
hawaiiensis, Leptoseris **2**:216-7
helianthoides, Favia **3**:110-1
Heliofungia **2**:254
 actiniformis **2**:254-5
Heliopora **3**:*399*,404-5
 coerulea **3**:404-5
heliopora, Diploastrea **3**:*84*,230-1,*433*
Helioporacea **3**:*399,404*
hemispherica, Montipora **1**:147
hemprichii, Acanthastrea **3**:22-3
hemprichii, Acropora **1**:194-5
hemprichii, Lobophyllia **3**:*2,30,43*,44-5,*50-1,429*
hermaphrodites *1:53*,**3**:417
hermatypic corals *3:465*
heronensis, Porites **3**:306-7
heronensis, Turbinaria **2**:404
Herpolitha *1:54*,**2**:291
 limax **2**:292-3
 weberi **2**:291
Heterocyathus **2**:412
 aequicostatus **2**:412-3
Heteropsammia **2**:407
 cochlea **2**:407
 eupsammides *2:407*
Hexacorallia *3:399,407*
hexagonalis, Cycloseris **2**:239
hexasepta, Cyphastrea **3**:245
high latitude
 corals **1**:28-9,**3**:414,441
 reefs **1**:28-9,**3**:414
hillae, Acanthastrea **3**:*13*,28-9
hirsuta, Montipora **1**:*155*,159
hirsutissima, Echinopora **3**:260-1
hispida, Montipora **1**:148-9
hispida, Mussismilia **3**:32-3
hodgsoni, Montipora **1**:72
hoeksemai, Acropora **1**:225,*305*
hoffmeisteri, Montipora **1**:123
Horastrea **2**:136
 indica **2**:136
horizontalata, Porites **3**:316-7
horrescens, Acrhelia *2:107*
horrescens, Galaxea **2**:107
horrida, Acropora **1**:*262*,266-7
horrida, Echinopora **3**:268
horrida, Fungia **2**:264-5

humilis, Acropora **1**:318-9,*325*
humilis, Agaricia **2**:177
hyacinthus, Acropora **1**:306-7,*3:432*
hyades, Solenastrea **3**:251
hybridisation **3**:419-20,438-9
Hydnophora: key to species **3**:455
Hydnophora *1:52*,**2**:364
 bonsai **2**:369
 exesa **2**:*362*,370-1
 grandis **2**:368
 microconos **2**:372-3
 pilosa **2**:364-5
 rigida **2**:366-7,*369*
Hydrozoa *3:399-403*
hystrix, Seriatopora **2**:*22,46,47*,48-9,*55,367*
identification guides **1**:7
illustrations **1**:6
immersa, Colangia *2:318*
immersed corallites **1**:178
inaequalis, Leptastrea **3**:233
incrassata, Montipora **1**:119
incrustans, Astreopora **1**:437
incrustans, Leptoseris **2**:218
Indian Ocean corals **3**:413
indiana, Acropora *1:345*
indiania, Pocillopora **2**:37
indica, Horastrea **2**:136
indonesia, Acropora **1**:224
Indo-Pacific corals **3**:412-3
Indophyllia **3**:81
 macassarensis **3**:81
inermis, Acropora **1**:281
inflata, Pocillopora **2**:41
informis, Montipora **1**:112-3
insignis, Acropora **1**:*161,279*,364-5
intermedia, Acropora *1:222*
internal fertilisation **3**:417-9
interrupta, Gyrosmilia **2**:*118*,128-9
intratentacular budding **1**:54
involuta, Pachyseris **2**:231
irregularis, Acropora *1:213*,**1**:214-5,*218*
irregularis, Anomastraea **2**:137
irregularis, Echinopora **3**:262
irregularis, Leptoria **3**:202-3
irregularis, Turbinaria **2**:398-9
ishigakiensis, Acanthastrea **3**:30-1
isolation
 genetic **3**:412-3,433,438-42
 geographic **3**:415,430,438-40
 reproductive **3**:412-3,425,433,441
Isophyllastrea *3:37,38*
 rigida *3:37*
Isophyllia *1:43*,**3**:36
 rigida **3**:37
 sinuosa **3**:36
Isopora, subgenus *1:180*
Isthmus of Panama **1**:42-3
jacquelineae, Acropora **1**:384-5
japonica, Acropora **1**:*29*,330
japonica, Alveopora **3**:392
japonica, Cyphastrea **3**:240
jardinei, Cataphyllia **2**:82-3
Javania exserta *3:476*
jebbi, Cantharellus **2**:253
Jukes, J.B. *1:12*
Jurassic
 corals **1**:36-7,40

reefs **1**:37
kelleheri, Pocillopora **2**:32
kellyi, Montipora **1**:164
kenti, Montigyra **2**:129
key to:
 Acanthastrea **3**:456
 Acropora **3**:447-9
 Acroporidae **3**:447
 Agariciidae **3**:451
 Alveopora **3**:457-8
 Anacropora **3**:449
 Astreopora **3**:450
 Astrocoeniidae **3**:451
 Caulastrea **3**:453
 Coscinaraea **3**:459
 Cycloseris **3**:455
 Cyphastrea **3**:453
 Dendrophylliidae **3**:452
 Echinophyllia **3**:456-7
 Echinopora **3**:453
 Euphyllia **3**:452
 Euphyllidae **3**:452
 Favia **3**:453
 Faviidae **3**:452-3
 Favites **3**:453
 Fungia **3**:455
 Fungiidae **3**:454
 Galaxea **3**:456
 Goniastrea **3**:453-4
 Goniopora **3**:458
 Hydnophora **3**:455
 Leptastrea **3**:454
 Leptoseris **3**:451
 Lobophyllia **3**:456
 Madracis **3**:452
 Meandrinidae **3**:455
 Merulinidae **3**:455
 Montastrea **3**:454
 Montipora **3**:450-1
 Mussidae **3**:455-6
 Mycedium **3**:457
 Oculinidae **3**:456
 Oxypora **3**:457
 Pavona **3**:451
 Pectinia **3**:457
 Pectiniidae **3**:456
 Platygyra **3**:454
 Pocillopora **3**:457
 Pocilloporidae **3**:457
 Porites **3**:458-9
 Poritidae **3**:457
 Psammocora **3**:459
 Seriatopora **3**:457
 Siderastrea **3**:459
 Siderastreidae **3**:459
 Stylophora **3**:457
 Symphyllia **3**:456
 Turbinaria **3**:452
keys *1:7*,**3**:447-59
kimbeensis, Acropora **1**:352-3
kirbyi, Madracis **2**:16
kirstyae, Acropora **1**:409
klunzingeri, Fungia **2**:266-7
kosurini, Acropora **1**:231
kuehlmanni, Stylophora **2**:62
Kuroshio **3**:414,441
labyrinthiformis, Diploria **3**:*87*,206-7

lacera, Oxypora **2:**336-7
lacera, Scolymia **3:***64,66*
lacrymalis, Cynarina **3:**82-3
lactuca, Pectinia **2:***349,*350-1
lacuna, Favia **3:**111
laddi, Barabattoia **3:**132
lamarcki, Acropora **1:**376-7
lamarcki, Agaricia **2:***viii,173,*176
lamarckiana, Mycetophyllia **3:**73
lamellina, Platygyra **3:**192-3
lamellosa, Echinopora **3:**254-5,*259*
laminar colonies **1:**56
lankaensis, Podabacia **2:**315
larvae **3:**417-20,*443*
    dispersal **3:**417-9
    endurance *1:43,***3:**420
    metamorphosis **3:**419
latistella, Acropora **1:**348-9
latistella, Porites **3:**320-1
latistellata, Moseleya **3:**269
latitudinal attenuation **3:**414
laxa, Favia **3:**105
Leeuwin Current **3:**414
Leptastrea: key to species **3:**454
Leptastrea **3:**232
    aequalis **3:**235
    bewickensis **3:**232
    bottae **3:**234
    inaequalis **3:**233
    pruinosa **3:**237
    purpurea **3:**236
    transversa **3:**238
leptophylla, Favia **3:**118
Leptoria **3:**202
    irregularis **3:**202-3
    phrygia **3:**204-5
Leptoseris: key to species **3:**451
Leptoseris **2:**202
    amitoriensis **2:**207
    cailleti **2:**205
    cucullata **2:**214-5
    explanata **2:**208-9
    foliosa **2:**219
    gardineri **2:**202-3,*206*
    hawaiiensis **2:**216-7
    incrustans **2:**218
    mycetoseroides **2:**213
    papyracea **2:**204-5
    scabra **2:**210
    solida **2:**211
    striata **2:**212
    tenuis **2:***219*
    tubulifera **2:**206
    yabei **2:**220
levis, Oulophyllia **3:**198-9
lianae, Acropora **1:**407
lichen, Porites **3:**304-5,*309*
lichtensteini, Physogyra **2:***86,*92-3
ligulata, Pocillopora **2:**38
limax, Herpolitha **2:**292-3
listeri, Acropora **1:**334
listeri, Astreopora **1:**439
Lithophaga lessepsiana **2:***407*
Lithophyllon **2:**306
    lobata **2:**307
    mokai **2:**306
    undulatum **2:**308-9

lizardensis, Favia **3:**120
lobata, Goniopora **3:***349,*354-5,*371*
lobata, Lithophyllon **2:**307
lobata, Porites **3:***279,*284-5,*432*
Lobophyllia: key to species **3:**456
Lobophyllia **1:***54,***3:**38
    corymbosa **3:**42-3
    dentatus **3:**46
    diminuta **3:**39
    flabelliformis **3:**48-9
    hataii **3:**47
    hemprichii **3:***2,30,43,*44-5,*50-1,429*
    pachysepta **3:**40
    robusta **3:**50-1,*67*
    serratus **3:**41
lobulata, Montipora **1:**95
loisetteae, Acropora **1:**346
lokani, Acropora **1:**379
long-distance dispersal **3:**416-20
longicyathus, Acropora **1:**430
longisepta, Galaxea **2:**116-7
Lophelia pertusa **1:***27*
lordhowensis, Acanthastrea **3:***13,*14-5
loripes, Acropora **1:**388-9
lovelli, Acropora **1:***189,*246-7
loyae, Blastomussa **3:***4*
luciphilia, Madracis **2:**12
lumping species **3:**426,431
lutea, Porites **3:***278,*287
lutkeni, Acropora **1:**252-3
macassarensis, Indophyllia **3:**81
macrostoma, Acropora **1:**375
macrostoma, Astreopora **1:**447
mactanensis, Montipora **1:**134-5
madagascarensis, Stylophora **2:**57
Madracis: key to species **3:**452
Madracis **2:**12
    asanoi **2:**17
    asperula **2:**18
    decactis **2:**18-9
    formosa **2:**14-5
    kirbyi **2:**16
    luciphilia **2:**12
    mirabilis **2:** 20-1
    pharensis **2:**12-3
    senaria **2:**14
Madrepora **2:***97*
magnistellata, Montastrea **3:**225
malampaya, Montipora **1:**163
maldivensis, Pavona **2:**192-3,**3:***427,440*
mamillata, Stylophora **2:**65
mammiformis, Echinopora **3:**266-7
mancaoi, Mycedium **2:**343
Manicina **3:**99
    areolata **3:**99
maps **1:**6
marenzelleri, Fungiacyathus **2:***235*
marionensis, Alveopora **3:**385
maritima, Favia **3:**130-1
marshae, Coscinaraea **2:**164
marshae, Favia **3:**122
maryae, Acropora **1:**392-3
mass extinctions **1:**37,40-1
mass spawning **3:**417-9
massawensis, Acropora **1:**336
massive colonies **1:**56,178
matthai, Anacropora **1:**174

matthaii, Favia **3:**106-7
maxima, Acanthastrea **3:**27
maxima, Echinophyllia **2:***349*
maxima, Favia **3:**126
maxima, Pectinia **2:**349
mayeri, Coeloseris **2:**221
mayeri, Porites **3:**289
mcneilli, Coscinaraea **2:**165
Meandrina **2:**120
    braziliensis **2:**122
    meandrites **1:***46,***2:**120-1,**3:***86*
meandrina, Montipora **1:**142-3
meandrina, Pocillopora **2:**30-1
Meandrinidae: key to species **3:**455
Meandrinidae **1:***42,49,51,***2:**119
meandrites, Meandrina **1:***46,***2:**120-1,**3:***86*
meandroid colonies 54-5
medusae **3:**400
meierae, Halomitra **2:**300
melicerum, Favites **3:***140*
memoralis, Goreaugyra **2:***119*
meridiana, Acropora **1:**280
merleti, Blastomussa **3:**4-5
Merulina **2:**376
    ampliata **2:***377,*378-9
    scabricula **2:**376-7,**3:***427*
    scheeri **2:**380-1
Merulinidae: key to species **3:**455
Merulinidae **2:**363
mesenteric filaments **1:**52
mesenteries **1:**48,52
mesenterina, Turbinaria **2:***389,*394-5,**3:***428*
mesoglea **1:**48,52
Mesozoic corals **1:**36-40
metamorphosis **3:**419
michelinii, Stephanocoenia **2:**9
micrantha, Tubastrea **2:***387,***3:***410*
micro-atoll **3:***287*
microclados, Acropora **1:**374
microconos, Hydnophora **2:**372-3
Micromussa: type species **3:**8
Micromussa **3:**8
    amakusensis **3:**10-1
    diminuta **3:**9
    minuta **3:**8
micropentagona, Favites **3:**137
microphthalma, Acropora **1:**258-9
microphthalma, Cyphastrea **3:**246-7
migrations **3:**416,420
Milankovitch cycles **3:**440
Millepora **3:**400-1
millepora, Acropora **1:**340-1
millepora, Montipora **1:**125
Milleporina **3:***399,*400-1
minor, Goniopora **3:**366-7
minuta, Acanthastrea **3:***8*
minuta, Acropora **1:**210
minuta, Alveopora **3:**396
minuta, Goniastrea **3:**158-9
minuta, Micromussa **3:**8
minuta, Pavona **2:**196
Miocene
    corals **1:**41-2
    reefs **1:**41
mirabilis, Acropora **1:**347
mirabilis, Madracis **2:**20-1
mokai, Lithophyllon **2:**306

mollis, Montipora **1**:117
molokensis, Pocillopora **2**:42
moluccensis, Fungia **2**:284-5
monasteriata, Montipora **1**:88-9
monile, Coscinaraea **2**:158-9
Montastrea: key to species **3**:454
Montastrea **3**:212
    annularis **3**:*87*,214-5
    annuligera **3**:220
    cavernosa **3**:*87*,222-3
    colemani **3**:219
    curta **3**:*212*,216-7
    faveolata **3**:*214*
    franksi **3**:*214*
    magnistellata **3**:225
    multipunctata **3**:221
    salebrosa **3**:218
    serageldini **3**:213
    valenciennesi **3**:*219*,224
monticules **1**:51-2,**2**:364
monticulosa, Acropora **1**:320-1,*325*
monticulosa, Porites **3**:314-5
Montigyra **2**:129
    kenti **2**:129
Montipora: corallite structure **1**:64-5
Montipora: diversity contours **1**:62
Montipora: growth-forms **1**:63
Montipora: key to species **3**:450-1
Montipora **1**:62-5
    aequituberculata **1**:*69*,76-7,**3**:*436*
    altasepta **1**:*151*,153
    angulata **1**:127
    aspergillus **1**:167
    australiensis **1**:152
    cactus **1**:150-1
    calcarea **1**:116
    caliculata **1**:128-9
    capitata **1**:144-5
    capricornis **1**:80-1
    cebuensis **1**:68-9,**3**:*468*
    circumvallata **1**:93
    cocosensis **1**:114-5
    confusa **1**:82-3
    corbettensis **1**:109
    crassituberculata **1**:78-9
    cryptus **1**:126
    danae **1**:140-1
    delicatula **1**:70-1
    digitata **1**:*151*,154-5,**3**:*429*
    dilatata **1**:98
    echinata **1**:166
    efflorescens **1**:104-5
    effusa **1**:108
    flabellata **1**:99
    florida **1**:74-5
    floweri **1**:124
    foliosa **1**:66-7
    foveolata **1**:131
    friabilis **1**:73
    gaimardi **1**:146
    grisea **1**:94
    hemispherica **1**:147
    hirsuta **1**:*155*,159
    hispida **1**:148-9
    hodgsoni **1**:72
    hoffmeisteri **1**:123
    incrassata **1**:119

    informis **1**:112-3
    kellyi **1**:164
    lobulata **1**:95
    mactanensis **1**:134-5
    malampaya **1**:163
    meandrina **1**:142-3
    millepora **1**:125
    mollis **1**:117
    monasteriata **1**:88-9
    niugini **1**:158
    nodosa **1**:110-1
    orientalis **1**:114
    pachytuberculata **1**:166
    palawanensis **1**:132-3
    patula **1**:106
    peltiformis **1**:100-1
    porites **1**:162
    pulcherrima **1**:*70*
    samarensis **1**:156-7
    saudii **1**:92
    setosa **1**:137
    spongiosa **1**:165
    spongodes **1**:122
    spumosa **1**:120-1
    stellata **1**:*155*,160-1
    stilosa **1**:102-3
    taiwanensis **1**:132
    tuberculosa **1**:90-1
    turgescens **1**:118
    turtlensis **1**:96-7
    undata **1**:86-7
    venosa **1**:130
    verrilli **1**:107
    verrucosa **1**:138-9
    verruculosus **1**:136
    vietnamensis **1**:84-5
mordax, Stylophora **2**:*58*
moretonensis, Astreopora **1**:436
morphological boundaries **3**:429-30
morphological variation **3**:426-9
Moseleya **3**:269
    latistellata **3**:269
mossambica, Acropora **1**:*322*
motuporensis, Podabacia **2**:312-3
mouth **1**:48,52,54
mucous **1**:52
mucous layer, *Porites* **3**:278
multiacuta, Acropora **1**:332
multipunctata, Montastrea **3**:221
muricata, Acropora **1**:*176*
murrayensis, Porites **3**:292
muscular system **1**:52-3
musica, Tubipora **3**:406-7
Mussa **3**:64
    angulosa **3**:64-5,*66*
Mussidae: key to species **3**:455-6
Mussidae **1**:*49,54*,**3**:3
Mussismilia **3**:32
    braziliensis **3**:34
    harttii **3**:35
    hispida **3**:32-3
mutations **3**:421
Mycedium: key to species **3**:457
Mycedium **2**:342
    elephantotus **2**:*320,342*,344-5
    mancaoi **2**:343
    robokaki **2**:346

    steeni **2**:347
    umbra **2**:342
Mycetophyllia **3**:72
    aliciae **3**:78-9
    danaana **3**:*73*,76-7
    ferox **3**:74
    lamarckiana **3**:73
    reesi **3**:75
mycetoseroides, Leptoseris **2**:213
myriophthalma, Astreopora **1**:*437*,442-3
myrmidonensis, Porites **3**:288
nana, Acropora **1**:354
napopora, Porites **3**:318
nariform corallites **1**:178
nasuta, Acropora **1**:400-1
natalensis, Acropora **1**:236-7
natans, Colpophyllia **3**:*73,207*,210-1
navini, Acropora **1**:431
negrosensis, Porites **3**:336-7
nematocysts **1**:48,52-3
Nemenzophyllia **2**:84
    turbida **2**:84-5
neoplasms **3**:421
nerve net **1**:52-3
nierstraszi, Psammocora **2**:153
nigrescens, Porites **3**:*330*,334-5
nishihirai, Echinomorpha **2**:333
nishihirai, Echinophyllia **2**:*333*
niugini, Montipora **1**:158
nobilis, Acropora **1**:222-3
nodifera, Porites **3**:296-7
nodosa, Montipora **1**:110-1
nominal species **1**:*11-2*,**3**:429
norfolkensis, Goniopora **3**:362-3
noumeae, Cantharellus **2**:252
novaehiberniae, Polyphyllia **2**:294
obtusangula, Psammocora **2**:145
ocean circulation patterns **3**:414-6,438-41
ocellata, Acropora **1**:312-3
ocellata, Alveopora **3**:397
ocellata, Astreopora **1**:446
ocellina, Cyphastrea **3**:244
Octocorallia **3**:*399*,404
Oculina **2**:98,**3**:*420*
    diffusa **2**:102
    patagonica **2**:99
    robusta **2**:101
    valenciennesi **2**:100
    varicosa **2**:98
Oculinidae: key to species **3**:456
Oculinidae **1**:*49,51*,**2**:95-7
okinawensis, Porites **3**:307
Oligocene corals **1**:41
oral cone **1**:48,52
oral disc **1**:48,52
orbicularis, Acropora **1**:244
orders of radial elements **1**:49-50
Ordovician reefs **1**:33-4
organ-pipe coral **3**:406-7
orientalis, Montipora **1**:114
origins
    of families **1**:36,*40-1*,**3**:411-2
    of species **3**:438-9,442
    of syngameons **3**:440
ornata, Porites **3**:340
orpheensis, Echinophyllia **2**:328-9
Oulastrea **3**:229

crispata **3**:229
Oulophyllia **3**:195
    bennettae **3**:200-1
    crispa **3**:*195*,196-7
    levis **3**:198-9
Oxypora: key to species **3**:457
Oxypora **2**:334
    convoluta **2**:340
    crassispinosa **2**:334-5
    egyptensis **2**:341
    glabra **2**:338-9
    lacera **2**:336-7
pachysepta, *Lobophyllia* **3**:40
Pachyseris **2**:224
    foliosa **2**:230
    gemmae **2**:224-5
    involuta **2**:231
    rugosa **2**:226-7
    speciosa **2**:228-9,*416*
pachytuberculata, *Montipora* **1**:166
Pacific Ocean corals **3**:413
pacificus, *Echinopora* **3**:252-3
paeonia, *Pectinia* **2**:*349*,354-5,*359*
Palaeogene corals **1**:41
Palaeozoic
    corals **1**:33-7
    reefs **1**:33-7
Palauastrea **2**:10
    ramosa **2**:10-1
palauensis, *Goniastrea* **3**:164-5
palawanensis, *Montipora* **1**:132-3
pali **1**:50
pali, *Porites* **3**:278
palifera, *Acropora* **1**:*27*,186-7
paliform crown **1**:50
paliform lobes **1**:48,50
paliformis, *Poritipora* **3**:347
pallida, *Favia* **3**:114-5
palmata, *Acropora* **1**:202-3
palmensis, *Goniopora* **3**:374
palmerae, *Acropora* **1**:211
panamensis, *Porites* **3**:283
pandoraensis, *Goniopora* **3**:*348-9,351*,370,*371*
paniculata, *Acropora* **1**:378
Panthalassa Ocean **1**:37,40
papillae **1**:64,*3:466*
papillare, *Acropora* **1**:345
papyracea, *Leptoseris* **2**:204-5
paraancora, *Euphyllia* **2**:74-5
Paraclavarina **2**:374
    triangularis **2**:374-5
Paraconotrochus **2**:*411*
Paracyathus **2**:*411*
paradivisa, *Euphyllia* **2**:*66*,73
paraflexuosa, *Favites* **3**:155
paraglabrescens, *Euphyllia* **2**:72
parahemprichii, *Acropora* **1**:274-5
parapharaonis, *Acropora* **1**:367
Parascolymia vitiensis **3**:*68*
Parasimplastrea **3**:239
    sheppardi **3**:239
    simplicitexta **3**:*239*
parilis, *Acropora* **1**:410-1
passive dispersal **3**:416
patagonica, *Oculina* **2**:99
patelliformis, *Cycloseris* **2**:246
patterns, geographic **1**:*6*,**3**:411-5,430

patula, *Echinophyllia* **2**:326-7
patula, *Montipora* **1**:106
patula, *Turbinaria* **2**:389
paucisepta, *Galaxea* **2**:112-3
paumotensis, *Fungia* **2**:282-3
Pavona: key to species **3**:451
Pavona **1**:*42*,**2**:178
    bipartita **2**:197
    cactus **2**:180-1,*183*
    clavus **2**:198-9
    danai **2**:179
    decussata **2**:*178*,194-5
    diffluens **2**:188
    duerdeni **2**:200-1
    explanulata **2**:*168*,184-5,*196*
    frondifera **2**:182-3
    gigantea **2**:189
    maldivensis **2**:192-3,**3**:*427*,*440*
    minuta **2**:196
    varians **2**:186-7
    venosa **2**:190-1
    xarifae **2**:*196*
pearsoni, *Goniopora* **3**:365
pectinata, *Echinophyllia* **2**:331
pectinata, *Goniastrea* **3**:*157*,174-5
pectinatus, *Acropora* **1**:264
Pectinia: key to species **3**:457
Pectinia **2**:348
    africanus **2**:353
    alcicornis **2**:356-7
    ayleni **2**:352
    elongata **2**:360
    lactuca **2**:*349*,350-1
    maxima **2**:349
    paeonia **2**:*349*,354-5,*359*
    pygmaeus **2**:361
    teres **2**:358-9
Pectiniidae: key to species **3**:456
Pectiniidae **1**:*49*,**2**:321
peltata, *Turbinaria* **2**:390-1
peltiformis, *Montipora* **1**:100-1
pendulus, *Goniopora* **3**:350
pentagona, *Favites* **3**:138-9
peresi, *Favites* **3**:*166*
peresi, *Goniastrea* **3**:166
pertusa, *Lophelia* **1**:*27*
petaloid septa **2**:*132*,**3**:*466*
Phacelocyathus **2**:*408*,*410*
phaceloid colonies **1**:54-5
pharaonis, *Acropora* **1**:296-7
pharensis, *Madracis* **2**:12-3
pharynx **1**:48,52
photographs **1**:6-7
phrygia, *Leptoria* **3**:204-5
Phyllangia **2**:*410*
    americana **2**:*410*
    consagensis **2**:*410*
Physogyra **1**:*52*,**2**:92
    lichtensteini **2**:*86*,92-3
Physophyllia ayleni **2**:*352*
pichoni, *Acropora* **1**:286
pileus, *Halomitra* **2**:*232*,302-3
pillai, *Anacropora* **1**:175
pilosa, *Hydnophora* **2**:364-5
pinguis, *Acropora* **1**:212-3
pini, *Platygyra* **3**:*87*,178-9
pistillata, *Stylophora* **2**:56,*57*,58-9

plana, *Acropora* **1**:302-3,**3**:*432*
plantaginea, *Acropora* **1**:391
planula larvae **3**:417-20,*443*
planulata, *Gardineroseris* **2**:222-3
planulata, *Goniopora* **3**:368-9
platform reefs **1**:*24*,*27*,**3**:*408-9*
Platygyra: key to species **3**:454
Platygyra **3**:176-7
    acuta **3**:190
    carnosus **3**:184
    contorta **3**:188-9
    crosslandi **3**:180
    daedalea **3**:*177*,191
    lamellina **3**:192-3
    pini **3**:*87*,178-9
    ryukyuensis **3**:182-3
    sinensis **3**:186-7
    verweyi **3**:*177*,181
    yaeyamaensis **3**:184-5
Plerogyra **1**:*52*,**2**:86
    discus **2**:86-7
    eurysepta **2**:*88*
    exerta **2**:*92*
    simplex **2**:90-1
    sinuosa **2**:*86*,88-9
Plesiastrea **3**:226
    devantieri **3**:228
    versipora **3**:226-7
plocoid colonies **1**:54-5
plumosa, *Acropora* **1**:288-9
Pocillopora: key to species **3**:457
Pocillopora **1**:*42,43*,**2**:24
    ankeli **2**:33
    capitata **2**:35
    damicornis **2**:26-7,*29*,**3**:*427*
    danae **2**:25
    effusus **2**:39
    elegans **2**:34-5
    eydouxi **2**:*31,43*,44-5
    fungiformis **2**:40
    indiania **2**:37
    inflata **2**:41
    kelleheri **2**:32
    ligulata **2**:38
    meandrina **2**:30-1
    molokensis **2**:42
    verrucosa **2**:28-9,*31,35-7,45,64*
    woodjonesi **2**:43
    zelli **2**:36
Pocilloporidae: key to species **3**:457
Pocilloporidae **1**:*36-7,42,49,50*,**2**:23
poculata, *Astrangia* **2**:319
Podabacia **2**:310
    crustacea **2**:310-1
    lankaensis **2**:315
    motuporensis **2**:312-3
    sinai **2**:314-5
Polycyathus **2**:*411*
polyformis, *Goniopora* **3**:360
polymorphic species **3**:428
polyp
    bail-out **3**:420
    expulsion **3**:420
    structure **1**:48-53
    tissues **1**:48,52-3
Polyphyllia **1**:*54*,**2**:294
    novaehiberniae **2**:294

talpina **2**:295
polyploidy **3**:421
polyps **1**:48-53
polystoma, Acropora **1**:335
populations *3:428-9*
Porites: corallite structure **3**:277-8
Porites: diversity contours **3**:276
Porites: key to species **3**:458-9
Porites: microatolls **3**:279
Porites: mucous layer: **3**:278
Porites **1**:*42,50,53,***3**:276-9
    annae **3**:310-1,*427*
    aranetai **3**:303
    arnaudi **3**:301
    astreoides **3**:280-1
    attenuata **3**:330
    australiensis **3**:*277,*286
    branneri **3**:325
    brighami **3**:295
    cocosensis **3**:311
    colonensis **3**:302
    columnaris **3**:300
    compressa **1**:*12,***3**:344-5
    cumulatus **3**:324
    cylindrica **3**:*274,314,330,332-3, 335*
    deformis **3**:323
    densa **3**:294
    desilveri **3**:308
    divaricata **3**:329
    duerdeni **3**:*344*
    echinulata **3**:290,*296*
    eridani **3**:322
    evermanni **3**:298
    flavus **3**:341
    furcata **3**:328
    harrisoni **3**:343
    heronensis **3**:306-7
    horizontalata **3**:316-7
    latistella **3**:320-1
    lichen **3**:304-5,*309*
    lobata **3**:*279,*284-5,*432*
    lutea **3**:*278,*287
    mayeri **3**:289
    monticulosa **3**:314-5
    murrayensis **3**:292
    myrmidonensis **3**:288
    napopora **3**:318
    negrosensis **3**:336-7
    nigrescens **3**:*330,*334-5
    nodifera **3**:296-7
    okinawensis **3**:307
    ornata **3**:340
    panamensis **3**:283
    porites **3**:326-7,*424*
    profundus **3**:338-9
    pukoensis **3**:299
    rugosa **3**:342
    rus **3**:*274,279,*312-3,*314,315*
    sillimaniana **3**:319
    solida **3**:282
    somaliensis **3**:291
    stephensoni **3**:293
    sverdrupi **3**:*283*
    tuberculosa **3**:331
    vaughani **3**:308-9
porites, Montipora **1**:162
porites, Porites **3**:326-7,*424*

Poritidae: diversity contours **3**:275
Poritidae: key to species **3**:457
Poritidae **1**:*49,***3**:275
Poritipora: type species **3**:347
Poritipora **3**:347
    paliformis **3**:347
pourtàles plan **1**:49-50
profundacella, Psammocora **2**:149
profundus, Porites **3**:338-9
prolifera, Acropora **1**:261
prostrata, Acropora **1**:338-9
prostrate colonies **1**:178
Proterozoic reefs **1**:*32*
Protoatlantic Ocean **1**:37
proximalis, Acropora **1**:278-9
pruinosa, Acropora **1**:270
pruinosa, Leptastrea **3**:237
Psammocora: key to species **3**:459
Psammocora **2**:144
    contigua **2**:146-7
    decussata **2**:144
    digitata **2**:154-5
    explanulata **2**:*132,*156
    haimeana **2**:152
    nierstraszi **2**:153
    obtusangula **2**:145
    profundacella **2**:149
    stellata **2**:148,*241*
    superficialis **2**:150-1
    vaughani **2**:157
    verrilli **2**:151
Pseudosiderastrea **2**:134
    tayami **2**:134-5
puertogalerae, Anacropora **1**:170-1
puishani, Fungia **2**:277
pukoensis, Porites **3**:299
pulcherrima, Montipora **1**:*70*
pulchra, Acropora **1**:344-5
punctata, Stylaraea **3**:346
purpurea, Leptastrea **3**:236
pygmaeus, Pectinia **2**:361
radians, Siderastrea **2**:142
radians, Symphyllia **3**:58-9
radicalis, Turbinaria **2**:397
radii, of *Porites* **3**:278
rafting **1**:*43,***3**:420
rambleri, Acropora **1**:298
ramosa, Goniastrea **3**:160
ramosa, Palauastrea **2**:10-1
randalli, Astreopora **1**:438
ranges, distribution **1**:6,**3**:430,432
rare events **3**:420
rates
    of evolution **3**:439-40
    of extinction **3**:439-42
recta, Symphyllia **3**:56-7
rediviva, Archohelia **2**:*96-7*
reef/reefs **1**:21-43
    aquaria **1**:14-5
    atolls **1**:23-6
    barrier **1**:23-7
    coral **1**:21-31,33-41
    Cretaceous **1**:37,40
    development **1**:23-7
    Devonian **1**:32-5
    distribution **1**:22
    Eocene **1**:41

    flats **1**:*24-6*
    fringing **1**:24-5
    high latitude **1**:28-9,**3**:414
    Jurassic **1**:37
    lagoons **1**:*23-6*
    Miocene **1**:41
    Ordovician **1**:33-4
    Palaeozoic **1**:33-7
    platform **1**:*24,27,***3**:*408-9*
    Proterozoic **1**:*32*
    Silurian **1**:*34*
    slopes **1**:*23-6*
    subtropical **3**:414
    temperate **1**:*28-9,***3**:414-5
    temperature of **1**:27-9
    Triassic **1**:37
    tropical **3**:414
    zonation **1**:24-7
reesi, Mycetophyllia **3**:75
references **1**:*7,***3**:469-73
    field guides **1**:*7,***3**:469-73
    taxonomic **1**:*7,***3**:469-73
regional endemism **3**:412-3
regional similarity **3**:413
regularis, Acanthastrea **3**:16
reniformis, Turbinaria **2**:*225,389,*396,**3**:*428*
repanda, Fungia **2**:258,272-3
repackaging, evolutionary **3**:438-43
reproductive isolation **3**:412-3,425,433,441
reproductive system **1**:53
reproduction **1**:*14-5,***3**:416-20
reticulata, Anacropora **1**:172
reticulate
    distributions **3**:416,438-41
    evolution **3**:438-43
    repackaging **3**:438-41
retiformis, Goniastrea **3**:162-3
retusa, Acropora **1**:322
Rhizangiidae **2**:317-8
Rhizotrochus typus **1**:*10*
rigida, Hydnophora **2**:366-7,*369*
rigida, Isophyllastrea **3**:*37*
rigida, Isophyllia **3**:37
robokaki, Mycedium **2**:346
robusta, Acropora **1**:216-7
robusta, Echinopora **3**:263
robusta, Lobophyllia **3**:50-1,*67*
robusta, Oculina **2**:101
robusta, Sandalolitha **2**:296-7
rosaria, Acropora **1**:394-5
rosaria, Favia **3**:119
roseni, Acropora **1**:218-9
rotumana, Favia **3**:121
rotundata, Favia **3**:124-5
rotundoflora, Acanthastrea **3**:20-1
rowleyensis, Australomussa **3**:80
rudis, Acropora **1**:201
rudist bivalves **1**:40-1
rufus, Acropora **1**:269
Rugosa **1**:*35,***3**:*399,*407
rugosa, Pachyseris **2**:226-7
rugosa, Porites **3**:342
rugose corals **1**:*35,***3**:*399,*407
rus, Porites **3**:*274,279,*312-3,*314,315*
russelli, Acropora **1**:420-1
russelli, Favites **3**:148-9
ryukyuensis, Platygyra **3**:182-3

salebrosa, Montastrea **3**:218
samarensis, Montipora **1**:156-7
samoensis, Acropora **1**:323
Sandalolitha **2**:296
    africana **2**:299
    dentata **2**:298
    robusta **2**:296-7
sarmentosa, Acropora **1**:326
satellite colonies **2**:82,**3**:352
saudii, Montipora **1**:92
savignyana, Siderastrea **2**:139
savignyi, Goniopora **3**:376
scabra, Astreopora **1**:441
scabra, Fungia **2**:274
scabra, Leptoseris **2**:210
scabricula, Merulina **2**:376-7,*3:427*
Scapophyllia **2**:383
    cylindrica **2**:383
scheeri, Merulina **2**:380-1
scherzeriana, Acropora **1**:200
Schizoculina **2**:104
    africana **2**:104
    fissipara **2**:105
schmitti, Acropora **1**:196-7
Scolymia: type species **3**:66
Scolymia **3**:66
    australis **3**:70-1
    cubensis **3**:66-7
    lacera *3:64,66*
    vitiensis *3:67,*68-9
    wellsi *3:66*
scruposa, Fungia **2**:259
scutaria, Fungia **2**:*258,*280-1
sea anemones *1:47,***3**:399
sea temperature **1**:*16,*27-9,**3**:414
sea-level change *1:23,***3**:441
secale, Acropora **1**:398-9
sekiseiensis, Acropora **1**:276
selago, Acropora **1**:362-3
self-fertilisation **3**:417
senaria, Madracis **2**:14
separately-sexed corals *1:53,***3**:417
septa **1**:48-50
    cycles of **1**:49-50
    of *Fungia* **2**:258
    of *Porites* **3**:277-8
septo-costae **1**:48-9
serageldini, Montastrea **3**:213
serailia, Cyphastrea **3**:*87,*242-3
seriata, Acropora **1**:254-5
Seriatopora: key to species **3**:457
Seriatopora **2**:46
    aculeata **2**:52
    caliendrum **2**:*47,48,*54-5
    dendritica **2**:46-7,*367*
    guttatus **2**:50-1
    hystrix **2**:*22,46,47,*48-9,*55,367*
    stellata **2**:53
serratus, Lobophyllia **3**:41
setosa, Montipora **1**:137
sex of coral **3**:417
seychellensis, Fungia **2**:279
sheppardi, Parasimplastrea **3**:239
sibling species **3**:428,430,439
Siderastrea: key to species **3**:459
Siderastrea **2**:138
    glynni **2**:138

radians **2**:142
savignyana **2**:139
siderea **2**:*139,*140-1,*142*
stellata **2**:143
Siderastreidae: key to species **3**:459
Siderastreidae *1:48-9,***2**:133
siderea, Siderastrea **2**:*139,*140-1,*142*
sillimaniana, Porites **3**:319
Silurian reefs *1:34*
similar species *1:4*
similarity, measures of **3**:413
Simplastrea **2**:103
    vesicularis **2**:103
simplex, Acropora **1**:284
simplex, Plerogyra **2**:90-1
simplicitexta, Parasimplastrea **3***:239*
sinai, Podabacia **2**:314-5
sinensis, Cycloseris **2**:238
sinensis, Platygyra **3**:186-7
sinuosa, Isophyllia **3**:36
sinuosa, Plerogyra **2**:*86,*88-9
sipunculids **2**:407
skeleton **1**:48-52
skeleton, lack of *1:15*
Solenastrea *1:43,***3**:250
    bournoni **3**:250
    hyades **3**:251
solida, Leptoseris **2**:211
solida, Porites **3**:282
solitary corals *1:49,*54
solitaryensis, Acropora **1**:234-5
somaliensis, Goniopora **3**:358-9
somaliensis, Porites **3**:291
somervillei, Cycloseris **2**:242-3
sordiensis, Acropora *1:397*
spat **3**:419
spathulata, Acropora *1:210*
spawning **3**:417-9
    control of **3**:418-9
    synchrony of **3**:418-9
species (see also 'distribution', 'evolution',
    'extinction','genetic','geographic','variation')
    abundance *1:4,***3**:442
    affinities **3**:413,430-3
    boundaries **3**:429-33
    cohesion **3**:430-3,440-2
    complexes **3**:429-33
    concepts **1**:11-3,**3**:425-6,441-3
    diversity **3**:412,415-6
    diversity contours **3**:412
    doubtful *1:6,***3**:429-32
    drop-out sequence **3**:414
    keys to **1**:7,**3**:447-59
    lumping of **3**:426,431
    nature of **3**:425-6
    nominal *1:11-2,***3**:429
    origination of **3**:438-9,442
    place of origin **3**:438-9,442
    polymorphic **3**:428
    sibling **3**:428,430,439
    similar *1:4*
    splitting of **3**:426,431
    time of origin **3**:438-9,442
    types of **3**:425-33,439
speciosa, Acropora **1**:424-5
speciosa, Favia **3**:108-9
speciosa, Pachyseris **2**:228-9,*416*

sperm **3**:417
spicifera, Acropora **1**:*210,*308-9,*344*
spinifer, Fungia **2**:275
spinosa, Anacropora **1**:173
spinosa, Favites **3**:142
splitting species **3**:426,431
spongiosa, Alveopora **3**:388-9
spongiosa, Montipora **1**:165
spongodes, Montipora **1**:122
spumosa, Anacropora **1**:171
spumosa, Montipora **1**:120-1
squarrosa, Acropora **1**:390
staghorn colonies **1**:178
steeni, Mycedium **2**:347
stellaris, Dichocoenia **2***:124*
stellata, Montipora **1**:*155,*160-1
stellata, Psammocora **2**:148,*241*
stellata, Seriatopora **2**:53
stellata, Siderastrea **2**:143
stelligera, Favia **3**:102-3
stellulata, Turbinaria **2**:400-1
Stephanocoenia **2**:9
    michelinii **2**:9
Stephanocyathus **2***:411*
stephanus, Fungiacyathus **2***:235*
stephensoni, Porites **3**:293
sterome **1**:48-9,51
stilosa, Montipora **1**:102-3
stoddarti, Acropora **1**:232-3
stokesi, Dichocoenia **2**:124-5
stokesi, Goniopora **3**:352-3
striata, Acropora **1**:272,*275*
striata, Leptoseris **2**:212
strigosa, Diploria **3**:208
stromatolites *1:33-4*
stromatoporoids *1:33-4*
structure **1**:47-57
stutchburyi, Goniopora **3**:377
Stylaraea **3**:346
    punctata **3**:346
Stylaster **3**:399,402-3
Stylasterina **3**:*399,402-3*
stylifera, Favites **3**:136
Stylocoeniella **2**:4
    armata **2**:*2,*4-5
    cocosensis **2**:8
    guentheri **2**:6-7
Stylophora: key to species **3**:457
Stylophora **1**:*41,42,***2**:56
    danae **2**:63
    kuehlmanni **2**:62
    madagascarensis **2**:57
    mamillata **2**:65
    mordax **2***:58*
    pistillata **2**:*56,57,*58-9
    subseriata **1***:393,***2**:60-1
    wellsi **2**:64
sub-, prefix **1**:56
subechinata, Acanthastrea **3**:13
subglabra, Acropora **1**:428
subseriata, Stylophora **1***:393,***2**:60-1
subspecies **3**:413,425,439
subtropical reefs **3**:414
subulata, Acropora **1**:350-1
suggesta, Astreopora **1**:440
suharsonoi, Acropora **1**:333
sukarnoi, Acropora **1***:214*

sultani, Goniopora **3**:355
superficialis, Psammocora **2**:150-1
surface circulation vicariance **3**:438-9
sverdrupi, Porites **3**:*283*
symbiosis **1**:*21,33,53*
symmetry **1**:49-50
sympatric origination **3**:439
sympatric patterns **3**:439
Symphyllia: key to species **3**:456
Symphyllia **1**:*54*,**3**:52
    agaricia **3**:60-1
    erythraea **3**:54-5
    hassi **3**:52
    radians **3**:58-9
    recta **3**:56-7
    valenciennesi **3**:62-3
    wilsoni **3**:53
synapticulae **1**:48-9
synapticular ring, *Porites* **3**:278
Synaraea subgenus **3**:279
synchrony of spawning **3**:418-9
syngameons **3**:426,428-33,439-40
    origins of **3**:440
synonymies **3**:429,432
Tabulata **1**:*34-5*,**3**:*399,407*
tabulate corals **1**:*34-5*,**3**:*399,407*
taiwanensis, Fungia **2**:278-9
taiwanensis, Montipora **1**:132
talpina, Polyphyllia **2**:295
tanegashimensis, Acropora **1**:310
taxonomic decisions **3**:429-32
taxonomic references **1**:*7*,**3**:469-73
taxonomists **1**:*11-14*,**3**:*425*,430-1
taxonomy **1**:*11-14*,**3**:425-33
tayami, Pseudosiderastrea **2**:134-5
taylorae, Echinophyllia **2**:327
temperate reefs **1**:*28-9*,**3**:414-5
temperature, ocean **1**:*16*,27-9,**3**:414
tenella, Acropora **1**:285
tenella, Goniopora **3**:360
tentacles **1**:48,52
tenuidens, Goniopora **3**:364
tenuifolia, Agaricia **2**:171,*172*
tenuis, Acropora **1**:360-1
tenuis, Cycloseris **2**:244
tenuis, Leptoseris **2**:*219*
teres, Acropora **1**:209
teres, Pectinia **2**:358-9
tethyan origins **3**:412
Tethys
    corals **1**:41-2
    Sea **1**:37,40-2
Thamnasteriidae **1**:*36-7*
thecata, Goniastrea **3**:169
tiranensis, Echinopora **3**:265
tizardi, Acropora **1**:277
tizardi, Alveopora **3**:392-3
togianensis, Acropora **1**:192
torihalimeda, Acropora **1**:421
torresiana, Acropora **1**:316
tortuosa, Acropora **1**:265
Trachyphyllia **3**:272
    geoffroyi **3**:*270*,272-3
Trachyphylliidae **3**:271
transversa, Leptastrea **3**:238
triangularis, Paraclavarina **2**:374-5
Triassic

corals **1**:36-9
reefs **1**:37
triplet, *Porites* **3**:278
Trochocyathus **2**:*411*
tropical reefs **3**:414
truncatus, Favia **3**:113
Tubastrea **1**:*44-5*,**2**:*384-7*,*414-5*
    faulkneri **2**:*386*
    micrantha **2**:*387*,**3**:*410*
tuberculae **1**:64
tuberculosa, Montipora **1**:90-1
tuberculosa, Porites **3**:331
Tubipora **3**:*399*,406-7
    musica **3**:406-7
tubular corallites **1**:178
tubulifera, Leptoseris **2**:206
tumida, Acropora **1**:271
tumida, Caulastrea **3**:94-5
turaki, Acropora **1**:429
turbida, Nemenzophyllia **2**:84-5
Turbinaria: key to species **3**:452
Turbinaria **2**:388
    bifrons **2**:402
    conspicua **2**:403
    frondens **2**:392-3
    heronensis **2**:404
    irregularis **2**:398-9
    mesenterina **2**:*389*,394-5,**3**:*428*
    patula **2**:389
    peltata **2**:390-1
    radicalis **2**:397
    reniformis **2**:*225*,389,396,**3**:*428*
    stellulata **2**:400-1
turgescens, Montipora **1**:118
turtlensis, Montipora **1**:96-7
tutuilensis, Acropora **1**:290-1
type localities **3**:*467*
type species **1**:176,**3**:*8,66,347,467*
typus, Rhizotrochus **1**:*10*
umbra, Mycedium **2**:342
undata, Agaricia **2**:175
undata, Montipora **1**:86-7
undulatum, Lithophyllon **2**:308-9
valenciennesi, Acropora **1**:226-7
valenciennesi, Montastrea **3**:*219*,224
valenciennesi, Oculina **2**:100
valenciennesi, Symphyllia **3**:62-3
valida, Acropora **1**:404-5
valley formation **1**:54
variabilis, Acropora **1**:406
varians, Pavona **2**:186-7
variation **3**:426-9
    environmental **3**:426-7
    genetic **3**:426-9
    geographic **3**:427-32,437
    morphological **3**:426-9
varicosa, Oculina **2**:98
variolosa, Acropora **1**:197
vasta, Favites **3**:*134*,152-3
vaughani, Acropora **1**:268-9
vaughani, Coscinaraea **2**:*157*
vaughani, Cycloseris **2**:244
vaughani, Porites **3**:308-9
vaughani, Psammocora **2**:157
venosa, Montipora **1**:130
venosa, Pavona **2**:190-1
vermiculata, Acropora **1**:359

veroni, Favia **3**:128-9
verrilli, Montipora **1**:107
verrilli, Psammocora **2**:151
verrilliana, Alveopora **3**:387
verrucae
    of *Montipora* **1**:64
    of *Pocillopora* **2**:24
verrucosa, Montipora **1**:138-9
verrucosa, Pocillopora **2**:28-9,*31,35-7,45,64*
verruculosus, Montipora **1**:136
versipora, Plesiastrea **3**:226-7
verweyi, Acropora **1**:386-7
verweyi, Platygyra **3**:*177*,181
vesicles **3**:*86,88,90,92*
vesicularis, Simplastrea **2**:103
vicariance biogeography **3**:438-9
vietnamensis, Favia **3**:127
vietnamensis, Montipora **1**:84-5
violacea, Distichopora **3**:*398*
viridis, Alveopora **3**:395
vitiensis, Parascolymia **3**:*68*
vitiensis, Scolymia **3**:*67*,68-9
walindii, Acropora **1**:287
wall of corallites **1**:48-9,51
    Porites **3**:278
wallaceae, Acropora **1**:245
wardii, Acropora **1**:*317*
weberi, Herpolitha **2**:291
wellsi, Blastomussa **3**:6-7
wellsi, Coscinaraea **2**:167
wellsi, Scolymia **3**:*66*
wellsi, Stylophora **2**:64
willisae, Acropora **1**:366
wilsoni, Symphyllia **3**:53
Wood-Jones, Fredric **1**:*12*
woodjonesi, Pocillopora **2**:43
woodsi, Astrangia **2**:*318*
xarifae, Pavona **2**:*196*
yabei, Leptoseris **2**:220
yaeyamaensis, Euphyllia **2**:76-7
yaeyamaensis, Platygyra **3**:184-5
yongei, Acropora **1**:262-3
zelli, Australogyra **3**:194
zelli, Pocillopora **2**:36
zonation, reef **1**:24-7
Zoopilus **2**:*234*,304
    echinatus **2**:*234*,304-5
zooxanthellae **1**:21,53
zooxanthellate corals **1**:53

The author Dr Veron took up diving in the mid 60's and has been working on corals ever since – work that has taken him to all the major coral reef regions of the world. He is principal author of 12 books and monographs on corals including the award winning *Corals in Space and Time*. Dr Veron has three higher degrees including a DSc for his work on coral taxonomy. For many years he has been Chief Scientist at the Australian Institute of Marine Science.

Corals of the World is an end product of thirty years of research. It is also very much the outcome of the author's love of nature and was written for the use and enjoyment of all who appreciate the beauty of coral reefs and who want to know more about the organisms that build them.

The scientific editor and producer Dr Stafford-Smith has worked for 20 years on coral reefs worldwide. She specialises in eco-physiological questions and management issues and has an extensive taxonomic knowledge of corals. Through this work she continued to expand her expertise in taxonomy and developed interests in other fields, including information technology and design.

Corals of the World was created through Dr Stafford-Smith's combined knowledge of coral taxonomy and computer technology. Like the author, it is also the outcome of a love of coral reefs and a desire to do her part for their long-term conservation.

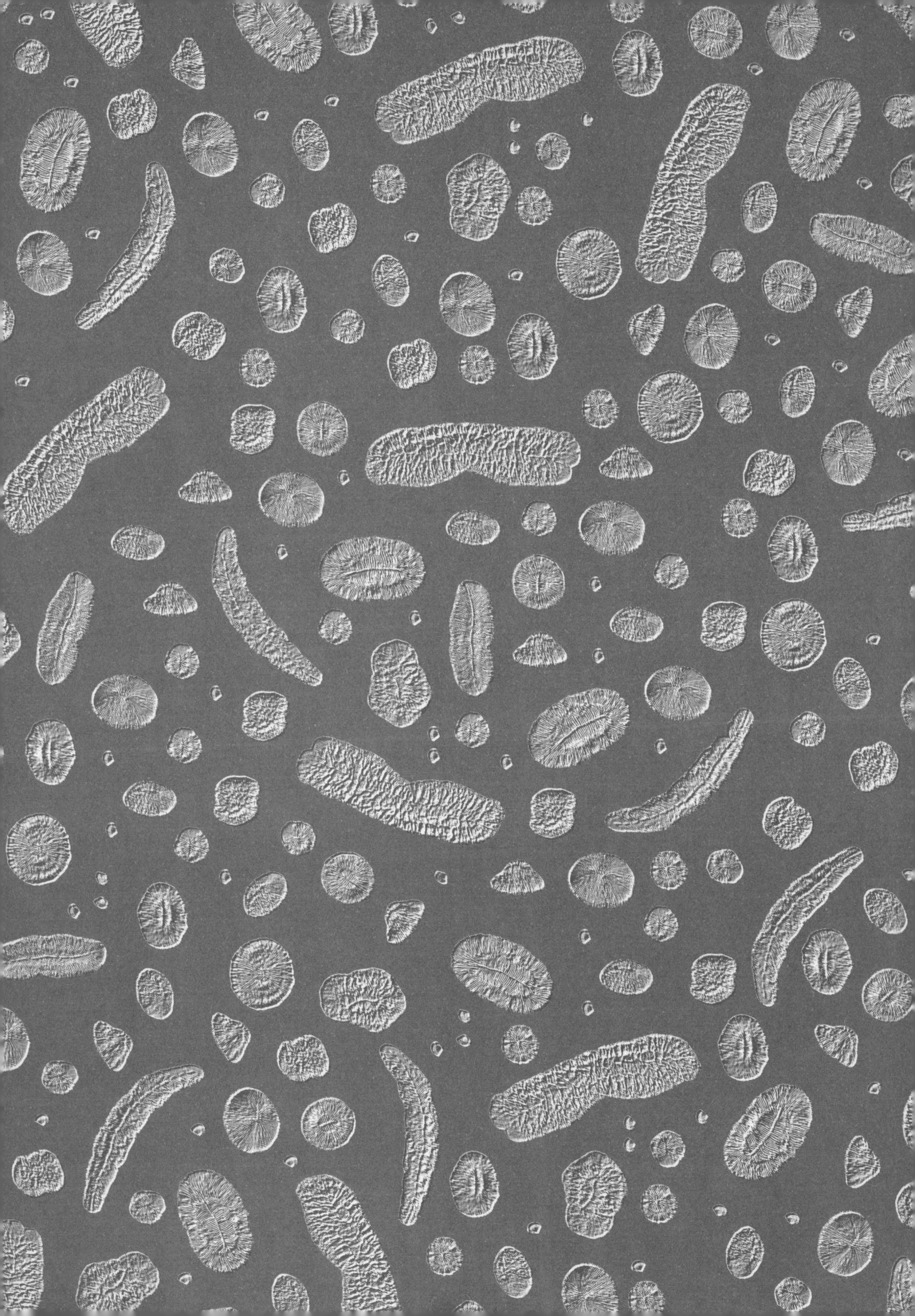